T0207522

Lecture Notes in Computer Science

Lecture Notes in Artificial Intelligence   13897

Founding Editor

Jörg Siekmann

Series Editors

Randy Goebel, *University of Alberta, Edmonton, Canada*
Wolfgang Wahlster, *DFKI, Berlin, Germany*
Zhi-Hua Zhou, *Nanjing University, Nanjing, China*

The series Lecture Notes in Artificial Intelligence (LNAI) was established in 1988 as a topical subseries of LNCS devoted to artificial intelligence.

The series publishes state-of-the-art research results at a high level. As with the LNCS mother series, the mission of the series is to serve the international R & D community by providing an invaluable service, mainly focused on the publication of conference and workshop proceedings and postproceedings.

Jose M. Juarez · Mar Marcos · Gregor Stiglic ·
Allan Tucker
Editors

Artificial Intelligence in Medicine

21st International Conference
on Artificial Intelligence in Medicine, AIME 2023
Portorož, Slovenia, June 12–15, 2023
Proceedings

 Springer

Editors
Jose M. Juarez 🄾
University of Murcia
Murcia, Spain

Mar Marcos 🄾
Universitat Jaume I
Castellón de la Plana, Spain

Gregor Stiglic 🄾
University of Maribor
Maribor, Slovenia

Allan Tucker 🄾
Brunel University London
Uxbridge, UK

ISSN 0302-9743   ISSN 1611-3349 (electronic)
Lecture Notes in Artificial Intelligence
ISBN 978-3-031-34343-8   ISBN 978-3-031-34344-5 (eBook)
https://doi.org/10.1007/978-3-031-34344-5

LNCS Sublibrary: SL7 – Artificial Intelligence

This Springer imprint is published by the registered company Springer Nature Switzerland AG
The registered company address is: Gewerbestrasse 11, 6330 Cham, Switzerland

Preface

The European Society for Artificial Intelligence in Medicine (AIME) was established in 1986 following a very successful workshop held in Pavia, Italy, the year before. The principal aims of AIME are to foster fundamental and applied research in the application of Artificial Intelligence (AI) techniques to medical care and medical research, and to provide a forum at conferences for discussing any progress made. The main activity of the society thus far has been the organization of a series of international conferences, held in Marseille, France (1987), London, UK (1989), Maastricht, Netherlands (1991), Munich, Germany (1993), Pavia, Italy (1995), Grenoble, France (1997), Aalborg, Denmark (1999), Cascais, Portugal (2001), Protaras, Cyprus (2003), Aberdeen, UK (2005), Amsterdam, Netherlands (2007), Verona, Italy (2009), Bled, Slovenia (2011), Murcia, Spain (2013), Pavia, Italy (2015), Vienna, Austria (2017), Poznan, Poland (2019), Minneapolis, USA (2020), Porto, Portugal (2021), and Halifax, Canada (2022).

The AIME 2023 goals were to present and consolidate the international state of the art of AI in biomedical research from the perspectives of theory, methodology, systems, and applications. The conference included two invited keynotes, presentations of full, short, and demonstration papers, tutorials, workshops, a doctoral consortium, and a panel discussion of editors-in-chief of prestigious journals.

AIME 2023 received 132 abstract submissions, 108 thereof were eventually submitted as complete papers. Submissions came from authors in 17 countries, including submissions from Europe, North and South America, and Asia.

All papers were carefully peer-reviewed by experts from the Program Committee, with the support of additional reviewers, and subsequently by members of the Senior Program Committee. Each submission was reviewed in most cases by three reviewers, and all papers by at least two reviewers. The reviewers judged the overall quality of the submitted papers, together with their relevance to the AIME conference, originality, impact, technical correctness, methodology, scholarship, and quality of presentation. In addition, the reviewers provided detailed written comments on each paper and stated their confidence in the subject area. One Senior Program Committee member was assigned to each paper, who wrote a meta-review and provided a recommendation to the Organizing Committee.

A small committee consisting of the AIME 2023 scientific co-chairs, Jose M. Juarez and Mar Marcos, the local organization chair, Gregor Štiglic, and the doctoral consortium chair, Allan Tucker, made the final decisions regarding the AIME 2023 conference program. This process was carried out with virtual meetings starting in March 2023.

As a result, 23 long papers (an acceptance rate of 21%), 21 short papers, and 3 demonstration papers were accepted. Each long paper was presented in a 15-minute oral presentation during the conference. Each regular short and demonstration paper was presented in a 5-minute presentation and by a poster or a demonstration session. The papers were organized according to their topics in the following main themes: (1) Machine Learning and Deep Learning; (2) Explainability and Transfer Learning; (3)

Natural Language Processing; (4) Image Analysis and Signal Analysis; (5) Data Analysis and Statistical Models; and (6) Knowledge Representation and Decision Support.

AIME 2023 had the privilege of hosting two invited keynote speakers: Zoran Obradovic, Distinguished Professor and center director at Temple University, giving the keynote entitled "Predictive Analytics for Clinical Decision Making", and Nada Lavrač, Research Councillor at Department of Knowledge Technologies, Jožef Stefan Institute, and Professor at University of Nova Gorica, Slovenia, describing "Learning representations for relational learning and literature-based discovery".

AIME 2023 also hosted an invited panel of editors-in-chief of prestigious high-impact journals. This panel focused on the current panorama of AI in Medicine, newly emerging research topics, and the publishing field. We were honored to have on this panel: Mor Peleg (Journal of Biomedical Informatics, Elsevier), Carlo Combi (Artificial Intelligence in Medicine Journal, Elsevier), Dimitrios I. Fotiadis (IEEE Journal of Biomedical and Health Informatics), and Zoran Obradovic (Big Data Journal, Mary Ann Liebert).

The doctoral consortium received 21 PhD proposals that were peer reviewed. AIME 2023 provided an opportunity for six of these PhD students to present their research goals, proposed methods, and preliminary results. A scientific panel, consisting of experienced researchers in the field, provided constructive feedback to the students in an informal atmosphere. The doctoral consortium was chaired by Dr. Allan Tucker (Brunel University London).

AIME 2023 invited researchers to submit proposals for workshops and tutorials. Five workshops were selected by the Organization Committee. These workshops were: the 2nd Workshop on Artificial Intelligence in Nursing (AINurse2023), the 3rd International Workshop on eXplainable Artificial Intelligence in Healthcare (XAI-Healthcare 2023); the 13th International Workshop on Knowledge Representation for Health Care (KR4HC 2023); and the 1st International Workshop on Process Mining Applications for Healthcare (PM4H 2023). In addition to the workshops, three interactive half-day tutorials were selected: (1) Evaluation of Prediction Models in Medicine, (2) pMineR & pMinShiny Tutorial, and (3) Mining and multimodal learning from complex medical data.

We would like to thank everyone who contributed to AIME 2023. First, we would like to thank the authors of the papers submitted and the members of the Program Committee together with the additional reviewers. Thank you to the Senior Program Committee for writing meta-reviews and for guidance. Thanks are also due to the invited speakers, the editor-in-chief panelists, as well as to the organizers of the tutorials, the workshops, and the doctoral consortium panel. Many thanks go to the local Organizing Committee members, who managed all the work making this conference possible: Lina Dvoršak, Nino Fijačko, Lucija Gosak, Primož Kocbek, Leon Kopitar, Aleksandra Lovrenčič, Ines Mlakar, Kasandra Musovič, and Marko Uršič. We would like to thank Springer for sponsoring the conference, and University of Maribor, Faculty of Health Sciences for all the support in organizing it. Finally, we thank the Springer team for helping us in the final preparation of this LNAI book.

April 2023

Jose M. Juarez
Mar Marcos
Gregor Stiglic
Allan Tucker

Organization

Scientific Chairs

Jose M. Juarez   University of Murcia, Spain
Mar Marcos   University Jaume I, Spain

Local Organizing Chair

Gregor Štiglic   University of Porto, Portugal

Doctoral Consortium Chair

Allan Tucker   Brunel University London, UK

Senior Program Committee

Ameen Abu-Hanna   University of Amsterdam, The Netherlands
Riccardo Bellazzi   University of Pavia, Italy
Carlo Combi   University of Verona, Italy
Arianna Dagliati   University of Pavia, Italy
Joseph Finkelstein   University of Utah, USA
Adelo Grando   Arizona State University, USA
Janna Hastings   University of Zurich, Switzerland
Milos Hauskrecht   University of Pittsburgh, USA
Pedro Henriques-Abreu   University of Coimbra, Portugal
John Holmes   University of Pennsylvania, USA
Elpida Keravnou-Papailiou   University of Cyprus, Cyprus
Nada Lavrač   Jozef Stefan Institute, Slovenia
Peter Lucas   Leiden University, The Netherlands
Martin Michalowski   University of Minnesota, USA
Stefania Montani   University of Piemonte Orientale, Italy
Robert Moskovitch   Ben-Gurion University, Israel
Barbara Oliboni   University of Verona, Italy
Enea Parimbelli   University of Pavia, Italy
Niels Peek   University of Manchester, UK

Mor Peleg	University of Haifa, Israel
Silvana Quaglini	University of Pavia, Italy
Syed Sibte Raza Abidi	Dalhousie University, Canada
Lucia Sacchi	University of Pavia, Italy
Yuval Shahar	Ben-Gurion University, Israel
Annette ten Teije	Vrije Universiteit Amsterdam, The Netherlands
Allan Tucker	Brunel University London, UK
Frank van Harmelen	Vrije Universiteit Amsterdam, The Netherlands
Szymon Wilk	Poznan University of Technology, Poland
Blaz Zupan	University of Ljubljana, Slovenia

Program Committee

Samina Abidi	Dalhousie University, Canada
Soon Chun	City University of New York, USA
Jose Luis Ambite	University of Southern California, USA
Lourdes Araujo	National Distance Education University, Spain
Isabelle Bichindaritz	State University of New York at Oswego, USA
Jerzy Błaszczyński	Poznań University of Technology, Poland
Henrik Boström	KTH Royal Institute of Technology, Sweden
Alessio Bottrighi	University of Piemonte Orientale, Italy
Marcos Luis de Paula Bueno	TU Eindhoven, The Netherlands
Iacer Calixto	University of Amsterdam, The Netherlands
Manuel Campos	University of Murcia, Spain
Bernardo Canovas-Segura	University of Murcia, Spain
Ricardo Cardoso	University of Coimbra, Portugal
Giovanni Cinà	University of Amsterdam, The Netherlands
Kerstin Denecke	Bern University of Applied Sciences, Switzerland
Barbara Di Camillo	University of Padova, Italy
Georg Dorffner	Medical University Vienna, Austria
Jan Egger	Graz University of Technology, Austria
Anna Fabijańska	Technical University of Lodz, Poland
Jesualdo T. Fernandez-Breis	University of Murcia, Spain
Daniela Ferreira-Santos	University of Porto, Portugal
Natalia Grabar	STL CNRS Université Lille 3, France
Bertha Guijarro-Berdiñas	University of A Coruña, Spain
Zhe He	Florida State University, USA
Jaakko Hollmen	University of Stockholm, Sweden
Arjen Hommersom	Open University of the Netherlands, The Netherlands
Zhengxing Huang	Zhejiang University, China

Guoqian Jiang	Mayo Clinic College of Medicine, USA
Rafał Jóźwiak	Warsaw University of Technology, Poland
Eleni Kaldoudi	Democritus University of Thrace, Greece
Aida Kamisalic	University of Maribor, Slovenia
Haridimos Kondylakis	Institute of Computer Science, FORTH, Greece
Jean B. Lamy	Université Paris 13, France
Pedro Larranaga	Technical University of Madrid, Spain
Giorgio Leonardi	University of Piemonte Orientale, Italy
Helena Lindgren	Umeå University, Sweden
Beatriz Lopez	University of Girona, Spain
Simone Marini	University of Florida, USA
Begoña Martinez-Salvador	University Jaume I, Spain
Ana Mendonça	University of Porto, Portugal
Silvia Miksch	Vienna University of Technology, Austria
Ioanna Miliou	Stockholm University, Sweden
Alina Miron	Brunel University London, UK
Laura Moss	University of Aberdeen, UK
Fleur Mougin	Université de Bordeaux, France
Loris Nanni	University of Padua, Italy
Østein Nytrø	Norwegian University of Science and Technology, Norway
Panagiotis Papapetrou	Stockholm University, Sweden
Pedro Pereira Rodrigues	University of Porto, Portugal
Luca Piovesan	DISIT, University of Piemonte Orientale, Italy
Jedrzej Potoniec	Poznan University of Technology, Poland
Lisiane Pruinelli	University of Minnesota, USA
Cédric Pruski	Luxembourg Institute of Science and Technology, Luxemburg
Nadav Rappoport	Ben-Gurion University of the Negev, Israel
Aleksander Sadikov	University of Ljubljana, Slovenia
Abeed Sarker	Emory University, USA
Isabel Sassoon	Brunel University London, UK
Michael Ignaz Schumacher	University of Applied Sciences Western Switzerland, Switzerland
Erez Shalom	Ben-Gurion University, Israel
Marco Spruit	Leiden University, The Netherlands
Manuel Striani	University of Piemonte Orientale, Italy
Samson Tu	Stanford University, USA
Ryan Urbanowicz	University of Pennsylvania, USA
Iacopo Vagliano	Amsterdam Universitair Medische Centra, The Netherlands
Alfredo Vellido	Technical University of Catalonia, Spain

Maria-Esther Vidal   Leibniz Information Centre for Science and
                             Technology University Library, Germany
Dongwen Wang   Arizona State University, USA
Jens Weber   University of Victoria, Canada
Jiayu Zhou   Michigan State University, USA
Pierre Zweigenbaum   Université Paris-Saclay, CNRS, France

External Reviewers

Agnieszka Onisko   Jana Fragemann
Alessandro Guazzo   Jens Kleesiek
Alexander Brehmer   Jianning Li
Andrea De Gobbis   Erica Tavazzi
Aref Smiley   Marcel Haas
Canlin Zhang   Marta Malavolta
Daniela Ferreira-Santo   Matthew Vowels
Enrico Longato   Natan Lubman
Giulia Cesaro   Niko Lukac
Grega Vrbančič   Sara Korat
Hamidreza Sadeghsalehi   Shubo Tian
Isotta Trescato   Vassilis Kilintzis

Sponsor

We would like to thank Springer for sponsoring the conference, and University of Maribor, Faculty of Health Sciences for all the support in organizing it.

Contents

Machine Learning and Deep Learning

Explainability and Transfer Learning

Natural Language Processing

Image Analysis and Signal Analysis

Data Analysis and Statistical Models

Knowledge Representation and Decision Support

Machine Learning and Deep Learning

Survival Hierarchical Agglomerative Clustering: A Semi-Supervised Clustering Method Incorporating Survival Data

Alexander Lacki[(✉)] and Antonio Martinez-Millana

ITACA Institute, Universitat Politecnica de Valencia, Valencia, Spain
alacki@upvnet.upv.es

Abstract. Heterogeneity in patient populations presents a significant challenge for healthcare professionals, as different sub-populations may require individualized therapeutic approaches. To address this issue, clustering algorithms are often employed that identify patient groups with homogeneous characteristics. Clustering algorithms are mainly unsupervised, resulting in clusters that are biologically meaningful, but not necessarily correlated with a clinical or therapeutical outcome of interest.

In this study we introduce survival hierarchical agglomerative clustering (S-HAC), a novel semi-supervised clustering method that extends the popular hierarchical agglomerative clustering algorithm. Our approach makes use of both patients' descriptive variables and survival times to form clusters that are homogeneous in the descriptive space and include cohesive survival times.

In a benchmark study, S-HAC outperformed several existing semi-supervised clustering algorithms. The algorithm was also evaluated on a critical care database of atrial fibrillation management, where it identified clusters that could readily be mapped to existing knowledge of treatment effects such as contraindications and adverse events following drug exposures. These results demonstrate the effectiveness of the algorithm in identifying clinically relevant sub-populations within heterogeneous patient cohorts.

S-HAC represents an attractive clustering method for the biomedical domain due to its interpretability and computational simplicity. Its application to different patient cohorts may enable healthcare professionals to tailor treatments and more effectively meet the needs of individual patients. We believe that this approach has the potential to greatly improve patient outcomes and enhance the efficiency of healthcare delivery.

Keywords: Semi-Supervised Clustering · Cluster Analysis · Survival Analysis · Phenotype Classification · Atrial Fibrillation

1 Introduction

Epidemiological research provides insights that augment and refine clinical decision-making processes. Studies often report generalized treatment effects on

J. M. Juarez et al. (Eds.): AIME 2023, LNAI 13897, pp. 3–12, 2023.
https://doi.org/10.1007/978-3-031-34344-5_1

a population, and, on occasion, on selected subgroups. Given the substantial heterogeneity within patient populations, it is not uncommon for a "heterogeneity of treatment effect" to manifest in analyses, with specific sub-groups of patients responding differently to evaluated treatments [11]. Such heterogeneous populations are increasingly analyzed using clustering algorithms which identify patient subgroups within cohorts.

Traditionally, clustering algorithms, such as the commonly employed K-Means and hierarchical agglomerative clustering (HAC), are unsupervised in nature, and only utilize patients' descriptive variables, not taking into account survival data. As a result, there is no assurance that the identified clusters correlate to survival times or provide clinical utility [3]. To address this limitation, previous studies have introduced semi-supervised clustering (SSC) algorithms that incorporate survival data into the clustering process. The objective of such algorithms is to identify clusters which are cohesive in the descriptive space, while also being predictive of a clinical or therapeutic outcome [3,15].

Clustering utilizing survival data remains relatively under-explored [2,6,15]. Early approaches involved the use of linear models for the identification of predictive variables, which are then used for clustering in unsupervised clustering algorithms [3,9]. More recent works [6,15] propose deep learning to form cohesive and predictive clusters. However, recent advancements in survival clustering have primarily been driven by the implementation of complex parametric models. This has led to increased computational complexity, and a reduction in algorithmic transparency and interpretability [18].

In this study, we introduce survival hierarchical agglomerative clustering (S-HAC), an extension to the popular HAC algorithm. S-HAC forms clusters using both, descriptive variables and survival data, identifying phenotypes with cohesive descriptive variables and distinct survival times. We compare S-HAC's performance to state-of-the-art SSC on several publicly available datasets, and apply it to a critical care cohort of patients with atrial fibrillation (AF), highlighting its ability to capture identifiable phenotypes with treatment effects corresponding to existing knowledge of adverse effects and contraindications of treatments.

2   Materials and Methods

2.1   Hierarchical Survival Agglomeration

Survival data often comes in triples (X_i, T_i, E_i), where X_i are a patient's descriptive covariates, T_i is the observation time, and E_i is the event indicator (1 if the event was observed, and 0 if the patient's outcome is censored). Unlike the original implementation of HAC, which exclusively uses the descriptive covariates X_i to compute the dissimilarity between observations, we propose the inclusion of the observation durations T_i and event indicators E_i into the dissimilarity computation as well as into a stopping criterion. The proposed method consists of several steps, which we introduce with help of a simulated example:

Consider a population of 1,250 patients described by two covariates, sampled from a bivariate Gaussian distribution with $\mu = (0.0, 0.0)$ and $\sigma = (1.0, 1.0)$.

Along the x-axis, the population is split into three phenotypes characterized by different survival distributions as depicted in Fig. 1A and B.

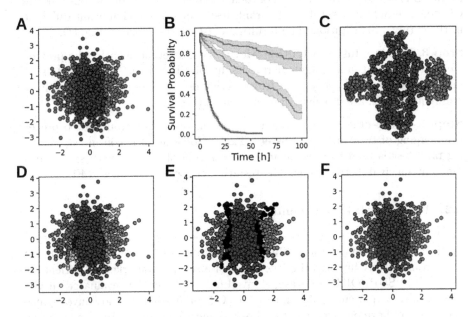

Fig. 1. Visualization of the clustering process. A) Simulated bivariate Gaussian distribution of subjects with three different phenotypes. B) Corresponding Kaplan-Meier curves for each phenotype. C) T-distributed stochastic neighbor embedding of the hybrid pairwise distance matrix. D) Clusters obtained from the pre-clustering step. E) Identified core clusters, with small clusters dissolved into singletons (black). F) Final clustering result.

Step 1 - Distance Matrix Computation: We include survival times and event indicators by representing each subject with a set of survival observations from its m nearest neighbors. That is, a given patient i can be represented by the set of partially censored survival times and event indicators $\{(t_{i,1}, e_{i,1}), (t_{i,2}, e_{i,2}), ..., (t_{i,m}, e_{i,m})\}$. Given two such sets of survival times, and event indicators, two patients i and j can be compared using a sample weighted log-rank test statistic $Z_{i,j}$. To preserve similarities in the covariate space, we define a distance matrix as a weighted sum of survival distance $Z_{i,j}$, and covariate distance $D_{i,j}$ (in the simulated example euclidean, even though other distance metrics can be used). The hybrid distance between the observations can be expressed as $H_{i,j} = \alpha * Z_{i,j} + (1 - \alpha) * D_{i,j}$. The effect of including the survival distance into the distance matrix computation is visualized in a t-distributed stochastic neighbor embedding (T-SNE) [14] of the resulting distance matrix (Fig. 1C). The distribution of subjects is no longer a 2-dimensional Gaussian, but three groups emerge in the form of high-density areas, corresponding to the simulated phenotypes.

Step 2 - Pre-clustering: The hybrid distance matrix H, is used to perform hierarchical agglomeration. During the agglomeration, any two clusters indicated for merging are compared in terms of their survival by means of the log-rank test. Clusters are only merged if the log-rank test shows a non-significant difference in the survival distribution. The resulting clusters are shown in Fig. 1D.

Step 3 - Core Cluster Identification: The clusters formed by the previous steps represent an under-agglomerated solution. To remove noisy clusters, we introduce a *minimum cluster size*. Clusters that do not meet the minimum cluster size requirement, are dissolved into singletons (Fig. 1E).

Step 4 - Post-clustering: A cannot-link constraint is introduced to prevent the merging of two singleton clusters. The agglomeration is continued, and the log-rank test is used to prevent an agglomeration of clusters with significantly different survival distributions. The resulting clusters are shown in Fig. 1F.

2.2 Benchmark Experiments

We benchmark S-HAC against previously proposed SSC algorithms using survival data. Three openly accessible real-world datasets are used: i) FLCHAIN, data from a study investigating the effect of a serum biomarker on patient survival [8], b) SUPPORT, a dataset derived from a study that evaluated survival in critically ill patients [13], and c) GBSG2, data from a study investigating the effects of hormone treatment on breast cancer recurrence [19]. The datasets' characteristics are shown in Table 2 in the appendix.

S-HAC is compared to HAC, SSC-Bair [3], and SSC-Gaynor [9]. All algorithms are tested using a 10-fold cross validation, and compared in terms of their log-rank scores and concordance indices. For each algorithm a clustering is performed on the training set, and Kaplan-Meier estimators are fitted to the identified clusters. The median survival time is used as a risk score in the concordance index computation. The log-rank test statistic is computed based on empirical distribution of subjects in the validation sets. SSC-Bair performs cluster predictions using a nearest centroid method. For the other algoritms cluster predictions are based on the nearest neighbor in the training set. For the algorithms HAC, SSC-Bair, and SSC-Gaynor, we evaluate k={2, 3, 4, 5, 6, 7, 8}, and select the best performing model. For S-HAC we perform a parameter sweep with the parameters shown in Table 3 in the appendix.

2.3 Case Study: Atrial Fibrillation Phenotypes

To qualitatively assess S-HAC, we apply it to a dataset of patients admitted to the intensive care unit [12], and evaluate the treatment effects of medical interventions in patients with AF. In short, patients are observed to have an AF episode and are exposed to an antiarrhythmic agent. Generate clusters based on the time until sinus rhythm is restored, the time until heart rate is controlled, and in-hospital mortality. Patients are divided into treatment groups based on an administration of beta blockers (BBs), potassium channel blockers (PCBs), calcium channel blockers (CCBs), and magnesium sulphate (MgS). We perform a parameter sweep to identify the the best set of hyperparameters using 10-fold cross-validation using the parameter combinations shown in Table 3 in the appendix. Treatment effects are evaluated in an intention-to-treat fashion using an inverse probability of treatment weighting and approximated using exponential decay models. We cluster patients and visualize the obtained phenotypes using T-SNE incorporating the cluster assignments as variables into the optimization to encourage cluster formation. We visualize treatment effects on the obtained embedding, and discuss the results in the context of existing literature.

3   Results and Discussion

3.1   Benchmark Experiments

The result of the benchmark experiments is presented in Table 1. S-HAC universally outperforms HAC highlighting the importance of including survival data into the clustering process, and highlighting the underlying problem of unsupervised algorithms in the identification of patient phenotypes. While HAC attempts to identify patient groups with similar covariates, the inclusion of survival data forms clusters with divergent survival times. This effect results in a higher concordance index, as well as a higher log-rank test statistic across all employed benchmark datasets.

Table 1. Benchmark metrics: Concordance index and log-rank test statistic with standard errors in parentheses

Dataset	Algorithm	Concordance Index	Log-Rank Statistic
FLCHAIN	HAC	0.676 (0.074)	27.9 (26.2)
	SSC-Bair	0.721 (0.037)	109 (47.7)
	SSC-Gaynor	0.653 (0.080)	52.7 (28.4)
	S-HAC (proposed)	**0.771 (0.014)**	**396 (81.9)**
SUPPORT	HAC	0.612 (0.008)	74.5 (18.9)
	SSC-Bair	0.614 (0.022)	89.0 (51.5)
	SSC-Gaynor	0.595 (0.016)	71.1 (23.6)
	S-HAC (proposed)	**0.776 (0.011)**	**648 (125)**
GBSG2	HAC	0.556 (0.059)	11.1 (9.33)
	SSC-Bair	0.544 (0.085)	5.52 (5.49)
	SSC-Gaynor	0.566 (0.080)	7.86 (3.81)
	S-HAC (proposed)	**0.602 (0.058)**	**12.1 (7.42)**

Compared to SSC-Bair and SSC-Gaynor, S-HAC demonstrated superior performance across all employed datasets. The concordance index and log-rank test statistic were found to be the highest for S-HAC. Notably, neither SSC-Bair nor SSC-Gaynor consistently outperformed unsupervised HAC on all datasets. A similar observation has been made by [6], who, among others, compared SSC-Bair to unsupervised K-Means. This observation may possibly be attributed to both algorithms' assumption of linear independence between covariates caused by the use of a linear Cox model, as well as inaccurate decision boundaries due to the use of the K-Means algorithm for cluster definition. Neither of these limitations holds true for S-HAC, resulting in superior test metrics.

3.2   Case Study: Atrial Fibrillation Phenotypes

S-HAC yielded a total of 26 patient clusters in the AF cohort, which are visualized in Fig. 2. Given the large amount of information generated during this process, we limit the evaluation to adjusted daily mortality rates across the different treatment groups (Fig. 3).

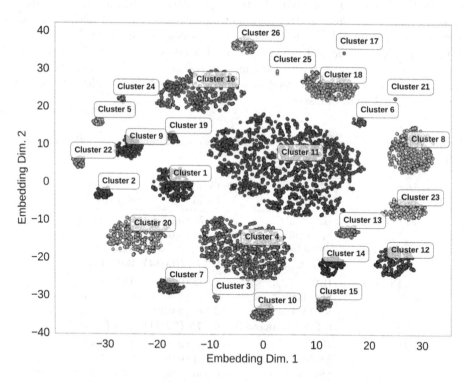

Fig. 2. T-distributed stochastic neighbor embedding of the obtained clusters

The use of PCBs has is associated with adverse effects on thyroid function, and is contraindicated in patients with thyroid disorders [4]. While patients in the PCB group generally had low mortality rates, patients in cluster 15 were observed to have the highest mortality rates when exposed to PCBs. Cluster 15 is primarily characterized by the highest rate of thyroid disorders (90.5%). Notably, this cluster was predominantly composed of female patients (88.0%), which is in line with the prevalence of thyroid disorders being approximately 10 times more common in women than in men [10]. In a similar fashion, PCB exposure has been associated with an increased risk of acute respiratory distress and mortality following the administration of PCBs [1], which is reflected in cluster 13 having the highest mortality when exposed to PCBs.

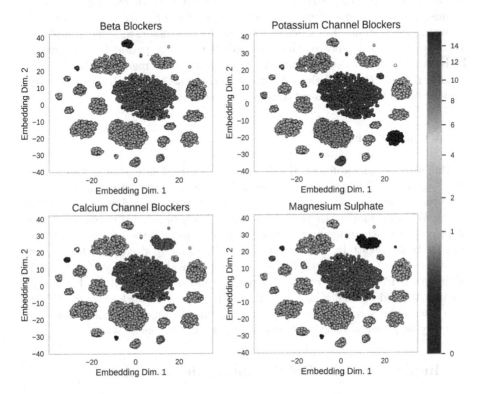

Fig. 3. Adjusted daily mortality rates for different treatment groups

MgS has been associated with reduced incidence of AF following cardiac surgery [7], and accelerated cardioversion rates as well as heart rate control rates [16]. The administration of MgS is, however, contraindicated in patients with severe renal failure, and, accordingly, shows an increased mortality in clusters characterized by renal failure (clusters 2, 7, 10, 16, 24, 26). A reduced table of cluster characteristics is presented in Table 4 in the appendix.

Within the case study, the phenotypes and treatment effects identified through the use of S-HAC align with existing knowledge on treatment effects in patients with AF. This lends support to the conclusion that, in addition to favorable quantitative performance, clusters formed by S-HAC are clinically useful. While several clusters correspond to existing clinical knowledge, the extent to which further results may provide insight into treatment effects, drive hypothesis generation, or provide a basis for clinical decision-making remains to be determined.

3.3 Limitations

While we believe that S-HAC is a valuable addition to existing SSC algorithms, it also has limitations. The requirement of computing pairwise distances between observations limits scalability due to the quadratic time complexity. While this obstacle may partially be addressed using paralellization, exceedingly large datasets may require the use of alternative methods. Additionally, S-HAC requires the definition of three hyperparameters, which results in additional computational overhead during hyperparameter tuning.

4   Conclusions

In this work we introduced S-HAC, a semi-supervised extension to the popular HAC algorithm. The proposed extension includes survival data into the clustering process, and forms clusters with cohesive descriptive variables and survival times. We benchmarked its performance in a study on several publicly available datasets, demonstrating superior performance to state of the art clustering algorithms. In a case study, we showed that S-HAC identified patient clusters with treatment effects that correspond to existing clinical knowledge. These results highlight S-HACs ability to identify patient phenotypes which are biologically meaningful and clinically useful. Further, the presented work supports the use of a phenotype stratification approach to support clinical decision-making by quantifying patients' treatment responses prior to treatment selection.

5   Implementation Details and Code Availability

To decrease computational time during the clustering process, we used the implementation of HAC from [17], and implemented the nearest neighbor chaining algorithm [5]. The implementation of S-HAC, together with the simulated example are available in a public repository: https://github.com/alexander-lacki/Survival-Hierarchical-Agglomerative-Clustering

Acknowledgements. This project has received funding from the European Union's Horizon 2020 research and innovation programme under the Marie Sklodowska-Curie grant agreement No 860974. This publication reflects only the authors' view, and the funding agencies are not responsible for any use that maybe made of the information it contains.

Appendix

Table 2. Dataset Characteristics

Dataset	No. of Subjects	Events (%)	No. of Covariates	Missing (%)	T_{max}
FLCHAIN	7,894	27.5	26	2.10	5,215
SUPPORT	9,105	68.1	59	12.6	2,029
GBSG2	686	43.6	10	0.00	2,659

Table 3. Tested hyperparameters for the benchmark and case studies

	Tested Values	
Hyperparameter	Benchmark Study	Case Study
No. of Neighbors	30, 60, 90, 120	20, 30, 45, 60, 90, 120, 160, 240, 320, 400
Min. Cluster Size	30, 50, 70, 90	20, 30, 50, 80, 160
Significance Level	0.05, 0.10, 0.20	0.05, 0.10, 0.20

Table 4. Selected cluster characteristics presented as counts and percentages

Variable	Clusters																									
	1	2	3	4	5	6	7	8	9	10	11	12	13	14	15	16	17	18	19	20	21	22	23	24	25	26
Thyroid Disorder	16 (3.82)	19 (13.8)	14 (29.8)	167 (14.0)	8 (10.7)	15 (20.6)	20 (10.6)	57 (9.93)	30 (12.1)	29 (15.2)	214 (8.21)	34 (8.88)	16 (10.5)	43 (18.9)	124 (90.5)	41 (5.03)	9 (31.0)	31 (7.08)	45 (42.5)	50 (8.26)	3 (7.32)	5 (4.35)	46 (14.1)	2 (3.28)	3 (10.7)	10 (5.41)
Renal Failure	107 (25.5)	114 (82.6)	4 (8.51)	695 (58.3)	25 (33.3)	23 (31.5)	139 (73.5)	504 (87.8)	89 (35.9)	51 (26.7)	243 (9.33)	21 (5.48)	135 (88.8)	168 (74.0)	12 (8.76)	35 (4.29)	19 (65.5)	229 (52.3)	40 (37.7)	409 (67.6)	31 (75.6)	10 (8.70)	66 (20.2)	50 (82.0)	24 (85.7)	18 (9.73)
Respiratory Failure	102 (24.3)	25 (18.1)	2 (4.26)	152 (12.7)	10 (13.3)	36 (49.3)	108 (57.1)	227 (39.6)	72 (29.0)	30 (15.7)	71 (2.73)	40 (10.4)	139 (91.5)	58 (25.6)	23 (16.8)	16 (1.96)	5 (17.2)	48 (11.0)	9 (8.49)	68 (11.2)	7 (17.1)	112 (97.4)	60 (18.4)	5 (8.20)	5 (17.9)	20 (10.8)
Male Sex	402 (95.9)	59 (42.8)	1 (2.13)	650 (54.5)	70 (93.3)	14 (19.2)	107 (56.6)	316 (55.1)	45 (18.2)	43 (22.5)	1622 (62.3)	215 (56.1)	108 (71.1)	104 (45.8)	15 (11.0)	264 (32.4)	1 (3.5)	399 (91.1)	14 (13.2)	542 (89.6)	10 (24.4)	71 (61.7)	110 (33.6)	22 (36.1)	17 (60.7)	143 (77.3)

References

1. Ashrafian, H., Davey, P.: Is amiodarone an underrecognized cause of acute respiratory failure in the ICU? Chest **120**, 275–282 (2001)
2. Bair, E.: Semi-supervised clustering methods. Wiley Interdisc. Rev.: Comput. Statist. **5**, 349–361 (2013)
3. Bair, E., Tibshirani, R.: Semi-supervised methods to predict patient survival from gene expression data. PLoS Biol. **2**, e108 (2004)
4. Basaria, S., Cooper, D.S.: Amiodarone and the thyroid. Am. J. Med. **118**, 706–714 (2005)
5. Chambers, J., Murtagh, F.: Multidimensional Clustering Algorithms. Physica-Verlag HD, Compstat lectures (1985)
6. Chapfuwa, P., Li, C., Mehta, N., Carin, L., Henao, R.: Survival cluster analysis. In: Proceedings of the ACM Conference on Health, Inference, and Learning, pp. 60–68 (2020)
7. Davey, M.J., Teubner, D.: A randomized controlled trial of magnesium sulfate, in addition to usual care, for rate control in atrial fibrillation. Annal. Emerg. Med. **45**, 347–353 (2005)
8. Dispenzieri, A., et al.: Use of nonclonal serum immunoglobulin free light chains to predict overall survival in the general population. Mayo Clinic Proceed. **87**, 517–523 (2012)
9. Gaynor, S., Bair, E.: Identification of biologically relevant subtypes via preweighted sparse clustering. CoRR abs/1304.3760 (2013)
10. Gessl, A., Lemmens-Gruber, R., Kautzky-Willer, A.: Thyroid Disorders, pp. 361–386 (2013)
11. Iwashyna, T.J., Burke, J.F., Sussman, J.B., Prescott, H.C., Hayward, R.A., Angus, D.C.: Implications of heterogeneity of treatment effect for reporting and analysis of randomized trials in critical care. Am. J. Resp. Crit. Care Med. **192**, 1045–1051 (2015)
12. Johnson, A.E., et al.: MIMIC-III, a freely accessible critical care database. Sci. Data **3**, 160035 (2016)
13. Knaus, W.A.: The support prognostic model: objective estimates of survival for seriously ill hospitalized adults. Annal. Internal Med. **122**, 191 (1995)
14. van der Maaten, L., Hinton, G.: Visualizing data using t-SNE. J. Mach. Learn. Res. **9**(86), 2579–2605 (2008)
15. Manduchi, L., et al.: A deep variational approach to clustering survival data. In: International Conference on Learning Representations (2022)
16. Miller, S.: Effects of magnesium on atrial fibrillation after cardiac surgery: a meta-analysis. Heart **91**, 618–623 (2005)
17. Müllner, D.: fastcluster: Fast hierarchical, agglomerative clustering routines for R and Python. J. Stat. Softw. **53**(9), 1–18 (2013)
18. Petch, J., Di, S., Nelson, W.: Opening the black box: The promise and limitations of explainable machine learning in cardiology. Canadian J. Cardiol. **38**, 204–213 (2022)
19. Schumacher, M., et al.: Randomized 2 x 2 trial evaluating hormonal treatment and the duration of chemotherapy in node-positive breast cancer patients. German breast cancer study group. J. Clin. Oncol. **12**, 2086–2093 (1994)

Boosted Random Forests for Predicting Treatment Failure of Chemotherapy Regimens

Muhammad Usamah Shahid[✉] [ID] and Muddassar Farooq

CureMD Research, 80 Pine St 21st Floor, New York, NY 10005, USA
{muhammad.usamah,muddassar.farooq}@curemd.com
https://www.curemd.com/

Abstract. Cancer patients may undergo lengthy and painful chemotherapy treatments, comprising several successive regimens or plans. Treatment inefficacy and other adverse events can lead to discontinuation (or failure) of these plans, or prematurely changing them, which results in a significant amount of physical, financial, and emotional toxicity to the patients and their families. In this work, we build treatment failure models based on the Real World Evidence (RWE) gathered from patients' profiles available in our oncology EMR/EHR system. We also describe our feature engineering pipeline, experimental methods, and valuable insights obtained about treatment failures from trained models. We report our findings on five primary cancer types with the most frequent treatment failures (or discontinuations) to build unique and novel feature vectors from the clinical notes, diagnoses, and medications that are available in our oncology EMR. After following a novel three axes - performance, complexity, and explainability - design exploration framework, boosted random forests are selected because they provide a baseline accuracy of 80% and an F1 score of 75%, with reduced model complexity, thus making them more interpretable to and usable by oncologists.

Keywords: Boosting · Random Forests · Chemotherapy · Treatment Failure · Feature Engineering

1 Introduction

Cancer patients undergo a lengthy and painful chemotherapy treatment that creates emotional toxicity, resulting in significant stress debt for the patients and their families. Unsuitable chemotherapy plans result in unpredictable and debilitating side effects that may lead to the death of patients in a worst-case scenario. When oncologists detect an ineffective plan, it is discontinued and often replaced with a new one; adding financial toxicity to patients, their families, and the healthcare system. To enhance patient-centered quality of care, an assistive tool – MedicalMind – for oncologists is developed that can predict, at the time of chemotherapy selection, the likelihood of failure of a chemotherapy plan based on the real-world evidence (RWE) that is collected from the historical EMR data. It also generates clinically relevant explanations that will empower oncologists

© The Author(s), under exclusive license to Springer Nature Switzerland AG 2023
J. M. Juarez et al. (Eds.): AIME 2023, LNAI 13897, pp. 13–24, 2023.
https://doi.org/10.1007/978-3-031-34344-5_2

to take informed decisions to not only reduce the number of preventable deaths due to the side effects of chemotherapy but also save patients and their families from emotional and financial toxicity when chemotherapy plans are changed.

Most of the patients' information - MRI scans, pet scans, pathology slides and genomics data - used in clinical trials or focused studies, is not available in the EMR systems of oncology providers in practice, since their major objective is to record clinical interventions - chemotherapy plans - and track their outcomes to gain reliable and valuable insights about their relative efficacy. Therefore, we are constrained to using the structured fields of EMR and providers' notes only, which poses unique challenges – noise, missing data fields, class imbalance – including providers' habit of not providing useful information like tumor size, toxicity, and staging even in the structured fields.

Historically, mathematical models based on survival analysis are used to predict treatment failure by modeling tumor dynamics [10]. Their well-known shortcomings are broad generalizations that are an outcome of often simple and linear relations that are asserted by the models that are developed from clinical studies of small cohort sizes. Within the context of EHR data, text-based models including Bag-of-Words representation [4] and Entropy-based categorization [7] are used. Sequential approaches that utilize states and events to find a correlation between two sequences [11] are also used. Generally speaking, RNN-based models outperform other sequential models, but their complexity and black-box nature make them unsuitable for use in medicine, as oncologists are clueless about the logic and reasoning of a decision.

Other direction of research focuses on predicting chemotherapy outcomes. The study presented in [11] uses toxicity as an outcome to predict chemotherapy treatment failure among breast cancer patients only. In [2], an instance-specific Bayesian model is reported for predicting survival in lung cancer patients. Tree ensemble methods have previously been used by researchers to predict treatment outcomes in cancer patients. The authors of [1] accurately predict 24-month progression-free survival in cervical cancer patients by using random forests; while the authors of [5] demonstrate the success of Xgboost in classifying treatment outcomes of pediatric patients suffering from acute lymphoblastic leukemia. A study by [9] uses Adaboost to predict chemotherapy response for breast cancer patients by using biomarkers from CT images.

In contrast, we work with simpler representations of oncology EMR data, where spatial snapshots of a patient's health and treatment profiles are taken at relevant time instances to look for hidden patterns. Therefore, after following a smart design exploration framework and extensive empirical evaluations, we discover that combining boosting and bagging of decision trees by using boosted random forests [8] is the optimal choice for treatment failure use-case in our multi-objective optimization landscape across three dimensions - performance, complexity, and explainability. To the best of our knowledge, our treatment failure models are novel and unique for the oncology use case, as they not only are relatively accurate but also explain the reasons for their inference to oncologists. Our models provide a baseline accuracy of 80% and an F1 score of 75% for the five most prevalent cancer types.

2   Real World Evidence and Oncology Data Overview

The medical records of cancer patients are collected from our Oncology EMR application. HIPAA guidelines are followed to anonymize and deidentify patient records to ensure the security and privacy of personally identifiable information. The registered patients have active treatments from 2015 to 2022. From a total of 21212 patients treated with chemotherapies, as shown in Table 1, we select the patients of top 5 cancer types that have the most frequent plan discontinuations.

Table 1. The total number of patients and patients with discontinued treatments

ICD10 code	Description	Patient Count	Treatment Failures
C18	Malignant neoplasm of colon	1034	332
C34	Malignant neoplasm of bronchia and lung	1547	489
C50	Malignant neoplasm of breast	2184	617
C61	Malignant neoplasm of prostate	1074	361
C90	Plasmocytoma and malignant plasma cell neoplasms	866	249

Table 2. Most frequent reasons for discontinuation of chemotherapy plans

Reason	% failures
Change in therapy (drugs, doses or schedule)	30.4
Progression of disease	18.0
Adverse effects and other health concerns	15.9
No comment	11.1
Ambiguous (e.g. "MD order")	10.3
Death or hospice	2.0

A chemotherapy plan comprises of multiple medications, dosages, and administration schedules. In our data, on average, patients are prescribed 3 plans over the course of their treatment, and approximately 40% of the patients undergo only one chemotherapy plan. Overall 16-18% of patients have at least one discontinued plan, and the most frequent reasons for discontinuations are tabulated in Table 2. It is clear that 34% of the discontinuations result due to poor treatment efficacy (disease progression or adverse effects) leading to the worst case outcome in 2% of the cases: death or hospice. We now discuss our design exploration framework, beginning with feature engineering, that leads to an optimal treatment failure model in terms of its performance, complexity, and explainability.

3   Feature Engineering

We aim for a minimalist and relevant feature set that captures dynamic changes in the diseases and treatment journeys of patients. Our feature engineering pipeline comprises of ingesting raw data of medical profiles of patients and then running autonomous transformation pipelines to convert it into processed information to enable feature engineering. Medical profiles of patients are grouped by cancer type which allows the creation of cancer-specific, unique features.

Feature Vector. For each patient, the data is ingested across multiple processed tables and arranged in chronological order. Chemotherapy plans are inferred from the orders by reviewing the medications present in them and by analyzing associated comments about therapy details and administration schedules. The designed features only consider a patient's profile to the date when the plan is being prescribed by oncologists. We only incorporate age and gender from a patient's demographics, as other information was either anonymized beforehand or scarcely available.

Table 3. Distribution of normal (successful) and abnormal (discontinued) plans)

ICD10 Code	Normal	Abnormal
C18	1024	706
C34	1466	945
C50	2130	1345
C61	914	685
C90	796	593

Medications. The EMR tracks, at length, different drugs and their doses that are planned, ordered, administered (on-site infusion), and/or prescribed for use at home. Medications are identified by their name and a universal National Drug Codes (NDC) system. To reduce the dimensionality of our drug dataset, we use the Generic Product Identifier (GPI) [6] - a 14-character hierarchical classification system that identifies drugs from their primary therapeutic use down to the unique interchangeable product. Depending on relevance to treatment, medications are abstracted to drug type (two digits), class (four digits) or sub-class (six digits). For a plan, each medication feature will have two occurrences:(1) total planned dosage if the proposed regimen were to be followed; and (2) total dose administered to a patient in the last six months to analyze the impact of a treatment or its side effects.

Diagnoses. International Classification of Diseases (ICD10) codes are used and collapsed into thirty different Elixhauser comorbidity groups [3]. Three types of features are then extracted from a list of the patient's active diagnoses at the time of selecting a plan: (1) certain ICD codes e.g. R97.2 - elevated levels of prostate-specific antigen, are used directly as features by searching whether they are present or not in the patient's records; (2) important comorbidity conditions e.g. renal failure are marked present or absent from the Elixhauser comorbidity groups derived from the ICD10 codes; and (3) Elixhauser readmission score, computed at the time of selecting a chemotherapy plan, and how it has changed in the last six months.

Vitals and Lab Results. At the time of selecting a plan, the last reading of a desired vital or lab test is used in the feature vector. Exponentially weighted moving averages are used for imputation and noise removal, complemented by substituting population means or clinically normal values for the patients with no relevant information stored in EMR. Currently, only body surface area and serum creatinine values are used as features, as they are recorded for approximately all of the patients in the EMR system.

Table 4. F1 scores (%) on the validation dataset, with top 3 highlighted

Cancer type	Log reg	Elastic net	KNN (k = 5)	SVM linear	SVM rbf	Tree (D = 10)	Naive Bayes	Ada-boost	Random forest	Xgb-oost	Boosted forest
C18	54.2	58.3	52.6	52.4	59.6	65.7	59.6	64.3	**75.3**	69.6	**76.4**
C34	55.3	60.9	52.0	50.0	55.4	64.6	54.3	62.5	**76.3**	**76.2**	**77.5**
C50	59.0	60.5	61.5	52.8	59.9	69.0	19.6	66.0	**81.9**	79.0	**82.4**
C61	49.7	57.1	57.7	15.5	51.1	60.8	61.8	63.1	**72.0**	**72.4**	**72.9**
C90	65.5	60.6	59.1	49.3	62.1	77.6	34.5	77.4	**84.6**	80.9	**85.9**

Notes. Critical information like staging, tumor information (size, grade, risk), ECOG or Karnofsky performance values, and biomarkers that are needed to select the type of chemotherapy are either completely missing or scarcely provided in the structured fields by oncologists. Luckily, they report most of these features in the clinical notes. We built a novel smart annotation engine that automatically annotates the oncologists' notes and subsequently extracts desired features from the annotated text. The detailed description of feature extraction from unstructured oncology notes is beyond the scope of this paper. We curate cancer-specific and relevant feature-set by filtering redundant and non-informative features.

Sampling. Plans are sampled for each patient before computing feature vectors or splitting the data for training and evaluation. The strategy is chosen to minimize class imbalance and remove bias in features. After ensuring the plan is prescribed for a chosen diagnosis, all unique chemotherapy plans, excluding the first one, are sampled for patients with no discontinuations. Abnormal samples are obtained by sampling discontinued plans and pairing them with the last unique plan prior to the discontinuation if any. The number of samples in the two classes - abnormal and normal - is reported in Table 3.

4   An Empirical Design Exploration Framework

Our design exploration framework consists of three steps:(1) select well-known machine learning classifiers from each learning paradigm and shortlist top 3 models that perform the best on our oncology dataset; (2) deep dive into the shortlisted models by exploring the best hyperparameter settings and select the best model; (3) build model insights for explainability to oncologists.

Step 1: Shortlist Top 3 Models. We select the following well-known machine learning classifiers to begin with for doing a comprehensive performance evaluation:(1) logistic regression, first with no penalties and then with combined l1 and l2 penalties (Elastic Net); (2) k Nearest neighbors (KNN); (3) SVM with linear and RBF kernels; (4) single decision tree, with a depth cut-off of 10; (5) Naive Bayes classifier; (6) Adaboost classifier with 100 decision stumps; and (7) the three tree ensemble methods – random forest, Xgboost, boosted forest – with 500 trees each.

Separate models are trained for each cancer type, with a three-way data split: 75% for training, 15% for validation and 10% for testing. The splits are random, but fixed, for each model over multiple iterations of training; keeping class distributions constant in all splits. To account for the randomness of some of the models, multiple instances of each model are trained and the model that performs the best on the validation set is selected. As is evident from Table 3, a significant imbalance exists between two classes; therefore, F1 score is more appropriate to benchmark the performance of different classifiers. The results are tabulated in Table 4. It is obvious that random forests (RF), Xgboost (XGB), and boosted forests (BF) - all ensembles of trees - are the best performing models for our treatment failure application.

Table 5. Hyperparameter settings for tree ensemble experiments

Identifier	Total trees	Forests	Trees per forest	Depth cut-off
HP1	500	10	50	None
HP2	100	10	10	None
HP3	100	10	10	10
HP4	50	5	10	10
HP5	25	5	5	10
HP6	10	2	5	10

Step 2: Hyperparameter Settings to Select the Best Model. We provide a description of the hyperparameters that are used for all three ensembles on the validation set, and how the respective boosted forest is constructed in Table 5. Table 6 tabulates the F1 scores of three ensemble methods on the validation set. Please note that the best results are marked in bold. It is obvious that a designer needs to trade performance for complexity and interpretability i.e. as he selects relatively less complex models that have better explainability (HP6 over HP1), the performance degrades from 3–7%. Generally speaking, the results validate the hypothesis that boosted forests outperform other ensemble methods for the majority of hyperparameter settings for all five cancer types on our oncology dataset.

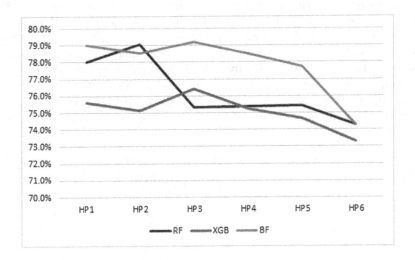

Fig. 1. Average F1 scores of 3 ensemble methods

To better visualize the impact of hyperparameter settings on three ensemble methods, we analyze the average F1 scores for each hyperparameter setting on the validation sets and plot the average in Fig. 1. It is interesting to note that once the tree depth is capped, the performance of random Forests is significantly degraded, whereas the performance of boosted ensembles slightly increase. Boosted forests provide superior performance compared with the other two except for the HP2 configuration. We also notice a common "elbow" on HP5, suggesting that these hyperparameter settings ($N = 25, D = 10$) are the best compromise between model interpretability and performance.

Table 6. F1 scores (%) for 3 ensemble methods on the validation set

Cancer type	HP1			HP2			HP3			HP4			HP5			HP6		
	RF	XGB	BF	RF	XGB	BF	RF	XGB	BF	RF	XGB	BF	RF	XGB	BF	RF	XGB	BF
C18	75.3	69.6	**76.4**	**76.3**	69.3	74.4	75.0	73.4	**75.4**	**75.3**	73.1	75.1	74.2	72.9	**75.5**	72.9	**73.5**	73.2
C34	76.3	76.2	**77.5**	**77.6**	75.5	76.7	72.1	77.0	**77.9**	73.4	75.4	**76.8**	72.7	74.6	**75.2**	69.7	**71.4**	70.0
C50	81.9	79.0	**82.4**	81.8	80.8	**82.1**	78.0	79.0	**82.0**	77.2	78.7	**81.1**	77.9	77.3	**80.1**	77.2	75.8	**75.9**
C61	72.0	72.4	**72.9**	73.8	68.4	**73.9**	70.8	70.1	**73.8**	70.0	68.4	**73.4**	70.3	69.7	**72.8**	69.8	68.7	**71.4**
C90	84.6	80.9	**85.9**	**85.9**	81.6	85.5	80.7	82.6	**87.0**	81.0	80.7	**86.0**	82.1	78.7	**85.1**	**81.8**	77.2	80.9

From Step 2, we select the boosted forests models and evaluate all of their performance metrics - accuracy, precision, recall, specificity, and F1 score - on the test dataset and tabulate the results in Table 7 on all five cancer datasets. It is safer to claim that the selected models maintain a baseline accuracy of 80% and F1 score of 75% to predict treatment failure of chemotherapy plans for five cancer types. Our models generally have much better specificity and precision than recall and this is a general symptom of imbalanced classes. But it also means that false positives are highly unlikely to occur although false negatives may happen, especially in our lung cancer model. The ROC curves in Fig. 2 suggest that explored models are very good at distinguishing between successful and failed chemotherapy plans. All five cancer models have an AUC value of 0.85 or more.

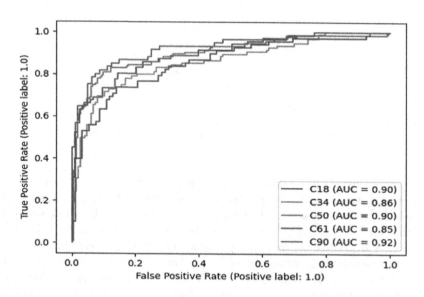

Fig. 2. ROC curves for the boosted forests calculated on the test dataset

Step 3: Model Insights and Explainability. The boosted forest models with the HP5 hyperparameter setting are the optimal models that are selected from Step 2 of our design exploration framework. We now provide model insights and discuss explainability.

Table 7. Results (in %) of the chosen boosted forest model on the test set

Cancer type	Accuracy	Precision	Recall	NPV	Specificity	F1 score	AUROC
C18	83.3	83.9	73.2	83.0	90.3	78.2	89.5
C34	81.0	85.5	62.1	79.2	93.2	72.0	85.8
C50	86.5	86.1	77.8	86.7	92.1	81.7	90.0
C61	80.6	79.4	73.6	81.4	85.9	76.3	85.5
C90	84.8	84.5	81.7	86.6	88.8	83.1	91.9

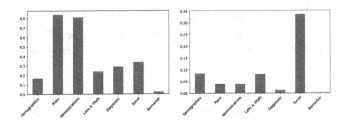

(a) Sum of MAD scores - how important is each category   **(b)** Average MAD scores of each feature in the category

Fig. 3. Feature importance by categories in the selected boosted forest model

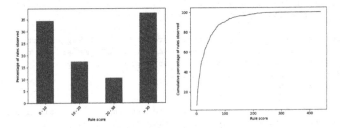

Fig. 4. Trends observed in rule scores of the 50 test samples

Feature Importance. Though we have engineered unique feature sets for each cancer type, still we can discuss important features by grouping them into broad categories as discussed in Sect. 3. We permute feature importance to calculate the mean accuracy decrease (MAD), if the value of a feature is randomly toggled. To compare the categories, we present the sum and the average of the MAD scores of all features belonging to each category, across all five models in Figs. 3a and b. On average, tumor features – tumor size, staging, grade, and risk – are among the most important features. Cumulatively, dosages of planned medications and past administrations, at the time of selecting a plan, have had the greatest impact on the predictions of our boosted forest models. Biomarkers - information extracted from notes about active mutations - make the least contributions to treatment failure decisions, perhaps due to a scarce availability even in the notes.

Explanations of Treatment Failures Models. One logical benefit of using boosted forests is that we can easily combine concepts of classifier weights, the probabilistic output of trees, and forward chaining to generate less number of simpler, yet powerful rules to explain the inference of our models. We use the following 3 step algorithm to shortlist rules:

1. Drop forests whose predictions contradict the prediction of the boosted ensemble model;
2. In the remaining forests, drop trees that predict an opposite outcome than the final prediction made by the model;
3. Rules, extracted by forward chaining on the remaining trees, are sorted by their importance or strength. We define a simple "rule score" as the product of the class probability determined by the leaf nodes of a tree and the residual number of samples at that leaf node.

Using this rule pruning algorithm, oncologists are presented with an ordered set of the most relevant and strong rules, for our treatment failure use case.

```
If: tumor size <=4.3 cm and Elixhauser readmission score < 39 and
    the plan comprises of <= 39.8 g of antimetabolites, <= 0.22 g
    of Opdivo, <= 0.07 g of corticosteroids and <= 1.35 g of anti-
    histamines, and in last six months patient has received
    <= 4.79 g of antimetabolites, <= 0.001 g of Topoisomerase I
    inhibitors and <= 0.021 g corticosteroids
Then Prediction: Treatment will fail
    Rule Score = 140

If: in the last six months patient has received <= 1.8 g of anti-
    metabolites and has not been diagnosed for adverse effect or
    underdosing of antineoplastic/immunosuppressive drugs
Then Prediction: Treatment will succeed
    Rule Score = 69

If: the patient is less than 36 years old and in the last six months
    patient has received > 1.8 g of antimetabolites
Then Prediction: Treatment will fail
    Rule Score: 11
```

Listing 1.1. Example rules from the boosted forest model

We examine 10 random test samples from each cancer type. On average, 4 out of 5 forests and 15 of the 25 individual trees are in agreement with the final prediction of the models. If we analyze the rule scores, only 6 per predictions have had a score above 30 (see Fig. 4). The trends observed in the rules are summarized in Fig. 4. The threshold to shortlist rules based on their importance or strength, computed from the real world evidence (RWE) in the data, may be arbitrarily chosen to meet the requirements of oncologists. The higher the value of the "rule score", the stronger the real world evidence that it represents. Listing 1.1 shows 3 handpicked rules from our models of varying strengths. A simple text matching and replacement strategy can translate feature names and thresholds to a natural language description of a rule that oncologists are accustomed to reading as clinical guidelines.

5   Conclusion

In this paper, we follow a customized design exploration framework to select optimal models for a novel use case of predicting the failure of chemotherapy regimens by using only the data that is available in an outpatient oncology EMR system. Our feature engineering pipeline is discussed to obtain distinctive feature sets for the five most prevalent primary cancers. Our studies demonstrate that boosted forests, which are little investigated in the prior art, provide the optimal models in our multi-objective exploration space of performance, complexity, and interpretability. The selected models provide a baseline accuracy of 85%, F1 score of 75%, and AUC of 0.85 on the test data. Finally, we explain models by discussing feature importance and the strength of shortlisted rules. In the future, we want to validate the system in controlled clinical trials.

References

1. Arezzo, F., et al.: A machine learning tool to predict the response to neoadjuvant chemotherapy in patients with locally advanced cervical cancer. Appl. Sci. **11**(2), 823 (2021)
2. Jabbari, F., Villaruz, L.C., Davis, M., Cooper, G.F.: Lung cancer survival prediction using instance-specific Bayesian networks. In: Michalowski, M., Moskovitch, R. (eds.) AIME 2020. LNCS (LNAI), vol. 12299, pp. 149–159. Springer, Cham (2020). https://doi.org/10.1007/978-3-030-59137-3_14
3. Elixhauser, A., Steiner, C., Harris, D.R., Coffey, R.M.: Comorbidity measures for use with administrative data. Medical Care, pp. 8–27 (1998)
4. French, J., et al.: Identification of patient prescribing predicting cancer diagnosis using boosted decision trees. In: Riaño, D., Wilk, S., ten Teije, A. (eds.) AIME 2019. LNCS (LNAI), vol. 11526, pp. 328–333. Springer, Cham (2019). https://doi.org/10.1007/978-3-030-21642-9_42
5. Kashef, A., Khatibi, T., Mehrvar, A.: Treatment outcome classification of pediatric acute lymphoblastic leukemia patients with clinical and medical data using machine learning: a case study at Mahak hospital. Inform. Med. Unlocked **20**, 100399 (2020)
6. Kluwer, W.: Medi-span Generic Product Identifier (GPI) (2019)
7. Malakouti, S., Hauskrecht, M.: Predicting patient's diagnoses and diagnostic categories from clinical-events in EHR data. In: Riaño, D., Wilk, S., ten Teije, A. (eds.) AIME 2019. LNCS (LNAI), vol. 11526, pp. 125–130. Springer, Cham (2019). https://doi.org/10.1007/978-3-030-21642-9_17
8. Mishina, Y., Murata, R., Yamauchi, Y., Yamashita, T., Fujiyoshi, H.: Boosted random forest. IEICE Trans. Inf. Syst. **98**(9), 1630–1636 (2015)
9. Moghadas-Dastjerdi, H., Sha-E-Tallat, H.R., Sannachi, L., Sadeghi-Naini, A., Czarnota, G.J.: A priori prediction of tumour response to neoadjuvant chemotherapy in breast cancer patients using quantitative CT and machine learning. Sci. Rep. **10**(1), 10936 (2020)

10. Ribba, B., et al.: A review of mixed-effects models of tumor growth and effects of anticancer drug treatment used in population analysis. CPT: Pharmacometr. Syst. Pharmacol. **3**(5), 1–10 (2014)
11. Silvina, A., Bowles, J., Hall, P.: On predicting the outcomes of chemotherapy treatments in breast cancer. In: Riaño, D., Wilk, S., ten Teije, A. (eds.) AIME 2019. LNCS (LNAI), vol. 11526, pp. 180–190. Springer, Cham (2019). https://doi.org/10.1007/978-3-030-21642-9_24

A Binning Approach for Predicting Long-Term Prognosis in Multiple Sclerosis

Robbe D'hondt[1,2]([✉]) [ID], Sinéad Moylett[3] [ID], An Goris[3] [ID],
and Celine Vens[1,2]([✉]) [ID]

[1] Department Public Health and Primary Care, KU Leuven, Kortrijk, Belgium
{robbe.dhondt,celine.vens}@kuleuven.be
[2] Itec, imec Research Group at KU Leuven, Kortrijk, Belgium
[3] Laboratory for Neuroimmunology, KU Leuven, Leuven, Belgium
{sinead.moylett,an.goris}@kuleuven.be

Abstract. Multiple sclerosis is a complex disease with a highly heterogeneous disease course. Early treatment of multiple sclerosis patients could delay or even prevent disease worsening, but selecting the right treatment is difficult due to the heterogeneity. To alleviate this decision-making process, predictions of the long-term prognosis of the individual patient are of interest (especially at diagnosis, when not much is known yet). However, most prognosis studies for multiple sclerosis currently focus on a short-term binary endpoint, answering questions like "will the patient significantly progress in 2 years". In this paper, we present a novel approach that provides a comprehensive perspective on the long-term prognosis of the individual patient, by dividing the years after diagnosis up into bins and predicting the level of disability in each of these bins. Our approach addresses several general issues in observational datasets, such as sporadic measurements at irregular time-intervals, widely varying lengths of follow-up, and unequal number of measurements even for the same follow-up. We evaluated our approach on real-world clinical data from an observational single-center cohort of multiple sclerosis patients in Belgium. On this dataset, a regressor chain of random forests achieved a Pearson correlation of 0.72 between its cross-validated test set predictions and the actual disability measurements assessed by a clinician.

Keywords: Machine learning for multiple sclerosis · Prognosis at diagnosis · Longitudinal data · Regressor chain

1 Introduction

Multiple sclerosis (MS) is a complex autoimmune disease with high patient-to-patient variation. Due to this variation, personalized prognosis predictions are of interest, both to patients as well as to clinicians. A prognosis at diagnosis would aid clinicians in initiating the right treatment (with the right efficacy) as soon as possible, hereby likely delaying or even preventing future disease worsening [4].

The level of (mostly ambulatory) disability in MS is most commonly measured in terms of the *expanded disability status scale* (EDSS) [6], an MS-specific

J. M. Juarez et al. (Eds.): AIME 2023, LNAI 13897, pp. 25–34, 2023.
https://doi.org/10.1007/978-3-031-34344-5_3

disability score ranging from 0 to 10 in steps of 0.5, indicating the impairment to several functional systems. Unless a patient is participating in a research study, years often go by between subsequent visits where an EDSS score is actually measured. The resulting observational datasets can contain sparse and irregularly sampled longitudinal signals, unsuited for traditional trajectory modeling approaches. Furthermore, the length of follow-up can vary widely between patients. Even when the length of follow-up is the same, the number of measurements can differ significantly, making it a challenge to evenly weigh the patients when training a model.

A recent review on the analysis of patient trajectories with artificial intelligence [1] indicates that current trajectory modeling approaches are unsuited for the task at hand. These current approaches are often focused on high-dimensional biomedical signals (e.g., from wearables or physiological measurements) and on using a history of targets or features. In contrast, we want to predict prognosis in terms of a sparse set of EDSS measurements and based only on static information available at diagnosis. Because of these challenges, the problem of patient trajectory modeling for multiple sclerosis is most often rephrased as a binary classification problem, with the aim to distinguish the group of stable or improving patients from the group of *worsening* patients at one specific time point in the future [3,7,8,10]. However, this disregards the temporal information inherently present in the data, and does not give any indication as to the degree of worsening.

In this work, we propose a machine learning approach to provide a long-term prognosis at diagnosis in terms of EDSS scores, based on real-world data. Our approach can incorporate temporal information and is modifiable into any application with sparse and irregularly sampled signals of unequal length. In particular, we recast the problem of patient trajectory modeling into a multi-target regression problem. We consider two training methods: one global method where all targets are predicted simultaneously, and one chaining method where the targets are predicted sequentially using predictions for previous targets as extra input features.

2   Dataset

To develop our approach, we used an observational single-center cohort of 174 multiple sclerosis patients from UZ Leuven, a university hospital in Belgium (approved by Ethics Committee Research UZ/KU Leuven S59940 & S60222). For these patients, the following data was used (all available at diagnosis): clinical features (gender, ethnicity, MS subtype, EDSS at diagnosis), temporal features (age at onset, age at diagnosis, time between onset and diagnosis, year of diagnosis), clinical markers (albumin, IgG index, OCB count+label, WBC count), and biomarkers (GPNMB, CCL18, CHIT1, NfL, sTREM2, YKL40). As expected from real-world data, values are sporadically missing, with 236 values missing out of 3306 values (174 patients × 19 variables) in total, or about 7% of all values missing. More details on the cohort characteristics are given in Table 1.

Table 1. Cohort characteristics of the dataset described in Sect. 2.

Feature	Distribution	Missingness
Gender	63% female, 37% male	0%
Ethnicity	98% Caucasian, 1% African	1%
MS subtype	83% RRMS, 15% PPMS	2%
Average age at onset	34.77 years (std: 12.37 years)	0%
Average age at diagnosis	39.45 years (std: 13.63 years)	0%

RRMS = relapsing-remitting MS, PPMS = primary progressive MS.

In terms of prognosis, each patient has EDSS scores for up to 5.8 years after diagnosis on average (with some patients up to 14 years), giving in total 382 disability measurements (2.2 per patient on average). To be able to estimate the disability for the first years after diagnosis, we assume for each patient an additional visit of EDSS = 0 (no disability) at disease onset. The date of onset is estimated by the attending clinician and is on average 4.7 years before diagnosis. This assumption results in 174 additional EDSS measurements, bringing the total up to 556. Of course, these additional measurements are ignored for all model evaluation purposes.

3   Methodology

In this section, we detail the multi-target binning approach presented in this paper. First, in Sect. 3.1, we explain how the EDSS measurements are converted into a set of targets for the machine learning model. Then, in Sect. 3.2, we describe the class of machine learning models used in this paper. Finally, in Sect. 3.3, we present the evaluation procedure used to evaluate the performance of the models.

3.1   Defining the Target

The main challenge in this context is to transform a set of sparse and sporadically measured signals of unequal length into a meaningful machine learning target. To this end, we divide the prediction window (i.e., the years after diagnosis) up into bins, in our case 10 bins each of width 1 year. Assuming that the EDSS progresses linearly in between visits, we can then compute (for each patient individually) the average disability attained in the extent of each bin (with missing values for the bins after the last visit for this patient). The collection of these 10 averages (one for each bin) then forms the target. This procedure is graphically represented in Fig. 1. For the dataset described in Sect. 2, this results in 1054 non-missing targets, obtained from just the 382 available EDSS measurements (and the 174 added visits). The remaining 686 missing targets represent a missingness of 39% out of the 1740 targets in total (1740 = 174 patients × 10 bins).

There are several advantages to this approach of defining the target. First of all, it can handle sparse and sporadically measured signals, such as EDSS

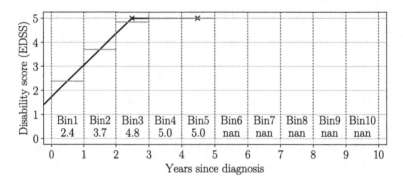

Fig. 1. Construction of the target for one particular patient as a piecewise average per bin. The crosses represent the actual visits. In bins 1, 2, 4, and 5, the target is simply the average value that a straight line obtained in this bin. In bin 3, a proportional average of 2 line segments must be taken, with as weights the fraction of the bin covered by each line segment. Finally, in bins 6 to 10, we have no more observations, so the target is set to missing.

measurements. Secondly, two patients with an equal length of follow-up will be weighted equally, even if there is a large difference in the number of measurements. This avoids a bias towards patients with many measurements, who might just be observed more intensely due to their disease course (i.e., informative sampling). Thirdly, this method can deal with patients having an unequal length of follow-up, as we make no extrapolations beyond what is actually measured. To intuitively see these advantages, a visual comparison of the patient trajectories and their binned target counterparts is shown in Fig. 2. As can be seen from this figure, the main disadvantage of this resampling approach is that we inevitably lose some information by averaging out closeby measurements. Therefore, the approach is most appropriate when the visits are sparse.

3.2   Machine Learning Model

In this work, we focused on (ensemble) tree-based models, which are able to capture complex relationships between the input and output variables while still offering some inherent transparency in the model predictions. Tree-based models recursively split the data based on a heuristic criterion. In particular, here we work with (random forests of) predictive clustering trees [2], a variant of regular decision trees which use a variance-based splitting criterion and an accompanying stopping criterion (namely an F-test assessing the statistical significance of a split). These models can inherently deal with missing values, both in input space (by proportionally distributing patients with missing values among both children of a split) and in output space (by forming leaf prototypes only based on observed targets). Furthermore, they are suited for both single as well as multi-target prediction tasks [5].

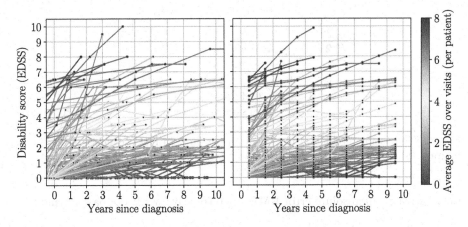

Fig. 2. Patient trajectories in terms of the visits (left) and in terms of the constructed targets, i.e., average disability per bin (right). The trajectories are colored by average EDSS score of the visits.

Two approaches are considered here. In the first, we train our model in a global multi-target regression setting, predicting the value for every bin simultaneously. As the bins are correlated, predicting them simultaneously instead of separately is expected to improve performance. In the second approach, we train a regressor chain of single-target models [9], each predicting the value for one bin. For each of these single-target models, the predictions of the models for the previous bins are added as additional features, as we expect these to be highly informative.

We did not perform any hyperparameter tuning, instead going with a rather standard set of default parameters. For the trees, we set the significance level for the F-test at $p = 1\%$ and require at least 5 patients per leaf node. For the random forests, we used 50 trees per forest and consider $\sqrt{k}$ features per split (where k is the number of features).

3.3   Evaluation Procedure

Due to the relatively small dataset of 174 patients, we used a 5-fold cross-validation setup. Any prediction results reported in this paper were obtained from the test folds (unless stated otherwise), and thus stem from data not seen by the model before. To see whether our models can identify a meaningful association between the input variables and the targets we constructed in Sect. 3.1, we took as baseline a model that predicts the mean over all the patients per bin. This baseline model is visualized in Fig. 3. In principle this would result in the same prediction for every patient, but due to the cross-validation setup there is some variability.

We used 3 evaluation metrics: the Pearson correlation $r_{\hat{y}y} \in [-1, 1]$ between the predictions $\hat{y}$ and the true values y, the mean absolute error MAE ≥ 0, and

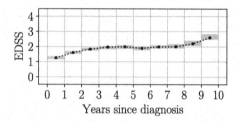

Fig. 3. The baseline prediction model (mean per bin). In black: over the full dataset. In gray: indication of variability in the 5-fold cross-validation.

the relative mean absolute error RMAE ≥ 0 (relative w.r.t. the baseline model):

$$r_{\hat{y}y} = \frac{\text{cov}(\hat{y}, y)}{\text{var}(\hat{y}) \cdot \text{var}(y)} \qquad \text{MAE} = \frac{1}{|y|} \sum_{i=1}^{|y|} |\hat{y}_i - y_i| \qquad \text{RMAE} = \frac{\text{MAE model}}{\text{MAE baseline}}$$

Note that for the perfect model $r_{\hat{y}y} = 1$ and RMAE $= 0$, and that a good model requires RMAE $\ll 1$. These metrics can be computed separately for every bin, or globally by aggregating all bins. Furthermore, they can be computed either w.r.t. the targets themselves, or w.r.t. the actual EDSS measurements. In the latter case, we need to extrapolate predictions to arbitrary time points, for which we use quadratic interpolation (assuming that predictions are made for the bin centers). The dotted line in Fig. 3 is an example of this interpolation procedure.

4   Results and Discussion

In Table 2, the performance of various models on the prognosis prediction task is evaluated. In particular, we consider the baseline model described in Sect. 3.3, a 'simple' decision tree (see below), a single global decision tree and a random forest and their regressor chain counterparts. The purpose of the simple decision tree was to provide a model small enough to be completely transparent and interpretable, and to reveal to a clinician the building block of our more complex models. This was achieved by increasing the minimum number of instances in a leaf from 5 to 15. When trained on the full dataset, this resulted in the clinically meaningful tree shown in Fig. 4, which also works as an interesting exploratory tool for the dataset at hand.

Based on Table 2, the best model is a chain of random forests, with a MAE on the actual measurements of 1.12 EDSS points. A t-test indicates that this average error of 1.12 is statistically significantly different from the average error of 1.67 EDSS points made by the baseline model ($p < 8 \times 10^{-8}$). Furthermore, for 71% of patients, the prediction of this model is closer to the true value than the prediction of the baseline model. This indicates that the significant difference in MAE is not just due to good performance on some patients far from the mean, but includes small improvements for many patients.

Table 2. Performance of various models, aggregated over all bins.

Model	Targets			Visits		
	$\rho_{\hat{y}y}$	MAE	RMAE	$\rho_{\hat{y}y}$	MAE	RMAE
Baseline	0.13	1.53	1.00	0.03	1.67	1.00
Small PCT	0.53	1.23	0.80	0.47	1.39	0.83
PCT global	0.55	1.19	0.78	0.51	1.34	0.80
RF global	0.69	1.15	0.75	0.66	1.32	0.79
PCT chain	0.61	1.12	0.73	0.64	1.18	0.71
RF chain	0.71	1.00	0.65	0.72	1.12	0.67

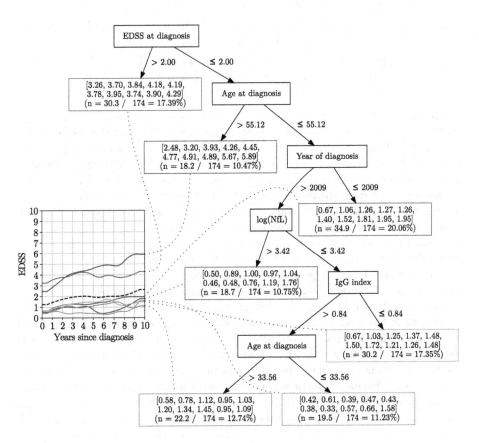

Fig. 4. A small decision tree (the building block of all our models) trained on the full dataset. In each leaf, the prototype is shown (average EDSS for each of the 10 bins), along with the fraction n of the total number of patients in that leaf. On the left, a plot of all these prototypes is shown (each connected with its respective leaf in the tree), along with the baseline model in a dashed black line.

Table 3. Performance of the chain of random forests per bin. For each bin, we also indicate the number n of instances in that bin (either number of non-missing targets for that bin or number of actual visits in the extent of the bin).

Bin	Extent	Targets				Visits			
		n	$r_{\hat{y}y}$	MAE	RMAE	n	$r_{\hat{y}y}$	MAE	RMAE
1	[0,1)	171	0.82	0.74	0.56	55	0.89	0.84	0.55
2	[1,2)	161	0.79	0.87	0.58	55	0.82	1.01	0.65
3	[2,3)	140	0.75	0.96	0.61	43	0.73	1.31	0.74
4	[3,4)	127	0.74	0.99	0.61	37	0.70	1.16	0.66
5	[4,5)	106	0.68	1.08	0.67	27	0.76	1.19	0.60
6	[5,6)	93	0.65	1.08	0.72	27	0.77	1.03	0.62
7	[6,7)	83	0.62	1.19	0.75	26	0.65	1.51	0.76
8	[7,8)	71	0.63	1.12	0.75	26	0.30	1.18	0.87
9	[8,9)	58	0.59	1.24	0.77	23	0.57	1.05	0.66
10	[9,10)	44	0.52	1.32	0.77	20	0.67	1.31	0.77
All	[0,10)	1054	0.71	1.00	0.65	339	0.72	1.12	0.67

Zooming in further on this chain of random forests, we see that this model achieves a Pearson correlation of 0.71 on the targets and 0.72 on the actual visits, and an RMAE of 0.65 on the targets and 0.67 on the actual visits. Of course, as expected, the error increases the further the bin is from diagnosis. In Table 3, we split up the metrics over the bins to show this decrease of performance. As we move further from diagnosis, the number of non-missing targets decreases, and naturally the MAE on the targets starts to increase, rising above the overall average of 1.00 EDSS points in the 4th year after diagnosis (bin 5). Despite this, the results indicate that the model is capable of providing every patient with a personalized prognosis, even on the level of the actual visits.

An example of such a personalized prognosis prediction is shown in Fig. 5. As we can see, at time of diagnosis the model is able to make predictions that are very close to the two actual EDSS measurements. However, we also see that predictions in the later years become unstable, because in the later bins there is less data to train our models on. More data could further improve the results, both in terms of number of patients and in terms of number of features (e.g., also incorporating other data modalities, such as genetic or radiomic markers). A preliminary feature importance analysis (not shown) indicates that the most important variables for the first random forest (i.e., in the first bin) are (in this order) the MS subtype, the baseline EDSS, the time between onset and diagnosis, and the baseline age. Further bins increasingly have predictions of previous models as the most important variables.

To verify the reliability and effectiveness of our approach, external validation would be necessary. The approach could also be improved with uncertainty quantification (especially for the individual patient predictions as in Fig. 5), by adding a statistically meaningful confidence interval to each bin prediction. This confidence interval could for example be based on the out of bag error of the random forest, or the variability in the predictions made by the individual trees. In further future work, the approach could be extended to other diseases such as amyotrophic lateral sclerosis (ALS), where it might also be appropriate to experiment with the number of bins and their size.

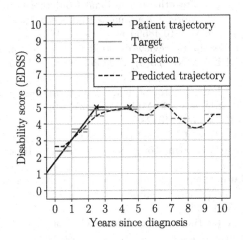

Fig. 5. Test fold prediction for an individual patient.

5   Conclusion

In this paper, we have presented an approach for modeling long-term disability progression in multiple sclerosis. The machine learning target for this approach was constructed by dividing the years after diagnosis up into bins, each of 1 year wide, and computing for each patient the average disability level in each bin (with missing values after the last measurement). Multi-target regression models (based on predictive clustering trees [2]) were then trained to predict this target. Models were evaluated on mean absolute error and Pearson correlation between true and predicted values, both based on the targets as well as on the actual EDSS measurements. The best model turned out to be a chain of random forests, where each forest predicts one bin and gets the predictions for the previous bins as extra input. With an overall mean absolute cross-validated error of 1.12 EDSS points on the visits, this model has shown to be capable of providing a long-term prognosis to each patient at diagnosis.

Acknowledgements. This work was funded by Research Fund Flanders (FWO fellowship 1S38023N) and supported by the Flemish government (through the AI Research Program) and Stichting MS Research (through a Monique Blom-de Wagt grant). We furthermore thank Professor Bénédicte Dubois, neurologist at UZ Leuven, for collecting the data that was used retrospectively in this work.

References

1. Allam, A., Feuerriegel, S., Rebhan, M., Krauthammer, M.: Analyzing patient trajectories with artificial intelligence. J. Med. Internet Res. **23**(12), e29812 (2021). https://doi.org/10.2196/29812
2. Blockeel, H., De Raedt, L., Ramon, J.: Top-down induction of clustering trees. In: Proceedings of the Fifteenth International Conference on Machine Learning. ICML 1998, pp. 55–63. Morgan Kaufmann Publishers Inc., San Francisco, CA, USA (1998). https://doi.org/10.48550/arxiv.cs/0011032
3. Brown, F., et al.: Systematic review of prediction models in relapsing remitting multiple sclerosis. PLoS ONE **15**(5), e0233575 (2020). https://doi.org/10.1371/journal.pone.0233575
4. Iaffaldano, P., et al.: Early treatment delays long-term disability accrual in RRMS: results from the BMSD network. Mult. Scler. **27**(10), 1543–1555 (2021). https://doi.org/10.1177/13524585211010128
5. Kocev, D., Vens, C., Struyf, J., Džeroski, S.: Tree ensembles for predicting structured outputs. Pattern Recogn. **46**(3), 817–833 (2013). https://doi.org/10.1016/j.patcog.2012.09.023
6. Kurtzke, J.F.: Rating neurologic impairment in multiple sclerosis: an expanded disability status scale (EDSS). Neurology **33**(11), 1444–1452 (1983). https://doi.org/10.1212/wnl.33.11.1444
7. Moazami, F., Lefevre-Utile, A., Papaloukas, C., Soumelis, V.: Machine learning approaches in study of multiple sclerosis disease through magnetic resonance images. Front. Immunol. **12**, 3205 (2021). https://doi.org/10.3389/fimmu.2021.700582
8. Seccia, R., Romano, S., Salvetti, M., Crisanti, A., Palagi, L., Grassi, F.: Machine learning use for prognostic purposes in multiple sclerosis. Life **11**(2), 122 (2021). https://doi.org/10.3390/life11020122
9. Spyromitros-Xioufis, E., Tsoumakas, G., Groves, W., Vlahavas, I.: Multi-target regression via input space expansion: treating targets as inputs. Mach. Learn. **104**(1), 55–98 (2016). https://doi.org/10.1007/s10994-016-5546-z
10. Stafford, I.S., Kellermann, M., Mossotto, E., Beattie, R.M., MacArthur, B.D., Ennis, S.: A systematic review of the applications of artificial intelligence and machine learning in autoimmune diseases. NPJ Digit. Med. **3**(1), 30 (2020). https://doi.org/10.1038/s41746-020-0229-3

Decision Tree Approaches to Select High Risk Patients for Lung Cancer Screening Based on the UK Primary Care Data

Teena Rai[1]([⊠])[iD], Yuan Shen[1][iD], Jaspreet Kaur[2][iD], Jun He[1][iD], Mufti Mahmud[1][iD], David J. Brown[1][iD], David R. Baldwin[2][iD], Emma O'Dowd[2][iD], and Richard Hubbard[2][iD]

[1] Department of Computer Science, Nottingham Trent University, Clifton Lane, Nottingham NG11 8NS, UK
teena.rai2022@my.ntu.ac.uk,
{yuan.shen,jun.he,mufti.mahmud,david.brown}@ntu.ac.uk
[2] Division of Epidemiology and Public Health, University of Nottingham, Nottingham NG5 1PB, UK
{jaspreet.kaur1,david.baldwin,emma.odowd,
richard.hubbard}@nottingham.ac.uk

Abstract. Lung cancer has the highest cancer mortality rate in the UK. Most patients are diagnosed at an advanced stage because common symptoms for lung cancer such as cough, pain, dyspnoea and anorexia are also present in other diseases. This partly attributes towards the low survival rate. Therefore, it is crucial to screen high risk patients for lung cancer at an early stage through computed tomography (CT) scans. As shown in a previous study, for patients who were screened for lung cancer and were identified with stage I lung cancer, the estimated survival rate was 88% compared to only 5% who have stage IV lung cancer. This paper aims to build tree-based machine learning models for predicting lung cancer risk by extracting significant factors associated with lung cancer. The Clinical Practice Research Datalink (CPRD) data was used in this study which are anonymised patient data collected from 945 general practices across the UK. Two tree-based models (decision trees and random forest) are developed and implemented. The performance of the two models is compared with a logistic regression model in terms of accuracy, Area Under the Receiver Operating Characteristic curve (AUROC), sensitivity and specificity, and both achieve better results. However, as for interpretability, it was found that, unlike coefficients in logistic regression, the default feature importance is non-negative in random forests and decision trees. This makes tree-based models less interpretable than logistic regression.

Keywords: lung cancer screening · CPRD · primary care · decision tree · random forest · lung cancer early detection · explainable AI

Supported by Nottingham Trent University Medical Technologies and Advanced Materials Strategic Research Theme. Teena Rai is funded by NTU VC PhD studentship.

J. M. Juarez et al. (Eds.): AIME 2023, LNAI 13897, pp. 35–39, 2023.
https://doi.org/10.1007/978-3-031-34344-5_4

1   Introduction

Two well known risk prediction models for lung cancer are commonly used: the Liverpool Lung Project (LLP) risk model [2] and PLCO$_{m2012}$ [10]. They both use logistic regression to build the risk prediction models. For the LLP risk model, various risk factors such as life-style characteristics, medical history and family history of cancer of participants were used to build the model which gave an AUROC of 0.70. For the PLCO$_{m2012}$ model, the AUROC for smokers in the control group was 0.803 and intervention group was 0.797.

Although the LLP risk model and PLCO$_{m2012}$ have been well validated [7,11], a lot of research has been undertaken to build non-linear machine learning models for risk prediction of different types of cancer with increased accuracy and AUROC. In [5], demographic characteristics, smoking information, BMI data and laboratory data were used to develop, extreme gradient boosting (XGBoost) and performed better than the PLCO$_{m2012}$, with an AUROC of 0.86 and had increased accuracy in identifying lung cancer among ever-smokers up to 12 months before clinical diagnosis (p-value $<$ 0.00001). Electronic Health Records (EHR) from the General Practice Research Database (GPRD) in the UK have been used to develop a risk prediction model with several machine learning classifiers for oseophago-gastric cancer [1] and performance compared with the current UK oesophago-gastric cancer risk-assessment tool (ogRAT). All predicitve models (Random Forest, linear SVM, Logistic Regression, Naive Bayes and XGBoost) used in this research outperformed ogRAT with an accuracy of 89%.

Therefore, machine learning models, and in particular tree based models, have been shown to develop risk prediction models with high accuracy and AUROC on tabular data [1,7]. The aim of this paper is to build two tree based models, decision tree and random forest, for predicting lung cancer risk by using CPRD data, a primary care EHR-based dataset.

2   Data and Methods

A subset of the CPRD data was used to develop the machine learning models in this study. This cohort comprised of patients who were between 50 and 80 years from 945 practices in the UK with continuous registration within the database from 01/01/2012 to 01/01/2014. The start of the follow up was 01/01/2014 and end of the follow up was the event which took place the earliest after the start of the follow up such as death date, diagnosis of lung cancer or end of the follow up period, 01/01/2020. The final data comprises of 867,670 healthy controls and 6,395 lung cancer cases after removing rows which have missing values.

We extracted 11 known factors associated with lung cancer [2,4,10] which are personal history of cancer, family history of cancer, current smoking status (smoking_status), smoking start date (start_date), smoking end date (end_date), number of cigarettes consumed per day (cig_per_day), number of cigars consumed per day (cigar_per_day), absence or presence of Chronic Obstructive Pulmonary Disease (COPD), Body Mass Index (BMI) and lastly, a binary predictor variable *lungcancer* to denote healthy controls and lung cancer cases. The `smoking_status` of a person is either ex, current or never. The duration of smoking and number of cigarettes smoked (smoking intensity) are two other related factors associated with lung cancer risk [3]. Therefore, we extract two further feature variables as follows:

1. smoking duration = (`start_date`) - (`end_date`)
2. smoking intensity = `cig_per_day` + 2 × `cigar_per_day`

As a result, we have 7 feature variables and a binary predictor variable to classify cases and controls. Additional feature engineering involved one-hot encoding of categorical feature `smoking_status` and normalization of training data with zero mean and unit variance.

The dataset was split into training and testing sets. Since we have a significantly larger number of controls compared to cases, we implement an undersampling approach in an ensemble setting [6] to overcome the class imbalance problem so that we have a one-to-one ratio of healthy controls and lung cancer cases to train over 100 random splits. We train each classifier (decision tree and random forest) independently in an ensemble setting which consists of 100 classifiers, where each classifier uses one of the training sets. The final prediction for testing data is then carried out by majority voting of the 100 predicted labels of each classifier.

Decision tree, random forests and logistic regression are used for the classification task in this paper. Their implementation is provided by scikit-learn library.

3   Results

The performance of the tree based models were evaluated on the basis of their overall accuracy, AUROC, sensitivity and specificity by comparing the results with logistic regression model. Both decision tree and random forest outperformed the logistic regression model in terms of accuracy. The mean accuracy of decision tree and random forests across 100 random splits was 73.17% and 73.19% respectively compared to 71.90% when the logistic regression approach was taken. The mean AUROC were also slightly better than the logistic regression model, with decision trees and random forests producing an AUROC of 0.7936 and 0.8005 respectively compared to 0.7933 for logistic regression. The sensitivity of decision tree and random forest were similar to that of logistic regression at 81%. The specificity of decision tree and random forest were slightly better with decision tree and random forest at 66% and 67% respectively compared to 62% for logisitic regression. Hence, random forest outperforms both decision tree and logistic regression in terms of accuracy and AUROC with a minor difference in terms of sensitivity.

Decision tree and random forests offer default feature importance based on mean decrease in impurity (MDI), i.e., impurity in this case is the gini index.

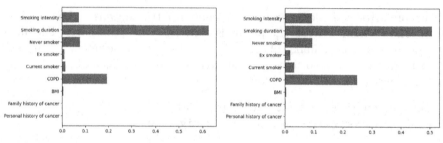

(a) Decision tree feature importance   (b) Random forest feature importance

Fig. 1. Feature importance by decision tree and random forest

The feature importance for decision tree and random forest are presented in Fig. 1 which shows that smoking duration and COPD are top two predictors.

Figure 1 shows feature importance from the perspective of machine learning. However, they are insufficient from a clinical decision point of view. Unlike logistic regression, which can distinguish positive or negative effects of features on lung cancer risk, MDI is always non-negative. Furthermore, unlike a single decision tree, the decision process of a random forest was difficult to interpret.

Note: Due to some changes in data preparation, the results of logistic regression in the current paper are different from our previous work [9]. All three models in this paper, including logistic regression, are slightly worse than [9]. However, for fair comparison, the same experimental conditions were used for all three models in current paper.

4   Conclusion

In this paper, we used CPRD data to develop risk prediction models on decision tree based approaches. We find that both random forest and decision tree outperform the logistic regression model, with random forest performing slightly better than decision tree both in terms of accuracy and AUROC. However, unlike positive and negative coefficients in logistic regression, the default feature importance is non-negative in tree-based models. This makes them less interpretable than logistic regression.

We only extracted a subset of features from the CPRD data to build the models. Therefore, our future work involves using more features to build a number of other explainable models. Decision trees are more explainable as we can easily plot a decision tree to see how the decision was made. However, this is not true for random forest as it makes its predictions on majority voting by the single decision trees built through bootstrap samples of the training data. Therefore, our aim is to work towards transforming a random forest into a single tree, as motivated by other research [8].

References

1. Briggs, E., de Kamps, M., Hamilton, W., Johnson, O., McInerney, C.D., Neal, R.D.: Machine learning for risk prediction of Oesophago-gastric cancer in primary care: comparison with existing risk-assessment tools. Cancers **14**(20), 5023 (2022). https://doi.org/10.3390/cancers14205023
2. Cassidy, A., et al.: The LLP risk model: an individual risk prediction model for lung cancer. Br. J. Cancer **98**(2), 270–276 (2008). https://doi.org/10.1038/sj.bjc.6604158
3. Doll, R., Peto, R., Boreham, J., Sutherland, I.: Mortality in relation to smoking: 50 years' observations on male British doctors. BMJ **328**(7455), 1519 (2004). https://doi.org/10.1136/bmj.38142.554479.AE
4. Durham, A.L., Adcock, I.M.: The Relationship between COPD and Lung Cancer. Lung Can. **90**(2), 121–127 (2015). https://doi.org/10.1016/j.lungcan.2015.08.017
5. Gould, M.K., Huang, B.Z., Tammemagi, M.C., Kinar, Y., Shiff, R.: Machine learning for early lung cancer identification using routine clinical and laboratory data. Am. J. Respir. Crit. Care Med. **204**(4), 445–453 (2021). https://doi.org/10.1164/rccm.202007-2791OC
6. Liu, X.-Y., Wu, J., Zhou, Z.-H.: Exploratory undersampling for class-imbalance learning. IEEE Trans. Syst. Man Cybern. Part B (Cybern.) **39**(2), 539–550 (2009). https://doi.org/10.1109/TSMCB.2008.2007853
7. Raji, O.Y., et al.: Predictive accuracy of the liverpool lung project risk model for stratifying patients for computed tomography screening for lung cancer. Ann. Int. Med. **157**(4), 242–250 (2012). https://doi.org/10.7326/0003-4819-157-4-201208210-00004
8. Sagi, O., Rokach, L.: Explainable decision forest: transforming a decision forest into an interpretable tree. Inf. Fusion **61**, 124–138 (2020). https://doi.org/10.1016/j.inffus.2020.03.013
9. Shen, Y., et al.: A logistic regression approach to a joint classification and feature selection in lung cancer screening using CPRD data. In: 2022 2nd International Conference on Trends in Electronics and Health Informatics (2022)
10. Tammemägi, M.C., et al.: Selection criteria for lung-cancer screening. N. Engl. J. Med. **368**(8), 728–736 (2013). https://doi.org/10.1056/NEJMoa1211776
11. Tammemägi, M.C., et al.: Development and validation of a multivariable lung cancer risk prediction model that includes low-dose computed tomography screening results: a secondary analysis of data from the national lung screening trial. JAMA Netw. Open **2**(3), e190204 (2019)

Causal Discovery with Missing Data in a Multicentric Clinical Study

Alessio Zanga[1,2]([✉]) [ID], Alice Bernasconi[1,3] [ID], Peter J.F. Lucas[4] [ID],
Hanny Pijnenborg[5] [ID], Casper Reijnen[5] [ID], Marco Scutari[6] [ID],
and Fabio Stella[1] [ID]

[1] DISCo, University of Milano - Bicocca, Milan, Italy
[2] F. Hoffmann - La Roche Ltd, Basel, Switzerland
a.zanga3@campus.unimib.it
[3] Fondazione IRCCS Istituto Nazionale dei Tumori, Milan, Italy
[4] University of Twente, Enschede, The Netherlands
[5] RadboudUMC, Nijmegen, The Netherlands
[6] IDSIA, Lugano, Switzerland

Abstract. Causal inference for testing clinical hypotheses from observational data presents many difficulties because the underlying data-generating model and the associated causal graph are not usually available. Furthermore, observational data may contain missing values, which impact the recovery of the causal graph by causal discovery algorithms: a crucial issue often ignored in clinical studies. In this work, we use data from a multi-centric study on endometrial cancer to analyze the impact of different missingness mechanisms on the recovered causal graph. This is achieved by extending state-of-the-art causal discovery algorithms to exploit expert knowledge without sacrificing theoretical soundness. We validate the recovered graph with expert physicians, showing that our approach finds clinically-relevant solutions. Finally, we discuss the goodness of fit of our graph and its consistency from a clinical decision-making perspective using graphical separation to validate causal pathways.

Keywords: Causal discovery · Causal graphs · Missing data

1 Introduction

Much of the data collected in clinical research is observational, collected as part of daily clinical practice. Correctly interpreting them requires a good understanding of their characteristics and of possible sources of bias. A common one is missing values, which may arise in three different ways [1]: *data missing completely at random* (MCAR), *data missing at random* (MAR), and *data missing not at random* (MNAR) that are neither MCAR nor MAR. MNAR is common in clinical observational data and thus interesting to study, as it is often possible to unravel the reason for the missingness: for instance, a laboratory test may be skipped in favour of a more precise ones available at a later stage.

© The Author(s), under exclusive license to Springer Nature Switzerland AG 2023
J. M. Juarez et al. (Eds.): AIME 2023, LNAI 13897, pp. 40–44, 2023.
https://doi.org/10.1007/978-3-031-34344-5_5

Missing values in clinical data are commonly imputed with heuristics or with single/multiple imputation. Such techniques assume that the data are MCAR or MAR; we cannot test whether these assumptions are valid without knowing the missingness mechanism but, at the same time, if these assumptions do not hold our clinical conclusions are likely to be biased [2]. A possible approach to this problem is *causal discovery*: modelling the missingness mechanism is to recover the underlying causal graph $\mathcal{G}^*$, given the data $\mathcal{D}$ and the prior knowledge $\mathcal{K}$ [3]. In our previous work on endometrial cancer [EC; 4], we proposed a new causal discovery approach based on bootstrapping for clinical data with low sample size and high missingness assuming MAR (*Bootstrap SEM*). Algorithms assuming MNAR were not available until recently when *HC-aIPW* [5] was introduced.

Our aim is to showcase how modern causal discovery techniques can model the biases in observational data, in particular for MNAR. For this purpose, we applied different causal discovery algorithms with different assumptions to data from a multicenter study on EC, highlighting the clinical implications of their biases on recovering the causal mechanisms behind the prognosis of EC.

2   Background

A causal graph $\mathcal{G} = (\mathbf{V}, \mathbf{E})$ is a directed acyclic graph (DAG) where for each directed edge $(X, Y) \in \mathbf{E}$, X is a direct cause of Y and Y is a direct effect of X. The vertex set $\mathbf{V}$ is usually split into two disjoint subsets $\mathbf{V} = \mathbf{O} \cup \mathbf{U}$, where $\mathbf{O}$ is the set of the *fully observed* variables (with no missing values), while $\mathbf{U}$ is the set of *fully unobserved* variables (the *latent* variables). A missingness graph $\mathcal{M} = (\mathbf{V}^*, \mathbf{E}^*)$ [6] is a causal graph where the vertices in $\mathbf{V}^*$ are partitioned into five disjoint subsets: $\mathbf{V}^* = \mathbf{O} \cup \mathbf{U} \cup \mathbf{M} \cup \mathbf{S} \cup \mathbf{R}$, where $\mathbf{M}$ is the set of the *partially observed* variables, that is, the variables with at least one missing value; $\mathbf{S}$ is the set of the proxy variables, that is, the variables that are actually observed; $\mathbf{R}$ is the set of the *missingness indicators*. Missingness graphs can be queried for independencies using *d-separation*. The set of variables $\mathbf{Z}$ d-separates X from Y, denoted by $X \perp\!\!\!\perp Y \mid \mathbf{Z}$, if it *blocks* every path π between X and Y. A path π is blocked by $\mathbf{Z}$ if and only if π contains: a fork $A \leftarrow B \rightarrow C$ or a chain $A \rightarrow B \rightarrow C$ so that B is in $\mathbf{Z}$, or, a collider $A \rightarrow B \leftarrow C$ so that B, or any descendant of it, is not in $\mathbf{Z}$. MCAR, MAR and MNAR result in different independence statements [1] which are linked to the independency statements implied by the missingness graph: MCAR implies $\mathbf{O} \cup \mathbf{U} \cup \mathbf{M} \perp\!\!\!\perp \mathbf{R}$, the missingness is random and independent from the fully observed and the partially observed variables; MAR implies $\mathbf{U} \cup \mathbf{M} \perp\!\!\!\perp \mathbf{R} \mid \mathbf{O}$, missingness is random only conditionally on the fully observed variables; MNAR if neither MCAR nor MAR. Since MCAR implies MAR, any method assuming MAR can be used on MCAR.

3   Multicentric Clinical Data on Endometrial Cancer

The observational data we explore in this paper comprise 763 patients with endometrial cancer from 10 gynecological oncological clinics in Europe that are

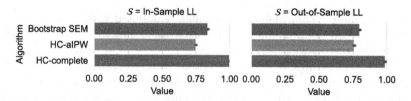

Fig. 1. In-sample and out-of-sample LL for each algorithm. Re-scaled by sample size and absolute maximum value. Lower values are better.

part of the European Network for Individualized Treatment of Endometrial Cancer (ENITEC). Clinical experts selected the variables that they considered most important for predicting the presence of lymph node metastases (LNMs) and survival [4]. The selected variables were: the cytology of the cervix uteri, the preoperative tumour grade, the postoperative tumour grade (after pathological examination of the tumour tissue obtained after surgical removal of the uterus), treatment by chemotherapy or radiatiotherapy, lymphvascular space invasion (that is, whether there is tumour growth into the lymph or blood vessels), the levels of estrogen and progesterone in blood, the presence of lymph node metastasis according to CT or MRI imaging, the CA125 tumour marker, L1CAM (an intracellular protein that promotes tumour cell motility), the p53 tumour suppressor gene, the number of platelets, presence of lymph node metastases, recurrence of the tumour, and lastly survival before and after 1, 3, and 5 years. The tumour markers (p53, CA125, L1CAM, estrogen and progesterone levels) are thought to offer causal prognostic information about tumour cell behaviour and thus tumour ingrowth, metastases, recurrence, and survival.

4 Experiments

We performed numerical experiments to compare the graphs recovered under MAR and MNAR by the Bootstrap SEM and HC-aIPW. For reference, we also reported the results for HC on data completed with single imputation, denoted *HC-complete*. Prior knowledge elicited from experts consists of forbidden and required edges (Survival1yr → Survival3yr, Survival3yr → Survival5yr).

Firstly, we evaluated the goodness of fit of the recovered graphs by computing the log-likelihood (LL) of both the data used to recover the graph (in-sample) and those held aside for validation (out-of-sample). The former allows us to see which algorithm fits a particular data set the best. The latter approximates the Kullback-Leibler divergence between the recovered causal graph and the unknown causal graph underlying the data, and allows us to see how close the two are and how well the recovered graph generalises to new data. We repeated causal discovery for 100 bootstrap replicates and computed the mean and the standard deviation of both in-sample and out-of-sample LL.

We observe (Fig. 1) that HC-aIPW, which assumes MNAR, dominates Bootstrap SEM and HC-complete, which assume MAR, for both in-sample and out-of-sample LL. In the case of in-sample LL, this may be attributed (at least in

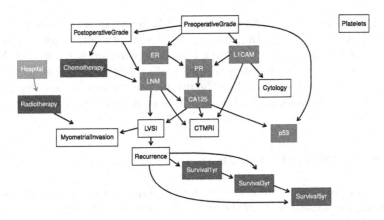

Fig. 2. Causal graph $\mathcal{M}_{\text{MNAR}}$ recovered by HC-aIPW under MNAR.

part) to making the correct missingness assumption: MCAR and MAR are too restrictive and limit how well the recovered graph fits the data; MNAR it is not strict enough when MCAR or MAR hold and let causal discovery algorithms overfit. This decreases the out-of-sample LL because overfitted models are too complex and do not generalize well. The fact that HC-aIPW, which assumes MNAR, outperforms Bootstrap SEM and HC-complete, which assume MAR, for both in-sample and out-of-sample LL suggests that the MNAR assumption is correct for the data and that it allows HC-aIPW to recover a causal graph that is close to the underlying model and that generalises better to new data.

Goodness-of-fit measures allow for a *quantitative* evaluation of the recovered graphs, but they say little about the *qualitative* information they encode in terms of *independence statements*. We denote the graph recovered by Bootstrap SEM as $\mathcal{M}_{\text{MAR}}$; Fig. 2 shows that recovered by HC-aIPW, $\mathcal{M}_{\text{MNAR}}$. For readability, we colored the vertices depending on their *semantic* interpretation: treatments (Radiotherapy and Chemotherapy) are colored in blue, outcomes (Survival1yr, etc.) in red, the event of interest (LNM) in orange, relevant biomarkers (ER, PR, CA125, etc.) in lightblue and the *context* variable Hospital in gray.

Focusing on the interactions of LNM, we observe using d-separation that LNM $\perp\!\!\!\perp$ {CA125, p53} | PostoperativeGrade is true in $\mathcal{M}_{\text{MAR}}$, but false in $\mathcal{M}_{\text{MNAR}}$, where CA125 and p53 are effects of LNM. This makes $\mathcal{M}_{\text{MNAR}}$ close to the clinical practice where both CA125 and p53 are considered relevant biomarkers linked to LNM, providing additional information on LNM even if PostoperativeGrade is observed. Indeed, a crucial difference between $\mathcal{M}_{\text{MAR}}$ and $\mathcal{M}_{\text{MNAR}}$ is that the biomarkers and LNM are d-separated from LNM in $\mathcal{M}_{\text{MAR}}$ if PostoperativeGrade is observed, but are descendants of LNM in $\mathcal{M}_{\text{MNAR}}$. Hence, if our goal is to detect the presence of LNM in EC patients, measuring CA125, p53 or any of their descendants is coherent with $\mathcal{M}_{\text{MNAR}}$.

Shifting the focus to the treatment variables Chemotherapy and Radiotherapy, LNM $\perp\!\!\!\perp$ Chemotherapy does not hold in either $\mathcal{M}_{\text{MAR}}$ or $\mathcal{M}_{\text{MNAR}}$, since Chemotherapy is a direct cause of LNM in both. Therefore, Chemotherapy is expected to influence the likelihood of LNM, which is exactly the reason why it is prescribed by clinicians. On the other hand, LNM ⊥⊥ Radiotherapy does hold in $\mathcal{M}_{\text{MNAR}}$ but not in $\mathcal{M}_{\text{MAR}}$, suggesting a spurious correlation induced by MAR. This is confirmed by the clinical literature: since Radiotherapy is aimed at local treatment of the tissue surrounding the uterus, and there is a clear dependence with MyometrialInvasion of the tumour (in both models), Radiotherapy effects on LNM are not expected.

5 Discussion and Conclusions

In this work we presented a systematic analysis of the impact of missingness assumptions using state-of-the-art causal discovery algorithms. We applied these methods to a real-world, observational multicentric study on EC patients, extending them to include expert prior knowledge without sacrificing theoretical soundness. Furthermore, we validated the obtained causal models with experienced physicians and clinical literature. We evaluated the goodness-of-fit of the recovered graphs with respect to the underlying data distribution, showing that stricter assumptions are associated to models that generalize poorly. Moreover, by leveraging the test for graphical separation, we explained how the missingness mechanism affects the causal pathways associated to the clinical decision-making perspective. Quantifying the bias due to missingness in other case studies and its overlap with the effects of hidden and selection variables are open problems.

Acknowledgements. Alessio Zanga is funded by F. Hoffmann-La Roche Ltd.

References

1. Rubin, D.B.: Inference and Missing Data. Biometrika **63**(3), 581–592 (1976). https://doi.org/10.1093/biomet/63.3.581
2. Stavseth, M.R., Clausen, T., Røislien, J.: How handling missing data may impact conclusions: a comparison of six different imputation methods for categorical questionnaire data. SAGE Open Med. **7**, 205 (2019). https://doi.org/10.1177/2050312118822912
3. Zanga, A., Ozkirimli, E., Stella, F.: A survey on causal discovery: theory and practice. Int. J. Approximate Reasoning **151**, 101–129 (2022). https://doi.org/10.1016/j.ijar.2022.09.004
4. Zanga, A., Bernasconi, A., Lucas, P.J.F., et al.: Risk assessment of lymph node metastases in endometrial cancer patients: a causal approach. In: Proceedings of the 1st Workshop on Artificial Intelligence For Healthcare (2022). https://ceur-ws.org/Vol-3307/
5. Liu, Y., Constantinou, A.C.: Greedy structure learning from data that contain systematic missing values. Mach. Learn. **111**(10), 3867–3896 (2022). https://doi.org/10.1007/S10994-022-06195-8
6. Mohan, K., Pearl, J.: Graphical models for processing missing data. J. Am. Stat. Assoc. **116**(534), 1023–1037 (2018). https://doi.org/10.1080/01621459.2021.1874961

Novel Approach for Phenotyping Based on Diverse Top-K Subgroup Lists

Antonio Lopez-Martinez-Carrasco[1]([⊠]) [iD], Hugo M. Proença[2] [iD],
Jose M. Juarez[1] [iD], Matthijs van Leeuwen[2] [iD], and Manuel Campos[1,3] [iD]

[1] MedAI-Lab, University of Murcia, Murcia, Spain
antoniolopezmc@um.es

[2] Leiden Institute of Advanced Computer Science, Leiden University,
Leiden, The Netherlands

[3] Murcian Bio-Health Institute (IMIB-Arrixaca), Murcia, Spain

Abstract. The discovery of phenotypes is useful to describe a population. Providing a set of diverse patient phenotypes with the same medical condition may help clinicians to understand it. In this paper, we approach this problem by defining the technical task of mining diverse top-k phenotypes and proposing an algorithm called DSLM to solve it. The phenotypes obtained are evaluated according to their quality and predictive capacity in a bacterial infection problem.

Keywords: Patient phenotyping · Subgroup Discovery · Subgroup List · the Minimum Description Length principle · Algorithm

1 Introduction

Phenotyping consists of finding a set of observable characteristics of patients with a medical condition [5]. Unfortunately, getting a single explanation of a medical phenomenon might be poor under the eyes of a clinical researcher, and observable descriptions could be ruled out. For this reason, multiple phenotypes might be provided, which implies that both the characteristics used and the patients represented must be diverse, thus providing the clinician with different explanations for the same medical phenomenon. From a machine learning perspective, phenotyping could be interpreted as the automatic generation of models describing a subset of the population with a specific feature. Therefore, we propose to use the Subgroup Discovery (SD) paradigm [1] as a building block of phenotypes. Moreover, the generation of multiple subgroup list models [3,4] is a suitable approach to generate diverse phenotypes. From a clinical perspective, a critical problem when creating subgroup lists is the large number of candidate subgroups that are mined by the SD algorithm. A feasible solution is to adopt the Minimum Description Length (MDL) principle, which is a method of inductive inference based on the idea that the model that best explains the data is the one that best compresses the data.

J. M. Juarez et al. (Eds.): AIME 2023, LNAI 13897, pp. 45–50, 2023.
https://doi.org/10.1007/978-3-031-34344-5_6

The contributions of this work are: (1) the definition of the new problem of mining diverse top-k phenotypes, and (2) a new algorithm called DSLM that uses SD and the MDL principle to solve this problem. These contributions provide clinicians with a method to obtain multiple and diverse explanations of a set of patients with statistically robust subgroups.

2   Problem Definition and Background

This section formally defines the novel problem of mining diverse top-k phenotypes and provides a short background of the methods used.

The following basic concepts are first introduced: (1) an attribute a is a relation between an object property and its value; (2) the domain of a ($dom(a)$) is the set of all unique values that a can take; (3) an instance i is a tuple $i = (a_1, \ldots, a_m)$ of attributes; and (4) a dataset d is a tuple $d = (i_1, \ldots, i_n)$ of instances. Note that an attribute can be nominal or numeric depending on its domain and that the notation $v_{x,y}$ is used to indicate the value of the x-th instance i_x and its y-th attribute a_y from a dataset d. Subsequently, the following definitions can be given. Given an attribute a_y from a dataset d, a binary *operator* $\in \{=, \neq, <, >, \leq, \geq\}$ and a value $w \in dom(a_y)$, then a selector e is defined as a 3-tuple of the form $(a_y, operator, w)$. In our problem, descriptions contained in a phenotype are related to a specific target value represented by a selector e. Given an instance i_x and a selector $e = (a_y, operator, w \in dom(a_y))$, then *selector_coverage*(i, e) function returns 1 if the binary expression "$v_{x,y}$ operator w" holds *true*, and returns 0 otherwise. A pattern p is a set of selectors of the form $\{e_1, \ldots, e_j\}$ that represents a conjunction of conditions describing a subset of the dataset, and in which all attributes of the selectors are different. Given an instance i and a pattern p, then *pattern_coverage*(i, p) function returns 1 if $\forall e_j \in p, selector\_coverage(i, e_j)$, and returns 0 otherwise. Additionally, we say that a pair (p, e) is positive if it exists an instance i covered by both p and e.

Given a dataset d and a target e, then a phenotype l is a list of patterns $< p_1, \ldots, p_z >$ such that $\forall p_z \in l, (p_z, e)$ is positive, and that cover dataset instances that are statistically different and interesting, compared to the dataset distribution. In the problem defined, we consider that multiple phenotypes must contain positive pairs pattern-target since the objective in this domain is to explain a specific target value from a dataset (e.g., *exitus = yes*). Moreover, when ensuring diversity, the outcome must provide clinicians with multiple phenotypes that are different and non-redundant. Finally, the problem of mining diverse top-k phenotypes is defined as follows: given a dataset d, a target e and the k maximum number of phenotypes to discover, the problem of mining diverse top-k phenotypes consists of generating a set of phenotypes $\{l_1, \ldots, l_k\}$ such that they only contain positive pairs pattern-target and are diverse.

Given a pattern p and a selector e, a subgroup s is a pair (p, e) in which the pattern is denominated as 'description' and the selector is denominated as 'target'. Additionally, given a subgroup s and a dataset d, a quality measure q is a function that computes one numeric value according to s and certain metrics from d (e.g., WRAcc). Finally, the SD problem consists of exploring the search space of a dataset d to mine subgroups whose quality value q is greater or equal to a given *threshold*.

A subgroup list [4] is a collection of ordered subgroups followed by a default subgroup, whose objective is to iteratively divide the input data into different subsets and to provide a description for each of them, except the last one, which corresponds to the default subgroup. While subgroups contained in a subgroup list cover the instances that are statistically different in comparison with the dataset distribution, the default subgroup represents the dataset average, covering the instances that are well described by the dataset distribution.

The authors of [4] used the MDL principle to build a single subgroup list. They defined the MDL encoding of the optimal subgroup list for a certain dataset and proposed a greedy algorithm called SDD++ that iteratively added one subgroup at a time to the subgroup list.

3   DSLM Algorithm

Diverse Subgroup Lists Miner[1] (DSLM) aims to generate diverse top-k subgroup lists by using SD and the MDL principle. This algorithm requires the following inputs: a dataset d, a collection C of subgroups, the number k of subgroup lists to generate, the maximum number of subgroups for each subgroup list (sl_max_size), and the maximum overlap permitted ($max\_overlap$). Note that subgroups contained in C can be generated by any SD algorithm and later filtered before executing this algorithm.

DSLM algorithm initially creates an empty collection $\mathcal{L}$. Next, it iterates k times and, in each loop, initializes an empty subgroup list, adds it to $\mathcal{L}$ and iterates over C to fill it. For each candidate from C, it is selected only if: (1) its score according to the MDL principle and the overlap factor is higher than the best so far, (2) it is positive (i.e., the pair pattern-target by which it is formed is positive), and (3) its overlap with the subgroups already contained in the subgroup list is less or equal to the maximum permitted. The overlap is controlled by the counter list that is obtained by *compute_overlap_factor* function. This function computes the proportion between the number of instances covered by the candidate c and the number of instances covered by the subgroup list sl (stored in a list called *overlap_counter*). Note that, for computing the overlap, subgroups are considered individually (i.e., not depending on their position in the subgroup list). When the best candidate is selected: (1) overlap counter is updated, (2) it is added to the subgroup list, (3) its refinements are deleted of C, and (4) all subgroups from C of which it is a refinement are deleted. Note that both the overlap computation and these last operations contribute to diversity. What is more, they also reduce the number of candidates to explore, which would

---

[1] Available at **subgroups** python library or https://github.com/antoniolopezmc/subgroups.

not be selected anyway due to their overlap. Finally, the collection $\mathcal{L}$ with top-k subgroup lists is returned.

4   Experiments and Discussion

The experiments aim to validate our proposal in the context of phenotyping antimicrobial resistances. We used real clinical data extracted from MIMIC-III public database using as target the patients infected by an Enteroccous Sp. bacterium resistant to Vancomycin. The final dataset had 9,240 instances and 12 attributes. Then, VLSD [2] exhaustive SD algorithm was applied (using WRAcc quality measure with a threshold of 0 and a maximum depth of 3), mining 473 subgroups. After that, the DSLM algorithm was executed using the previous subgroups and with $k = 2$, $sl\_max\_size = 6$ and $max\_overlap = 0.06$, obtaining the subgroup lists depicted in Table 1.

Table 1. Top-2 phenotypes from our dataset.

	Subgroup description (Pattern)	Subgroup coverage Pos-Neg	Contribution Pos-Neg	Phenotype coverage Pos-Neg
s1	culture_type = 'SWAB', icu_when_culture = 'SICU'	466-168	466-168	466-168
s2	culture_type = 'URINE', previous_vancomycin = 'yes'	145-85	145-85	611-253
s3	days_admitted_before_ICU = 'OneDayOrMore', discharge_location = 'DEAD/EXPIRED', patient_age = 'ADULT'	167-100	147-96	758-349
s4	culture_type = 'BLOOD_CULTURE', previous_vancomycin = 'yes'	112-111	98-110	856-459
s5	admission_location = 'PHYS_REFERRAL/NORMAL_DELI', readmission = 'yes', service_when_culture = 'SURG'	124-86	81-80	937-539
s6	culture_type = 'SWAB', previous_vancomycin = 'yes'	114-97	76-89	1013-628
	Default subgroup	-	1113-6486	2126-7114
	Subgroup description (Pattern)	Subgroup coverage Pos-Neg	Contribution Pos-Neg	Phenotype coverage Pos-Neg
s1	culture_type = 'SWAB', service_when_culture = 'SURG'	380-241	380-241	380-241
s2	days_admitted_before_ICU = 'OneDayOrMore', previous_vancomycin = 'yes'	166-136	150-126	530-367
s3	service_when_culture = 'OMED'	115-95	85-92	615-459
s4	days_admitted_before_ICU = 'ZeroDays', previous_vancomycin = 'yes', service_when_culture = 'SURG'	110-77	76-64	691-523
s5	culture_type = 'SWAB', service_when_culture = 'MED'	205-321	193-320	884-843
s6	days_admitted_before_ICU = 'OneDayOrMore', discharge_location = 'DISTINCT_PART_HOSP', service_when_culture = 'SURG'	125-121	76-81	960-924
	Default subgroup	-	1166-6190	2126-7114

The first phenotype describes patients admitted to an ICU ward, adults, and different types of culture (swab, urine, or blood), with treatment by vancomycin in previous admissions, and that could die. The second phenotype represents patients from emergencies or medical wards (SURG, OMED or MED), that could or not be complicated and moved to ICU, and the cultures were swab mainly.

Moreover, it also had vancomycin in previous admissions. Finally, to evaluate the phenotypes from an objective point of view, we show their predictive capacity by using them as dummy variables in a classification algorithm that predicts the target variable of the phenotype. Two classification models were fitted: (1) with the original dataset, and (2) adding two attributes to the original dataset, indicating whether an instance is covered or not by each phenotype. Random Forest, Gradient Boosting Classifier and Logistic Regression were used and the second classification model always obtained a statistically higher accuracy than the first one, demonstrating that both phenotypes have predictive capacity.

5   Conclusions

In this research, we proposed a novel approach for phenotyping based on the task of mining diverse top-k subgroup lists. The aim was to provide clinicians with few descriptions about a specific target value of interest that are diverse both in coverage and in descriptions. Moreover, we also proposed the Diverse Subgroup Lists Miner algorithm (DSLM) algorithm, which generates subgroup lists based on the subgroup discovery paradigm and the minimum description length principle. We carried out the experiments about phenotyping antimicrobial resistances in the MIMIC-III database. The results showed that the top-2 phenotypes represented by the subgroup list model have valuable properties: they are statistically robust, they are legible by the clinical experts, they are diverse, and they have few selectors. Finally, the predictive capacity of the phenotypes obtained has been probed by significantly increasing the accuracy of all the classification algorithms used to predict the same outcome of the phenotype after including the phenotypes as new independent variables in the dataset.

Acknowledgements. this work was partially funded by the CONFAINCE project (Ref: PID2021-122194OB-I00) by MCIN/AEI/10.13039/501100011033 and, as appropriate, by "ERDF A way of making Europe", by the "European Union", and by the GRALENIA project (Ref: 2021/C005/00150055) supported by the Spanish Ministry of Economic Affairs and Digital Transformation, the Spanish Secretariat of State for Digitization and Artificial Intelligence, Red.es and by the NextGenerationEU funding. This research was also partially funded by a national grant (Ref: FPU18/02220), of the Spanish Ministry of Science, Innovation and Universities (MCIU) and by a mobility grant (Ref: R-933/2021), of the University of Murcia.

References

1. Lavrac, N., Kavsek, B., Flach, P.A., Todorovski, L.: Subgroup discovery with CN2-SD. J. Mach. Learn. Res. **5**, 153–188 (2004)
2. Lopez-Martinez-Carrasco, A., Juarez, J.M., Campos, M., Canovas-Segura, B.: VLSD - an efficient subgroup discovery algorithm based on equivalence classes and optimistic estimate. Knowledge and Information Systems. Status: submitted
3. Lopez-Martinez-Carrasco, A., Proença, H.M., Juarez, J.M., van Leeuwen, M., Campos, M.: Discovering diverse top-k characteristic lists. In: 21th Symposium on Intelligent Data Analysis (IDA 2023) (2023)

4. Proença, H.M., Grünwald, P., Bäck, T., van Leeuwen, M.: Robust subgroup discovery. Data Mining and Knowledge Discovery (2022)
5. Wojczynski, M.K., Tiwari, H.K.: Definition of phenotype. Genetic dissection of complex traits. Adv. Genetics **60**, 75–105 (2008)

Patient Event Sequences for Predicting Hospitalization Length of Stay

Emil Riis Hansen[1][(✉)], Thomas Dyhre Nielsen[1], Thomas Mulvad[2],
Mads Nibe Strausholm[2], Tomer Sagi[1], and Katja Hose[1,3]

[1] Department of Computer Science, Aalborg University, Aalborg, Denmark
{emilrh,tdn,tsagi,khose}@cs.aau.dk
[2] Unit of Business Intelligence, North Denmark Region, Aalborg, Denmark
{tml,mns}@rn.dk
[3] TU Wien, Vienna, Austria
katja.hose@tuwien.ac.at

Abstract. Predicting patients' hospital length of stay (LOS) is essential for improving resource allocation and supporting decision-making in healthcare organizations. This paper proposes a novel transformer-based model, termed Medic-BERT (M-BERT), for predicting LOS by modeling patient information as sequences of events. We performed empirical experiments on a cohort of $48k$ emergency care patients from a large Danish hospital. Experimental results show that M-BERT can achieve high accuracy on a variety of LOS problems and outperforms traditional non-sequence-based machine learning approaches.

Keywords: length of stay prediction · transformers · sequence models

1 Introduction

Increasingly scarce hospital resources challenge (often oversaturated) hospital wards, negatively impacting the quality of health care [1]. Models for predicting the remaining patient hospitalization time, i.e., patient length of stay (LOS), is a valuable tool for healthcare facilities in resource availability planning of, e.g., beds and staff. For instance, prediction of discharge time can be used to preemptively free in-hospital resources to alleviate hospital ward oversaturation [11]. However, LOS prediction is a challenging problem, requiring methods for handling missing data [6] and temporal event dependencies integration.

Previous work on LOS prediction often models patient hospitalizations using standard ML models, such as RFs, GBs, and ANNs, relying on imputation techniques for replacing missing values [2]. However, missing observations in healthcare data are often not missing at random (NMAR), and the mere fact that observations are missing is essential information [6].

To alleviate this and other shortcomings, attention-based models have recently been investigated for sequence structured Electronic Health Record (EHR) data [8,10]. Attention models address the inefficiency of recurrent networks for long sequences [10] while still capturing significant sequential information by learning from the order of tokens in a sequence. However, in medical

J. M. Juarez et al. (Eds.): AIME 2023, LNAI 13897, pp. 51–56, 2023.
https://doi.org/10.1007/978-3-031-34344-5_7

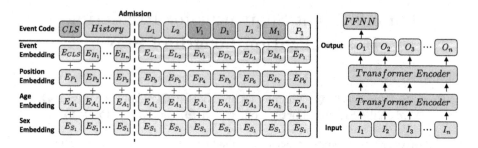

Fig. 1. Medical event sequence pre-pended with the patient's medical history.

data, observations are often grouped with the same timestamp. For example, a blood panel drawn from a patient contains several measurements whose order is undefined. Based on layers of transformer encoders, we propose the Medic-BERT (M-BERT) model inspired by the original BERT model [3]. Using sequences of in-hospital medical events exhibiting event concurrences common in EHR data, we employ M-BERT for LOS prediction. We evaluate M-BERT on a cohort of 48k patient admissions from a large Danish hospital with information on diverse medical events, such as measurements of vital parameters, medication adminis-tration, laboratory tests, and conducted procedures.

2   Transformer Models for EHR

Patient hospitalizations can naturally be modeled as sequences of medical codes for determining, measuring, or diagnosing the patients' conditions. To standard-ize how procedures are described, medical facilities code concepts using accepted taxonomies. Hence, hospitalization can be described as a sequence of concept tokens detailing the medical procedures pertaining to a patient and coded using taxonomical concepts. For some procedures, such as vitals and lab tests, a numer-ical measurement value accompanies the procedure. These measurement values are mapped into normal, abnormal-low, or abnormal-high tokens.

Furthermore, as patient history is essential for correct treatment, we pre-pend the patient's medical history to the hospitalization sequence as a tokenized vector consisting of 38 tokens. These include comorbidities (Charlson Index [12]), five years of prescription history grouped by the first level of the ATC hierarchy [9], and the mode, time, and initial hospitalization triage category [13].

In this paper, we propose Medic-BERT (M-BERT), see Fig. 1, an EHR-data modification of the Bidirectional Encoder Representations from Transformers (BERT) model [3]. BERT is an NLP model based on a stack of encoder lay-ers. The transformer encoder naturally handles complex long-term dependen-cies that occur between medical concepts through its utilization of multi-head self-attention. Furthermore, BERT naturally operates in domains with irregular intervals between events, as is the case with EHR data. BERT can also naturally

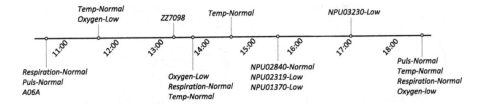

Fig. 2. Patient event sequence illustrating events grouping together.

integrate disparate input, such as diagnostic and therapeutic events, encoding each event as an n-dimensional vector.

M-BERT learns embeddings for the demographic features and the medical event tokens. Together with positional embeddings, the model can learn temporal dependencies within a sequence. We use a static positional embedding modified for usage on medical event sequences: medical events with no natural chronological ordering (e.g., multiple lab tests done on the same sample as illustrated in Fig. 2) are assigned the same positional embedding. Finally, we use a classification (CLS) token as a final aggregate representation fed to a linear output layer for LOS classification and regression tasks.

3   Empirical Evaluation and Results

3.1   Data and Experimental Setting

Table 1. Concept types with occurrences.

Event Type	#Tokens	#Events
Lab Tests	748	2,774,790
Vitals	22	837,931
Medication	1,441	376,591
Procedures	2049	247,924
History	81	1,880,580

We compiled a dataset of hospital emergency care admissions in northern Jutland (Denmark) between 2018–2021. The dataset comprises 48,177 admissions (>one day). Figure 3 presents the remaining length of stay distribution. Table 1 summarizes the event types present in the data.

Due to our focus on a single prediction task, we directly train model parameters and token embeddings toward the downstream task of LOS prediction without any unsupervised pre-training.[1] Using the data gathered within 24 h of

---

[1] https://github.com/dkw-aau/medic_transformer.

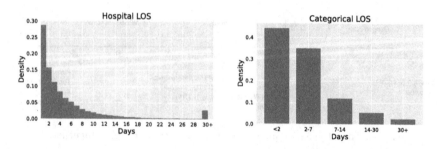

Fig. 3. Length of stay distribution over patient hospitalizations.

admission, we evaluate M-BERT on three LOS prediction tasks: **Binary** classi-fication of LOS > 2 days, a three-class **Category** task of LOS > 2, $2 \leq$ LOS ≤ 7, and LOS > 7 days with class balances as shown in Fig. 3, and the **Real** regression task of predicting LOS in days (decimal).

We compare our approach with an RF, ANN, and SVM model as imple-mented in the Sklearn library [7] using default hyper-parameters. The last mea-sured value for each event type (within 24 h) defined the input features [5]. For features with missing values, we relied on imputation using the mean of the fea-tures, and subsequently scaled the values to be between 0 and 1. A chi^2 test was finally used for extracting the 50 most relevant features.

M-BERT was trained with a 1e−5 learning rate on a 80/10/10 random split of patients. We use the evaluation loss for early stopping within ten epochs. The model architecture has six hidden layers with an intermediate layer size of 288, eight attention heads, and an input token embedding size of 288. We truncate sequences to 256 tokens, as most sequences adhere to this limit. To counter overfitting, we add a 10% dropout layer after the final encoder layer, a 10% attention dropout, and a weight decay of 0.003. Further details are available in the extended version of this paper [4].

3.2 Results

Table 2 presents AUROC and F1 scores for the **Binary** and **Category** exper-iments and Mean Absolute Error (MAE) and Mean Squared Error (MSE) for the regression task. Model performance is stable for different age groups. The results indicate that M-BERT can leverage temporal dependencies inherent to EHR data for increased predictive accuracy. Being a transformer-based model, M-BERT overcomes the challenge of missing data as patient sequences are not required to contain the same events or be of the same length.

Table 2. Experimental results.

	Binary		Category		Real	
	AUROC	F1	AUROC	F1	MAE	MSE
RF	0.72	0.70	0.66	0.45	4.18	39.08
ANN	0.67	0.68	0.63	0.43	4.09	38.10
SVM	0.70	0.70	0.65	0.38	3.56	43.36
M-BERT	**0.78**	**0.77**	**0.74**	**0.54**	**3.42**	**37.48**

4 Conclusion

We have proposed a novel approach for predicting LOS by modeling patient information as event sequences. We adapt the transformer-based architecture to sequence prediction over grouped events of varying data types as typically found in medical event sequences. Our empirical evaluation on a large cohort of emergency care patients from a Danish hospital demonstrates high accuracy on various LOS problems, while also outperforming traditional non-sequence-based approaches. Future work includes model pre-training as well as evaluation of the predictive uncertainty offer by the model. Overall, the proposed approach has the potential to improve resource allocation in healthcare organizations by providing accurate and reliable predictions of LOS.

Acknowledgments. This study received funding from the Region North Denmark Health Innovation Foundation. This study is also supported by the Poul Due Jensen Foundation.

References

1. af Ugglas, B., Djärv, T., Ljungman, P.L., Holzmann, M.J.: Association between hospital bed occupancy and outcomes in emergency care: a cohort study in stockholm region, sweden, 2012 to 2016. Ann. Emerg. Med. 76(2), 179–190 (2020)
2. Bacchi, S., Tan, Y., Oakden-Rayner, L., Jannes, J., Kleinig, T., Koblar, S.: Machine learning in the prediction of medical inpatient length of stay. Intern. Med. J. 52(2), 176–185 (2022)
3. Devlin, J., Chang, M., Lee, K., Toutanova, K.: BERT: pre-training of deep bidirectional transformers for language understanding. In: NAACL-HLT 2019, pp. 4171–4186 (2019). https://doi.org/10.18653/v1/n19-1423
4. Hansen, E.R., Nielsen, T.D., Mulvad, T., Strausholm, M.N., Sagi, T., Hose, K.: Hospitalization length of stay prediction using patient event sequences (2023)
5. Iwase, S., Nakada, T.A., Shimada, T., Oami, T., Shimazui, T., Takahashi, N., et al.: Prediction algorithm for ICU mortality and length of stay using machine learning. Sci. Rep. 12(1), 1–9 (2022)
6. Li, J., Yan, X.S., Chaudhary, D., Avula, V., Mudiganti, S., Husby, H., et al.: Imputation of missing values for electronic health record laboratory data. NPJ Digital Med. 4(1), 1–14 (2021)

7. Pedregosa, F., Varoquaux, G., Gramfort, A., Michel, V., Thirion, B., Grisel, O., et al.: Scikit-learn: machine learning in Python. JMLR **12**, 2825–2830 (2011)
8. Rasmy, L., Xiang, Y., Xie, Z., Tao, C., Zhi, D.: Med-BERT: pretrained contextualized embeddings on large-scale structured electronic health records for disease prediction. NPJ Digital Med. **4**(1), 1–13 (2021)
9. Ronning, M.: A historical overview of the ATC/DDD methodology. WHO Drug Inf. **16**(3), 233 (2002)
10. Song, H., Rajan, D., Thiagarajan, J., Spanias, A.: Attend and diagnose: clinical time series analysis using attention models. In: AAAI 2018, vol. 32 (2018)
11. Stone, K., Zwiggelaar, R., Jones, P., Mac Parthaláin, N.: A systematic review of the prediction of hospital length of stay: towards a unified framework. PLOS Digital Health **1**(4), e0000017 (2022)
12. Sundararajan, V., Henderson, T., Perry, C., Muggivan, A., Quan, H., Ghali, W.A.: New ICD-10 version of the charlson comorbidity index predicted in-hospital mortality. J. Clin. Epidemiol. **57**(12), 1288–1294 (2004)
13. Wireklint, S.C., Elmqvist, C., Göransson, K.E.: An updated national survey of triage and triage related work in sweden: a cross-sectional descriptive and comparative study. In: SJTREM '21 29(1), pp. 1–8 (2021)

Autoencoder-Based Prediction of ICU Clinical Codes

Tsvetan R. Yordanov[1,2(✉)] [ID], Ameen Abu-Hanna[1,2] [ID], Anita CJ. Ravelli[1,2] [ID], and Iacopo Vagliano[1,2] [ID]

[1] Department Medical Informatics, Amsterdam UMC, University of Amsterdam, Amsterdam, Netherlands
{t.yordanov,a.abu-hanna,a.c.ravelli,i.vagliano}@amsterdamumc.nl
[2] Amsterdam Public Health Research Institute, Meibergdreef 9, Amsterdam, Netherlands

Abstract. Availability of diagnostic codes in Electronic Health Records (EHRs) is crucial for patient care as well as reimbursement purposes. However, entering them in the EHR is tedious, and some clinical codes may be overlooked. Given an incomplete list of clinical codes, we investigate the performance of ML methods on predicting the complete ones, and assess the added predictive value of including other clinical patient data in this task. We used the MIMIC-III dataset and frame the task of completing the clinical codes as a recommendation problem. We consider various autoencoder approaches plus two strong baselines; item co-occurrence and Singular Value Decomposition (SVD). Inputs are 1) a record's known clinical codes, 2) the codes plus variables. The co-occurrence-based approach performed slightly better (F1 score = 0.26, Mean Average Precision [MAP] = 0.19) than the SVD (F1 = 0.24, MAP = 0.18). However, the adversarial autoencoder achieved the best performance when using the codes plus variables (F1 = 0.32, MAP = 0.25). Adversarial autoencoders performed best in terms of F1 and were equal to vanilla and denoising autoencoders in term of MAP. Using clinical variables in addition to the incomplete codes list, improves the predictive performance of the models.

Keywords: Prediction · Medical codes · Recommender Systems · Autoencoders

1 Introduction

During a patient's stay at an Intensive Care Unit (ICU) their record will often have multiple diagnostic and procedure codes entered. The clinical codes for a patient may be incomplete or missing for various reasons. To help automate data entry, we investigate the task of predicting clinical diagnostic and procedure codes for ICU patients. We consider missing clinical codes as a recommendation problem, and develop a Recommender System (RS) that suggests clinical codes for a patient stay given what is already present in their EHR record. We focus on AutoEncoders (AEs) as they are known to perform well at recommendation problems [1, 2]. In the current experiments we investigate different AE types and apply them to the problem of clinical code recommendation.

© The Author(s), under exclusive license to Springer Nature Switzerland AG 2023
J. M. Juarez et al. (Eds.): AIME 2023, LNAI 13897, pp. 57–62, 2023.
https://doi.org/10.1007/978-3-031-34344-5_8

We aimed to evaluate the predictive performance of different types of AEs (vanilla, denoising, variational, and adversarial) using two types of input data: 1. The known clinical codes for a record, and 2. The codes and variables [structured data]). We compare their results against two baseline models.

2   Methods

Problem statement. We define a set of m ICU patient records P and a set of n clinical codes C, where we attempt to model the spanned space of $P \times C$. A sparse matrix $X \in \{0, 1\}^{m \times n}$ is used to represent the codes of each patient ICU stay where X_{ij} indicates the presence of clinical code j in patient record i. Additional information from a record is held in the supplementary information matrix S [2]. As supplementary information we considered the measurements of patient vitals during ICU stay (e.g., heart rate) along with demographics (e.g., age).

Dataset. The MIMIC-III dataset is a large freely-available dataset of 57,786 ICU hospital admissions of 46,476 patients from one US hospital between 2001–2012 [3].

For structured data, each record holds laboratory measurements and vital signs, patient demographics, and a list of clinical diagnostic and procedure codes. Additionally, there are semi-structured data like the textual descriptions of the clinical codes, and unstructured data like clinical notes.

The International Statistical Classification of Diseases and Related Health Problems version 9 (ICD9) was used for coding patient conditions and the procedures received. ICD9 consists of about 13,000 unique codes in varying levels of detail.

Inclusion/Exclusion Criteria. We consider codes which appear at least 50 times in the dataset [4] and only include admissions of patients older than 18 years of age.

Features. We used the list of known diagnostic and procedure codes per patient ICU stay. For the various AEs we also included the supplementary information of the structured patient data.

We considered all numeric and categorical features available during the whole ICU stay. Time-series were aggregated with the mean and the difference between the last and first measurement. Time-series were imputed by a sliding-window average of length one, where a missing measurement was set to the average of the two non-missing nearest measurements.

Models. We compare performances between four types of autoencoders – 1) vanilla AEs, 2) Denoising Autoencoders (DAEs) [5], 3) Variational Autoencoders (VAEs) [6], and 4) Adversarial Autoencoders (AAEs) [7]. We evaluated each AE type using the two different input sets described earlier.

We considered the following two baseline models: 1) a simple co-occurrence based method and 2) the Singular Value Decomposition (SVD) matrix factorization method. Since these baselines do not support the provision of supplementary information, only the lists of ICD codes were used as input.

Model Training and Evaluation Strategy. We follow the pre-processing steps described in [4]. We used 5-fold Cross-Validation (CV) with a 80/10/10 split of each fold into training (P_{train}), validation (P_{val}) and test sets (P_{test}). Hyperparameter values obtained from using grid-search in the first fold were then used and not re-computed for the remaining folds.

During training, models were provided with the full list of clinical codes, and where applicable supplementary data, from a patient record, while during evaluation on P_{val} or P_{test} they received a randomly-sampled 50% subset of the codes. The models were evaluated on predicting the missing codes.

For tuning the count-based and SVD models, we used correlation order and number of dimensions, respectively. For AEs, we optimized the learning rate, number of epochs, batch size, number of neurons per hidden layer, and for the latent space.

Performance Metrics. We report on the F1 score for measuring the harmonic mean of the precision and recall of the recommendations, as well as the Mean Average Precision (MAP) for measuring the relevance of the recommendations.

3 Results

Table 1. Averaged performance metric results from 5-fold cross-validation on the test sets.

		Only codes	Codes + TD
Item co-occurrence	F1 (SD)	0.26 (0.0019)	-
	MAP (SD)	0.19 (0.0018)	-
SVD	F1 (SD)	0.24 (0.0014)	-
	MAP (SD)	0.18 (0.0012)	-
Vanilla AE	F1 (SD)	0.31 (0.0035)	0.31 (0.0034)
	MAP (SD)	0.25 (0.0037)	0.25 (0.0035)
DAE	F1 (SD)	0.30 (0.0035)	0.31 (0.0039)
	MAP (SD)	0.24 (0.0034)	0.25 (0.0043)
VAE	F1 (SD)	0.20 (0.0029)	0.23 (0.0033)
	MAP (SD)	0.13 (0.0025)	0.16 (0.0026)
AAE	F1 (SD)	0.31 (0.0059)	**0.32 (0.0013)**
	MAP (SD)	0.25 (0.0055)	0.25 (0.0019)

Types of input data considered: only clinical codes ('**only codes**'); clinical codes and Tabular Data ('**codes + TD**'). SVD – Singular Value Decomposition; DAE – Denoising Autoencoder; VAE - Variational Autoencoder; AAE – Adversarial Autoencoder.

Dataset. The pre-processed dataset contained 49,002 admissions of 38,402 patients. 822,047 codes were recorded with more than 50 occurrences, of which 1,581 unique codes (1,208 diagnostic and 372 procedural). The median number of codes given per record was 15 with an Inter-Quartile-Range (IQR) of 10–21.

Performance. All AE's apart from VAEs achieved a higher F1 score and MAP compared to the baselines (Table 1). The count-based predictor achieved a slightly higher F1 (0.26) than the SVD (0.24), as well as a slightly better MAP (0.19 versus 0.18). Overall, the best model was the AAE using supplementary data (F1 0.32; MAP 0.25), followed by the vanilla AE and DAE using the same supplementary information (F1 0.31; MAP 0.25).

4 Discussion

In our experiments we demonstrated that for an ICU patient record using supplementary information in addition to the incomplete clinical code list leads to an improvement in the F1 score and MAP metric in 3 out of 4 and 2 out of 4 types of AEs respectively.

Related Work. The challenge of predicting clinical codes in ICU EHR patients has received attention in recent works [8, 9]. In all comparative literature the authors formulated the problem as a Multilabel Classification (MC) task, and thus their performance metrics reported should not be directly compared to those from our experiments where we phrased the problem as a RS task.

Xu et al. developed an ensemble ML method for predicting ICD codes in the MIMIC-III dataset [9]. They used the 32 most frequently-occurring ICD codes. Their ensemble model achieved an F1-micro of 0.76, but that work addressed a simpler problem than ours by using only the most common ICD codes, and the architecture of the proposed solution involved a rather complex ensemble of ML methods.

Bao et al. [8] developed a hybrid capsule network model om MIMIC-III using only the clinical notes as inputs. Their best performance F1-micro was 0.67. Unlike our approach, Bao et al. used rolled-up representation of the clinical codes up to the first three characters and only a selection of 344 codes of three characters for their experiments.

Strengths and Limitations. Strengths of the current study include its use of a large well-known freely-available EHR dataset, the adoption of pre-processing steps described from previous studies, the employment of cross-validation for optimism correction in the results, and the battery of different types of autoencoders considered. Compared to the previous approaches for clinical code prediction, we adopt a re-phrasing of the task as a recommender system problem and propose a novel application using autoencoders to solve it. Our code is available at https://github.com/tsryo/aae-recommender.

Our study has some limitations. Due to resource constraints, hyperparameter optimization was only done for the first fold of each cross-fold validation and the selected parameters were used in subsequent folds. We only considered autoencoders in the current experiments as they were promising for recommendation tasks. It is unclear whether other deep learning approaches could have performed even better.

Future Work. In the current approach we did not use semi-structured or unstructured data as inputs, where we suspect there could be further gains in performance possible. Future work should investigate the possibility of incorporating additional sources of information to the ones presented here and report on their impact on model performance. We reported on overall performance across all clinical diagnostic and procedure codes, whilst the question of how that performance would translate into subgroups of codes and code hierarchies was left open. It would be interesting to see how performance varies between the different categories of clinical codes.

We only considered one value for the proportion of codes to hide from a record for evaluating the models' performance (50%). It would be valuable to investigate its impact on model performance.

We excluded clinical codes that occurred less than 50 times in the dataset. Future research should evaluate the impact on performance when including rare codes.

The small gain in performance from using structured supplementary data could potentially increase with more sophisticated encoding techniques, such as using Long-Short Term Memory network [10] or Time2Vec [11] to encode time series, and merits further investigation.

Our system could be used by EHR administrators when performing clinical code entry- where they obtain recommendations for which codes to enter in a record in order to decrease manual workload for instance in case of an applying for reimbursement. Before that can happen, however the utility of introducing such a tool in practice must be evaluated.

References

1. Vagliano, I., Galke, L., Scherp, A.: Recommendations for item set completion: on the semantics of item co-occurrence with data sparsity, input size, and input modalities. Information Retrieval Journal **25**, 269–305 (2022)
2. Galke, L., Mai, F., Vagliano, I., Scherp, A.: Multi-Modal Adversarial Autoencoders for Recommendations of Citations and Subject Labels. Proceedings 26th Conference on User Modeling, Adaptation and Personalization, pp. 197–205. ACM (2018)
3. Johnson, A.E.W., Pollard, T.J., Shen, L., Lehman, L.-w.H., Feng, M., Ghassemi, M., Moody, B., Szolovits, P., Anthony Celi, L., Mark, R.G.: MIMIC-III, a freely accessible critical care database. Scientific Data 3, 160035 (2016)
4. Lovelace, J., Hurley, N.C., Haimovich, A.D., Mortazavi, B.J.: Dynamically Extracting Outcome-Specific Problem Lists from Clinical Notes with Guided Multi-Headed Attention. In: Finale, D.-V., Jim, F., Ken, J., David, K., Rajesh, R., Byron, W., Jenna, W. (eds.) Proceedings of the 5th Machine Learning for Healthcare Conference, vol. 126, pp. 245--270. PMLR, Proceedings of Machine Learning Research (2020)
5. Vincent, P., Larochelle, H., Lajoie, I., Bengio, Y., Manzagol, P.-A.: Stacked Denoising Autoencoders: Learning Useful Representations in a Deep Network with a Local Denoising Criterion. JMLR **11**, 3371–3408 (2010)
6. Kingma, D.P., Welling, M.: Auto-Encoding Variational Bayes. CoRR abs/1312.6114, (2013)
7. Makhzani, A., Shlens, J., Jaitly, N., Goodfellow, I., Frey, B.: Adversarial Autoencoders. pp. arXiv:1511.05644 (2015)
8. Bao, W., Lin, H., Zhang, Y., Wang, J., Zhang, S.: Medical code prediction via capsule networks and ICD knowledge. BMC Med. Inform. Decis. Mak. **21**, 55 (2021)
9. Xu, K., Lam, M., Pang, J., Gao, X., Band, C., Mathur, P., Papay, F., Khanna, A.K., Cywinski, J.B., Maheshwari, K., Xie, P., Xing, E.P.: Multimodal Machine Learning for Automated ICD Coding. In: Finale, D.-V., Jim, F., Ken, J., David, K., Rajesh, R., Byron, W., Jenna, W. (eds.) Proceedings of the 4th Machine Learning for Healthcare Conference, vol. 106, pp. 197--215. Cambridge MA: JMLR, Ann Arbor, Michigan (2019)
10. Hochreiter, S., Schmidhuber, J.: Long Short-Term Memory. Neural Comput. **9**, 1735–1780 (1997)

11. Kazemi, S.M., Goel, R., Eghbali, S., Ramanan, J., Sahota, J., Thakur, S., Wu, S., Smyth, C., Poupart, P., Brubaker, M.A.: Time2Vec: Learning a Vector Representation of Time. CoRR abs/1907.05321, (2019)

Explainability and Transfer Learning

Hospital Length of Stay Prediction Based on Multi-modal Data Towards Trustworthy Human-AI Collaboration in Radiomics

Hubert Baniecki[1,2(✉)] ⓘ, Bartlomiej Sobieski[2], Przemysław Bombiński[2,3],
Patryk Szatkowski[2,3], and Przemysław Biecek[1,2] ⓘ

[1] MI2.AI, University of Warsaw, Warsaw, Poland
h.baniecki@uw.edu.pl
[2] MI2.AI, Warsaw University of Technology, Warsaw, Poland
przemyslaw.biecek@pw.edu.pl
[3] Medical University of Warsaw, Warsaw, Poland

Abstract. *To what extent can the patient's length of stay in a hospital be predicted using only an X-ray image?* We answer this question by comparing the performance of machine learning survival models on a novel multi-modal dataset created from 1235 images with textual radiology reports annotated by humans. Although black-box models predict better on average than interpretable ones, like Cox proportional hazards, they are not inherently understandable. To overcome this trust issue, we introduce time-dependent model explanations into the human-AI decision making process. Explaining models built on both: human-annotated and algorithm-extracted radiomics features provides valuable insights for physicians working in a hospital. We believe the presented approach to be general and widely applicable to other time-to-event medical use cases. For reproducibility, we open-source code and the TLOS dataset at https://github.com/mi2datalab/xlungs-trustworthy-los-prediction.

Keywords: explainable AI · survival analysis · healthcare · radiology · interpretable machine learning

1 Introduction

Predicting patients' hospital length of stay (LoS) is a challenging task supporting the day-to-day decisions of medical doctors and nurses [6]. For example, accurate LoS prediction can increase hospital service efficiency, cutting costs and improving patient care. Historically, white-box statistical learning methods were used to estimate the anticipated LoS [2]. These provide clear reasoning behind the prediction, which is especially important in medical applications requiring stakeholders to comprehend "Why?" [11]. Nowadays, advancements in machine and deep learning for healthcare provide valuable improvements in the performance of predicting LoS [8,17,18]. The natural drawback of using not inherently

© The Author(s), under exclusive license to Springer Nature Switzerland AG 2023
J. M. Juarez et al. (Eds.): AIME 2023, LNAI 13897, pp. 65–74, 2023.
https://doi.org/10.1007/978-3-031-34344-5_9

interpretable black-box models is their complex nature [1,11]. Indeed, a recent systematic review on the exact topic of hospital LoS prediction concludes with a concrete statement that there are no studies on the explainability of black-box models predicting LoS [14], a matter of high importance for diverse stakeholders involved in this healthcare process. Therefore in this paper, we demonstrate the applicability of explainable machine learning methods [1,7] in the LoS prediction task as an enabler towards trustworthy human-AI collaboration.

Contribution. We summarize our contributions as follows. In Sect. 2, we introduce a novel task of hospital LoS prediction based on multi-modal X-ray data and benchmark on it machine learning survival models. To achieve this, we create the TLOS dataset by manually annotating 1235 X-ray textual radiology reports from one of the Polish hospitals resulting in 17 interpretable features. Moreover, we include the state-of-the-art radiomics features extracted from images and critically evaluate their predictive performance. In Sect. 3, we put recent advancements in time-dependent explainable machine learning to practice. In that, we explain the best-performing models to gain insights into the importance of features and their effects on LoS prediction. Analysing complementary explanations leads to an improved human understanding of AI, e.g. allows discovering bias in a model, increasing trust. We conclude with a discussion on the limitations of our study and potential future work in Sect. 4.

Related Work. Applying AI through machine learning to predict hospital length of stay from data is broadly studied as it has a high potential to support decision making in healthcare [14]. In [6], LoS is predicted based on information from clinical treatment processes, i.e. sequences of time-point hospital events. In [2], hospital events like the number of tests and time of arrival are aggregated to quantify their importance in patient discharge. We base our analysis on clinical features impacting physician understanding of the patient's severity instead. In [8], machine learning models predict LoS after brain surgery using clinical features in a single-value regression task. This results in simple feature importance and effects explanations without the valuable time dimension. In [18], deep learning models classify the patient staying in a hospital for longer than 7 days based on multi-modal data combining unstructured notes and tabular features. In [17], various machine and deep learning survival models are benchmarked for predicting LoS of COVID-19 patients. Both studies focus on predictive performance without explaining the models. For a broader overview of works on LoS prediction, we refer the reader to [14].

Contrary to the above-mentioned studies, we specifically use raw X-ray images to analyze the predictive power of radiomics features and explain the prediction in a time-dependent manner. To achieve it, we rely on `pyradiomics` – the state-of-the-art radiomics feature extraction tool [15] and `survex` – a toolbox for explaining machine learning survival models [13] implementing among others the SurvSHAP(t) explanation method [7]. Related to this are works introducing alternative explanation methods to interpret time-to-event models predicting patients' survival [9,16]. We explain survival models predicting LoS instead.

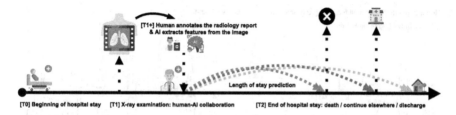

Fig. 1. Human-AI collaboration in radiomics. Shortly after the X-ray examination, a physician annotates the radiology report and AI extracts features from the image. Based on this data, a time-to-event model is used to predict the hospital length of stay.

2 Predicting Hospital Length of Stay Using X-ray Images

We consider a setting where LoS is predicted using time-to-event survival models instead of single-value time estimation with regression models or time-span classification. Besides giving a more holistic prediction, survival analysis naturally allows for censored observations in data, e.g. a patient was discharged from one hospital and moved to another with further information missing (see Fig. 1).

A unique multi-modal dataset used in this study is created based on image, text and tabular data of 1235 patients from one of the Polish hospitals. The *target feature* is the time between the patient's radiological examination and hospital discharge (in days, $min = 1$, $median = 7$, $mean = 13.73$, $max = 330$). Due to the high skewness of the time distribution, we model the logarithm of time in practice. About 20% of outcomes are right-censored, e.g. due to death.

Each radiologic exam consists of an X-ray image with a written report stating observable features, e.g. pathological signs, lung lesions and pleural abnormalities, but also the occurrence of medical devices on the image, e.g. tubing and electrocardiographic leads. We manually annotate each report into 17 interpretable binary features informing whether the pathology occurs or is absent. Note that we sampled patients at random and capped their quantity after reaching the reasonable capacity of human annotators. Moreover, we automatically extract 76 numerical features from the image using the `pyradiomics` tool [15]. It computes various statistics based on an image and a lung segmentation mask. e.g. various aggregations of the gray-level co-occurrence matrix.[1] A pretrained CE-Net [3] model was used to obtain lung segmentation masks inputted to `pyradiomics`. We treat it as a reasonable baseline approach while acknowledging that segmentation errors will inevitably contribute to errors in algorithm-extracted features.

The described procedure leads to obtaining four feature sets referred to as:

- *baseline* (number of features: $d = 2$) – includes the patient's age and sex,
- *human-annotated* ($d = 2 + 17$) – includes *baseline* and pathology occurrences,
- *algorithm-extracted* ($d = 2 + 76$) – includes *baseline* and image statistics,

---

[1] A detailed description of algorithm-extracted features from `pyradiomics` is available at https://pyradiomics.readthedocs.io/en/v3.0.1/features.html.

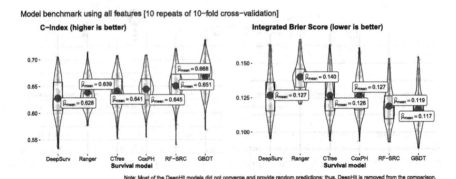

Fig. 2. Benchmark of machine learning survival models predicting LoS using features from X-ray images. Based on 10 repeats of 10-fold cross-validation, the GBDT algorithm performs best on average (marked with $\hat{\mu}_{mean}$), i.e. achieves 0.668 C-index and 0.117 IBS. In contrast to Fig. 3, we omit reporting p values for significant differences as there are too many.

– *all features* ($d = 2 + 17 + 76$).

We use the dataset to answer the question of interest: *To what extent can the patient's length of stay in a hospital be predicted using only an X-ray image?*

Models. We first use all features to compare various machine learning survival models in predicting LoS using X-ray images and then evaluate the impact of particular feature sets on the best models' predictive performance.

We perform a comprehensive benchmark of relevant learning algorithms available in the `mlr3proba` toolbox [12]: a decision tree (CTree), gradient boosting decision trees (GBDT), two implementations of random survival forest (Ranger & RF-SRC), Cox proportional hazards (CoxPH), and two implementations of neural networks (DeepSurv & DeepHit). We use 10 repeats of 10-fold cross-validation and assess the predictive performance of survival models with two measures: C-index where higher means better performance and the baseline value of a random model equals 0.5, and Integrated Brier Score (IBS) where lower is better and the baseline value of a random model equals 0.25. The evaluation protocol mimics a benchmark of survival prediction methods described in [5].

Figure 2 presents the results where models are sorted by an average C-index. First, note that most DeepHit models did not converge and provide random predictions; thus, DeepHit is removed from the comparison. We observe that, on average, the best algorithm is a gradient boosting decision tree (0.668 C-index, 0.117 IBS) with a random survival forest in second. Interestingly, the interpretable and widely-used CoxPH model performs worse (0.645 C-index, 0.127 IBS). Overall, neural networks have a hard time learning meaningful models.

Comparing the raw performance values with results from related work leads to the conclusion that *predicting the patient's LoS from an X-ray image is indeed*

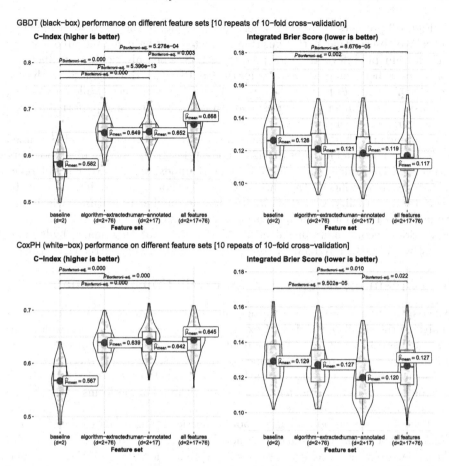

Fig. 3. Benchmark of feature sets in predicting LoS using GBDT black-box and CoxPH white-box model. The p values mark significant differences between the averages $\hat{\mu}_{mean}$.

possible, but challenging. Possible future improvements include improving data quality and tuning the hyperparameters of learning algorithms (see Sect. 4).

Feature Performance. We now tackle the performance-interpretability trade-off in machine learning for medicine [11]. Based on the benchmark results reported in Fig. 2, we choose the best black-box algorithm (GBDT) to compare with CoxPH – the widely-adopted interpretable approach to time-to-event analysis. Note that one can consider the human-annotated features as interpretable and algorithm-extracted features as a black-box approach, i.e. training the CoxPH model on algorithm-extracted features is not necessarily a white-box model. We test the two algorithms on four feature sets using the same cross-validation scheme and performance measures as the previous benchmark.

Figure 3 presents the results where feature sets are sorted by an average C-index. We observe that both algorithm-extracted and human-annotated features include valuable information for predicting LoS. The only significant difference (on average) between the two sets is for the CoxPH algorithm evaluated with IBS. For GBDT, using all features results in the best performance, while for CoxPH, increasing the number of features leads to the same or worse performance due to the curse of dimensionality.

In summary, the best-performing interpretable algorithm is CoxPH trained on human-annotated features (0.642 C-index, 0.120 IBS), and the black-box approach is GBDT trained on all features (0.668 C-index, 0.117 IBS). The difference in C-index is significant (p value < 0.001). Although this difference may be neglectable in reality, CoxPH is limited by the number of features, which now rapidly increases in medical applications, e.g. in radiology where tools for feature extractions become more prevalent [2]. Moreover, annotating images by humans is costly, and GBDT remains more efficient with algorithm-extracted features.

3   Explaining Length of Stay Predictions to Humans

A classic approach to the explanatory analysis of time-to-event models involves analysing the significance of the CoxPH model's coefficients. For a broader context, we use all observations in data to fit CoxPH models to the four feature sets. Table 1 reports features with significant coefficients, effectively serving as a list of features important to predicting LoS. The main limitation of this approach is explaining only a particular learning algorithm that performs worse in the predictive task, as in our case.

Therefore, we propose to use model-agnostic explanations to interpret the predictions of any black-box survival model predicting LoS in general. We extend the explanatory model analysis framework [1] to include the SurvSHAP(t) explanation [7], time-dependent What-if and Partial dependence plots, all implemented in **survex** [13]. For a concrete example, we use all observations in data and fit a GBDT model to the human-annotated feature set.

Figure 4 presents four complementary local and global explanations providing a multi-faceted understanding of the black-box model. It is accompanied by exemplary X-ray images of lung disease and healthy lungs with medical devices. We first interpret the particular prediction for a 78-year-old male patient with parenchymal opacification in the lungs and possibly also pleural effusion. SurvSHAP(t) attributes high importance to the occurrence of parenchymal opacification, but also the absence of medical devices on the chest X-ray. In fact, the latter decreases the predicted probability of longer LoS as X-rays with observable medical devices usually indicate a severe patient condition. This image feature is a potential bias in data that later propagates to a predictive model. Next, we perform a What-if analysis for the ambiguous pleural effusion feature to explain the uncertainty in LoS prediction conditioned on this feature. Over the 60 days since the X-ray examination, there is an increased probability of staying in a hospital by 0.125 when a pleural effusion occurred (for this patient).

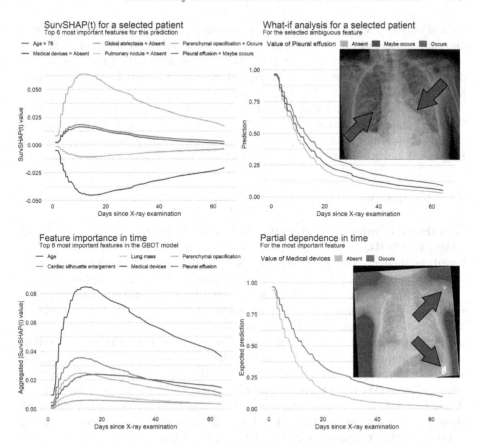

Fig. 4. Complementary time-dependent explanations of the GBDT model trained on age, sex, and human-annotated radiomics features. These are accompanied by exemplary X-ray images of (**top**) lung disease in an adult and (**bottom**) healthy children's lungs with medical devices. **Top left**: SurvSHAP(t) local explanation for a selected patient informing about the 6 most important features and their effect on the predicted length of stay at each day since X-ray examination. **Top right**: What-if analysis for the same patient and the selected ambiguous feature, informing about how the predicted length of stay would change upon the change in feature value. **Bottom left**: Feature importance global explanation based on aggregated SurvSHAP(t) values for a subset of patients informing about the 6 most important features overall. **Bottom right**: Partial dependence global explanation for the most important feature informing about its effect on the predicted length of stay on average.

The bottom of Fig. 4 presents global explanations of the model's behaviour: feature importance and effects, also referred to as Partial dependence. We obtain time-dependent feature importance by aggregating absolute SurvSHAP(t) values for a representative subset of patients. The visualization indicates that the model finds the occurrence of medical devices as a proxy for LoS, which may be correlated with the patient's condition. Other important radiomics features include pleural effusion, age, parenchymal opacification, cardiac silhouette enlargement and lung mass. For each of these pathologies, a Partial dependence plot explains its aggregated effect on the LoS prediction. Specifically, when medical devices occur on an image, there is an increased probability by 0.25 of staying in a hospital 20 days after the X-ray examination.

As illustrated here, incorporating time-dependent explanations of machine learning models predicting LoS into the existing decision support systems can provide useful information for physicians. We believe the presented approach to be general and widely applicable to other time-to-event medical use cases.

4   Discussion

Details of Human Annotation. The original raw dataset includes X-ray images and textual radiology reports, which were manually annotated by two board-certified radiologists. First, we developed a project-specific ontology of chest pathologies that can be observed on X-ray images. It compiles information from the selected relevant radiology literature [4], existing ontologies like RadLex [10], and popular X-ray databases available online (CheXpert, MIMIC, VinBigData challenge). Second, we identified problems in our reports related to, e.g. ambiguous interpretations of phrases in the radiological reports, differences in the quality and length of descriptions prepared by different physicians and for different types of examinations (like follow-up or comparison to previous examinations). We fixed them by updating the ontology in accordance with the domain knowledge. As a result, a set of 35 classes was obtained. Finally, we chose the 17 most common features, i.e. with more than 3% occurrence among patients, for further analysis in this study. Although we are yet unable to share X-ray images and textual reports due to privacy concerns, the preprocessed TLOS dataset with further documentation and code to reproduce our results is available at https://github.com/mi2datalab/xlungs-trustworthy-los-prediction. We hope this resource can spark further research on trustworthy LoS prediction.

Limitations. First, the segmentation model used to obtain masks for feature extraction propagates errors to LoS predictions, and it can be improved if the final goal of a study would be a definitive evaluation of those features. Tuning the hyperparameters of survival models is also an option. Moreover, physicians and other stakeholders interpreting predictions need to understand that the presented explanations are only an approximation of the black-box model [11]. One needs to be sure that stakeholders properly interpret these visualizations.

Table 1. Coefficients of only significant features (p value < 0.05) in the CoxPH models fitted to all observations and the four feature sets. For example, the CoxPH model fitted on age and sex of 1235 patients relies on age, not sex to make the prediction. A detailed description of algorithm-extracted features with their acronyms is available at https://pyradiomics.readthedocs.io/en/v3.0.1/features.html.

Feature set	Feature name	Estimate	p value
baseline ($d = 2$)	Age	$0.007_{\pm 0.001}$	0.000
baseline with human-annotated features ($d = 2 + 17$)	Parenchymal opacification	$-0.194_{\pm 0.081}$	0.016
	Pleural effusion	$-0.334_{\pm 0.084}$	0.000
	Medical devices	$-0.628_{\pm 0.083}$	0.000
	Lung mass	$-0.311_{\pm 0.142}$	0.029
	Reticular pattern	$-0.397_{\pm 0.156}$	0.011
	Global atelectasis	$-0.430_{\pm 0.216}$	0.046
baseline with algorithm-extracted features ($d = 2 + 76$)	Shape–Maximum2DDiameterSlice	$-1.547_{\pm 0.512}$	0.003
	Firstorder–Entropy	$422.9_{\pm 161.1}$	0.009
	GLCM–ID	$-325.9_{\pm 140.0}$	0.020
	GLCM–IDN	$1000_{\pm 425.7}$	0.019
	GLCM–IMC1	$296.4_{\pm 85.14}$	0.000
	GLCM–InverseVariance	$93.89_{\pm 43.39}$	0.030
	GLCM–JointEntropy	$-533.5_{\pm 177.5}$	0.003
	GLDM–DependenceEntropy	$156.7_{\pm 57.70}$	0.007
	GLDM–LGLE	$-120.9_{\pm 49.07}$	0.014
	GLRLM–GLNN	$-497.3_{\pm 207.2}$	0.016
	GLRLM–LRE	$6.213_{\pm 2.180}$	0.004
	GLSZM–SALGLE	$-694.2_{\pm 283.6}$	0.014
all features ($d = 2 + 17 + 76$)	Pleural effusion	$-0.385_{\pm 0.095}$	0.000
	Medical devices	$-0.389_{\pm 0.101}$	0.000
	Chest wall subcutaneous emphysema	$0.354_{\pm 0.180}$	0.049
	Shape–Maximum2DDiameterRow	$0.730_{\pm 0.313}$	0.020
	Shape–Maximum2DDiameterSlice	$-1.531_{\pm 0.536}$	0.004
	Firstorder–Entropy	$407.0_{\pm 165.3}$	0.014
	Firstorder–Range	$-4.361_{\pm 1.858}$	0.019
	GLCM–ID	$-358.4_{\pm 142.8}$	0.012
	GLCM–IDN	$926.6_{\pm 438.3}$	0.034
	GLCM–IMC1	$274.2_{\pm 87.89}$	0.002
	GLCM–InverseVariance	$94.82_{\pm 44.86}$	0.035
	GLCM–JointEntropy	$-517.9_{\pm 182.2}$	0.004
	GLDM–DependenceEntropy	$185.9_{\pm 59.69}$	0.002
	GLDM–LGLE	$-136.8_{\pm 50.56}$	0.007
	GLRLM–GLNN	$-606.7_{\pm 211.5}$	0.004
	GLRLM–LRE	$6.261_{\pm 2.176}$	0.004
	GLSZM–SALGLE	$-581.1_{\pm 288.6}$	0.044
	NGTDM–Busyness	$1.252_{\pm 0.606}$	0.039
	NGTDM–Complexity	$1.406_{\pm 0.625}$	0.024

Future Work. A natural future direction can be to improve the dataset and models for a more accurate LoS prediction. While this paper proposes explaining LoS predictions for knowledge discovery, e.g. importance of features, future work could address external evaluation of human-AI collaboration in a hospital.

Acknowledgements. This work was financially supported by the Polish National Center for Research and Development grant number INFOSTRATEG-I/0022/2021-00, and carried out with the support of the Laboratory of Bioinformatics and Computational Genomics and the High Performance Computing Center of the Faculty of Mathematics and Information Science, Warsaw University of Technology.

References

1. Biecek, P., Burzykowski, T.: Explanatory model analysis. Chapman and Hall/CRC (2021)
2. Chaou, C.H., et al.: Predicting length of stay among patients discharged from the emergency department-using an accelerated failure time model. PLoS ONE **12**(1), e0165756 (2017)
3. Gu, Z., et al.: CE-Net: context encoder network for 2D medical image segmentation. IEEE Trans. Med. Imaging **38**(10), 2281–2292 (2019)
4. Hansell, D.M., et al.: Fleischner society: glossary of terms for thoracic imaging. Radiology **246**(3), 697 (2008)
5. Herrmann, M., et al.: Large-scale benchmark study of survival prediction methods using multi-omics data. Brief. Bioinform. **22**(3), bbaa167 (2021)
6. Huang, Z., et al.: Length of stay prediction for clinical treatment process using temporal similarity. Expert Syst. Appl. **40**(16), 6330–6339 (2013)
7. Krzyziński, M., Spytek, M., Baniecki, H., Biecek, P.: SurvSHAP(t): time-dependent explanations of machine learning survival models. Knowl.-Based Syst. **262**, 110234 (2023)
8. Muhlestein, W.E., et al.: Predicting inpatient length of stay after brain tumor surgery: developing machine learning ensembles to improve predictive performance. Neurosurgery **85**(3), 384–393 (2019)
9. Rad, J., et al.: Extracting surrogate decision trees from black-box models to explain the temporal importance of clinical features in predicting kidney graft survival. In: AIME, pp. 88–98 (2022)
10. Radiological Society of North America: Radiology Lexicon. https://radlex.org
11. Rudin, C.: Stop explaining black box machine learning models for high stakes decisions and use interpretable models instead. Nat. Mach. Intell. **1**, 206–215 (2019)
12. Sonabend, R., et al.: mlr3proba: an R package for machine learning in survival analysis. Bioinformatics **37**, 2789–2791 (2021)
13. Spytek, M., Krzyziński, M., Baniecki, H., Biecek, P.: survex: explainable machine learning in survival analysis. R package version 0.2.2 (2022). https://github.com/modeloriented/survex
14. Stone, K., et al.: A systematic review of the prediction of hospital length of stay: towards a unified framework. PLOS Digital Health **1**(4), e0000017 (2022)
15. Van Griethuysen, J.J., et al.: Computational radiomics system to decode the radiographic phenotype. Cancer Res. **77**(21), e104–e107 (2017). https://github.com/aim-harvard/pyradiomics
16. Wang, Z., et al.: Counterfactual explanations for survival prediction of cardiovascular ICU patients. In: AIME, pp. 338–348 (2021)
17. Wen, Y., et al.: Time-to-event modeling for hospital length of stay prediction for COVID-19 patients. Mach. Learn. Appl. **9**, 100365 (2022)
18. Zhang, D., et al.: Combining structured and unstructured data for predictive models: a deep learning approach. BMC Med. Inform. Decis. Mak. **20**(1), 1–11 (2020)

Explainable Artificial Intelligence for Cytological Image Analysis

Stefan Röhrl[1]([✉])[iD], Hendrik Maier[1][iD], Manuel Lengl[1][iD], Christian Klenk[2][iD], Dominik Heim[2][iD], Martin Knopp[1,2][iD], Simon Schumann[1][iD], Oliver Hayden[2][iD], and Klaus Diepold[1][iD]

[1] Chair of Data Processing, Technical University of Munich, Munich, Germany
`stefan.roehrl@tum.de`
[2] Heinz-Nixdorf Chair of Biomedical Electronics, Technical University of Munich, Munich, Germany

Abstract. Emerging new technologies are entering the medical market. Among them, the use of Machine Learning (ML) is becoming more common. This work explores the associated Explainable Artificial Intelligence (XAI) approaches, which should help to provide insight into the often opaque methods and thus gain trust of users and patients as well as facilitate interdisciplinary work. Using the differentiation of white blood cells with the aid of a high throughput quantitative phase microscope as an example, we developed a web-based XAI dashboard to assess the effect of different XAI methods on the perception and the judgment of our users. Therefore, we conducted a study with two user groups of data scientists and biomedical researchers and evaluated their interaction with our XAI modules, with respect to the aspects of behavioral understanding of the algorithm, its ability to detect biases and its trustworthiness. The results of the user tests show considerable improvement achieved through the XAI dashboard on the measured set of aspects. A deep dive analysis aggregated on the different user groups compares the five implemented modules. Furthermore, the results reveal that using a combination of modules achieves higher appreciation than the individual modules. Finally, one observes a user's tendency of overestimating the trustworthiness of the algorithm compared to their perceived abilities to understand the behavior of the algorithm and to detect biases.

Keywords: XAI · Quantitative Phase Imaging · Blood Cell Analysis

1 Introduction

The current gold standard sending and presenting hematological laboratory results is in tabular form with numbers and benchmarks. There are neither detailed insights provided on the methodology nor the possibility to interpret or question the results. With current medical analysis, this information it is not relevant for physicians and patients, as they are mainly interested in the plain results. However, as machine learning (ML) comes into place, the need for additional information increases. The workflow will also change for laboratory personnel and pathologists who are directly interacting with the algorithms

© The Author(s), under exclusive license to Springer Nature Switzerland AG 2023
J. M. Juarez et al. (Eds.): AIME 2023, LNAI 13897, pp. 75–85, 2023.
https://doi.org/10.1007/978-3-031-34344-5_10

and responsible for the correctness of the result. Since one decade, deep learning models have pushed the boundaries in various fields of ML [11] and have shown a superior performance also in computer vision [9]. However, the increasing model complexity comes at the cost of interpretability. This lack of transparency is a problem that has been recognized broadly in legislation as well as academia in the recent years [6,18,20].

After the GDPR was introduced, in 2018, an independent expert group was set up by the European Commission to further evaluate implications for the research and deployment of AI, resulting in the *Ethics Guidelines For Trustworthy AI* [5]. They note that especially transparency, which state of the art models are missing, is the key to establish trust and accordingly they demand traceability, explainability and adequate communication.

To demonstrate our work, we investigate the use case of hematological analysis, which is one of the most common laboratory tests [7], as it delivers comprehensive information about the health status of an organism. In contrast to conventional blood analysis via molecular labeling [16] or the gold standard blood smear [2], we focus on the ascending *quantitative phase imaging* approach combined with *microfluidics*. As a quantitative phase microscope offers a higher dynamic range than an unstained bright-field microscope, this technology works label-free and, therefore, needs no time-consuming sample preparation. The sample presentation via a microfluidics channel leverages the approach to a high statistical power, while keeping the cells near *in vivo* conditions. Various publications demonstrate its diverse potentials and versatility in the domains of oncology [10,13], hematology [15,19] and beyond [14].

Alongside with all its advantages, this new platform technology comes with several challenges. Besides the inexactness of viscoelastic focusing and orientation, the problem changes from classical cell sorting to a computer vision problem which is preferably solved by machine learning [9,14].

2   Background

2.1   Differential Blood Count Using Quantitative Phase Imaging

Blood contains a vast amount of information about the state of human health. Especially the composition of white blood cells (WBCs) and their functions form the basis for the detection of hematological, oncological or immunological diseases. However, many biomarkers remain hidden due to the technical limitations of conventional analyzers. This may be due to volatility of some biomarkers, the insufficient contrast or resolution provided by optical methods or the lack of suitable antibodies for fluorescent staining [16]. Using a quantitative phase microscope, the problem transitions into the domain of object detection, pattern recognition and classification. Skipping the tedious sample preparation, the measurement can be performed within 15 min after blood draw, which paves the way to the analysis of the kinetics of intra-cellular changes closer to the point of care [2,8]. Though, before pursuing the analysis of internal cell structures and morphological changes, the *Five-Part Differential* of WBCs has to be established using computer vision and machine learning techniques. For healthy individuals, Neutrophils (62%) make up the biggest proportion, followed by Lymphocytes (30%), Monocytes (5.3%), Eosinophils (2.3%) and Basophils (0.4%) [1]. Manual staining or costly molecular labeling are currently employed to solve this problem and have coined the biomedical community [2]. In contrast, phase images are largely unfamiliar and new visualizations and interpretations must be found to support the clinical decision-making process. Figure 1 shows typical examples of WBCs inside a phase microscope. In these grayscale images the brighter parts correspond to higher optical phase shifts $\Delta\phi$.

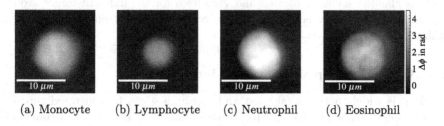

(a) Monocyte (b) Lymphocyte (c) Neutrophil (d) Eosinophil

Fig. 1. Quantitative phase images of WBC subtypes (Brightness $\sim$ Phase Shift)

2.2 Image Acquisition and Data Processing Pipeline

For our experiments, we use a custom-made differential holographic microscope by *Ovizio Imaging Systems* like [8,19]. It is equipped with a microfluidics chip to align the diluted blood sample stream in the focal plane of the microscope. Starting on the left of Fig. 2, the raw phase images, with a size of 512×384 pixels, undergo a simple background subtraction and threshold segmentation, before feeding the individual images of single cells into the next stages. Here, the path splits into several possibilities. The most transparent one is the extraction of handcrafted morphological features, which describe e.g. the `optical volume` of a cell or its `granularity` [15,19]. A subsequent interpretable classifier like a *naive Bayes* or *Random Forest* can use these features to predict the cell's class affiliation. On the other hand, we can pass the image to a deep (convolutional) neural network, which can learn important features to optimize cell classification without prior expert knowledge. Considering solely the classification accuracy, we noticed that the data driven black box models outperformed the classical approaches [11,14]. Therefore, we trained an *AlexNet* [9] architecture to perform the described WBC classification task on 7706 cells, which were balanced according to their class label. Due to their rare occurrence, we were not able to obtain a decent number of basophil cells near *in vivo* conditions. Therefore, this group of WBCs is not part of the data set. The fine tuned *AlexNet* classifier achieved an F_1-score of 0.963 and an accuracy of 96.7% on the unknown test set of 1000 cells with 250 cells per class.

Note that this work is not intended to improve the existing approaches regarding their accuracy. We investigate how the established methods and their

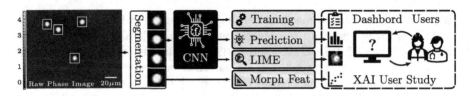

Fig. 2. The XAI dashboard needs to provide different means of explanations to optimally communicate the data processing pipeline to the domain experts.

outputs should be communicated and visualized for the different target groups to maximize the knowledge gain and acceptance of this emergent platform technology.

2.3 Evaluating Explainability and Interpretability

While it is widely agreed that trust and acceptance can be generated through transparency, explainability and robustness, there is still a broad discourse on how to define and how to determine these indicators [12]. Possible dimensions include (1) measuring the quality that is subjectively perceived by an individual or (2) measuring proxies for the sufficiency of the model. For the first dimension, we apply a *human-grounded metric*, where an evaluation should depend only on the "quality of the explanation, regardless of whether the explanation is the model itself or a post hoc interpretation of a Black Box model, and regardless of the correctness of the associated prediction" [3]. Here, we implemented an adapted form of the *binary forced choice*, where different visualizations are rated by humans according to a selection of questions. At the end, the users need to decide for their personal favorite. The second dimension focuses on the opacity of the chosen classifier and therefore evaluates its proxy model in the following aspects: First, the completeness compared to the original model, i.e., how closely it approximates the model to be explained. Second, the ability of the model to detect biases in the original model. And third, the ability of humans to "evaluate explanations for reasonableness, that is how well an explanation matches human expectations" [4].

3 XAI Dashboard Prototype

Based on preceded expert interviews and taking into account the technologies' boundary conditions, we identified four suitable interpretation approaches which we implemented as so-called modules in the prototype of an XAI dashboard.

3.1 Module 1: General Information on Training and Validation

The first XAI module displayed in Fig. 3a provides background information on the algorithm that was deployed for the prediction. On the left, a table lists

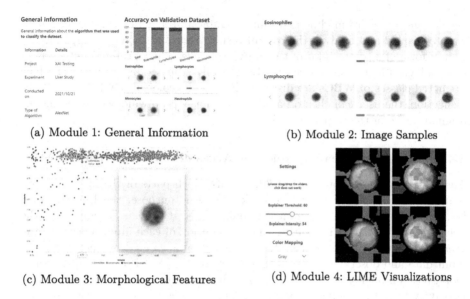

(a) Module 1: General Information

(b) Module 2: Image Samples

(c) Module 3: Morphological Features

(d) Module 4: LIME Visualizations

Fig. 3. Screenshots of the modules in the XAI Dashboard Prototype (Color figure online)

general information about the algorithm, where on the right, a barplot shows the performance of the algorithm on a validation data set. This offers the user an impression of the overall capability of the algorithm after the training has been completed. Finally, this module summarizes information about the training data set. The user has the option to view sample images for each class of WBCs specifically requested by interviewees with a biomedical background.

3.2   Module 2: Image Samples of Classified Cells

The second module shows cell samples from the actual prediction results. The cells are grouped by their predicted class, what can be seen in Fig. 3b. Since the underlying data are phase images (see Sect. 2.1), the images are not colorized by default. However, as a suitable color map can reveal more of the inner structure of the cell, the user has the option to chose a custom coloring. This element is based on a significant need of biomedical users to be able to have a look at cells. As it was identified in the preceding interviews, it would allow this target group to visually double-check if the classifications are meaningful.

3.3   Module 3: Morphological Features in a Scatter Plot

Many publications and interviews report the importance of morphological features for differentiating cells [8,14,15,19]. Furthermore, they are often used as input features for computer vision algorithms or for dimensionality reduction techniques. In contrast to the second module, which displays only individual predictions, the third module takes into account all predictions and gives a neat way to display and analyze the overall result. Figure 3c shows its implementation in form of a scatter plot of the individual cells. The axes represent selected morphological features, that can be dynamically adjusted by the user. The four predicted classes of WBCs are distinguished by the color of the dots. For closer inspection, the user can click on a dot to open a pop-up window containing the original image of the respective cell.

3.4   Module 4: Revealing Relevant Areas of an Image Using LIME

The fourth module, shown in Fig. 3d, reveals which parts of the image are relevant for the employed neural networks. For this purpose, the LIME library [17,21] has become one of the most popular tools. It provides insights into the behavior of a model by measuring the contribution of each input feature to the overall prediction of the sample. The visualization is created by perturbing the input data and observing the resulting effects on the model prediction. Furthermore, it is a model-agnostic linear proxy which identifies areas that contributed positively to the predicted class (green) and the ones which opposed that decision (red). Users interact with the method by being able to view only a minimum effect strength, tuning the overlays transparency to inspect the corresponding cell structures and finally focusing only on explanations concerning a specific class. This module might be the closest to human perception but also the most complex one in our prototype.

4   User Study Results

In total, the study cohort consists of 57 people from different scientific backgrounds and comprises students as well as researchers from various local institutes. Their distribution is displayed in Fig. 4a. The demographic composition of the cohort shows as shift towards the younger generation, as most of the participants are younger than 35. To simplify the observations we split the participants into a **biomedical (bm)** group and a **data science (ds)** group, in which people had to state a ML experience ≥ 5 on a scale from 1 to 8 and not being accounted

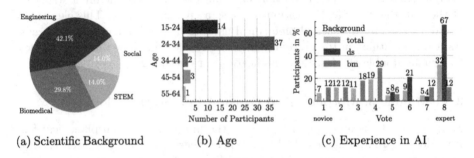

(a) Scientific Background   (b) Age   (c) Experience in AI

Fig. 4. Demographics of the participants

to the other group. This leads to Fig. 4c, which shows the distribution of participants' experience with machine learning algorithms. Naturally, data scientists have the most experience, with two-thirds of them reporting the highest score. They are therefore expected to provide the reference values for the study. In the biomedical group, the full range of experience is represented.

4.1   Evaluation of the Overall XAI Dashboard

For the estimation of the overall impact of the dashboard, a set of three questions was asked a first time when the users where confronted with the bare classification results and a second time when they had interacted with the XAI dashboard. The questions relate to the user's (a) understanding of the algorithm's behavior, (b) ability to detect biases, and (c) impression of the trustworthiness. We recorded their answers on an evenly distributed Lickert scale from 1 to 8. As the subplots in Fig. 5 show, the first take away is the dashboard's positive effect on all three aspects. This confirms the assumption that both user groups have difficulties to judge and understand the ML algorithm without the XAI spyhole. When looking at the **behavioral understanding**, the data scientist experience the highest improvement whereas the biomedical group states a slightly higher overall value even without the dashboard. When it comes to the **detection of biases** the user groups show an inverted influence. Another aspect to bear in mind is that the ability to judge the **trustworthiness** is rated higher than the understanding of the algorithm as well as the ability to detect biases.

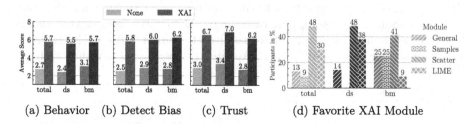

(a) Behavior   (b) Detect Bias   (c) Trust   (d) Favorite XAI Module

Fig. 5. Overall improvements through XAI dashboards

At the end of the survey, the participants were asked to pick their **favorite module** of the dashboard. The results in Fig. 5d expose module 3 as the preferred one for both user groups. The highly complex LIME explanation ranks second best at the total group and among the data scientists. In contrast, it ranks last for the biomedical group. This is to be expected, since LIME is a method focusing on the needs of data scientist rather than biomedical people. Those are more interested in the representation of the familiar morphological features and the modules about the training/validation of the algorithm as well as the cell samples. Note that the data science group could not profit from cell images.

4.2   Comparison of the XAI Modules

In the survey, each module is examined by five aspects. In addition to the three questions introduced in the previous section, we asked (d) if the shown module is relevant for the task and (e) if the users understand the displayed information. These additional questions elaborate whether the previous are influenced by other factors and to ensure that the modules used are comprehensible in themselves. Concerning the perceived **relevancy**, Fig. 6d shows that all modules score relatively high. We partly attribute this to the fact that all modules were designed based on the needs of the various user groups. When comparing the modules, it is noticeable that the modules that refer globally to the algorithm (1, 3) are more relevant than the modules that refer to a local explanation (2, 4). Users are more interested in general information than investing time in individual examples. A good **understanding** of what is being shown is paramount to any data presentation and, in this case, a prerequisite for all other aspects. Equally to the relevancy, the modules are understood by the users pretty well as indicated by Fig. 6e. Data scientists generally exhibit a little higher understanding compared to biomedical, which is comprehensible considering this being a dashboard about ML. An exception is the module on showing classified cell samples. Surprisingly, across all modules biomedical users indicate a higher level of **understanding of the algorithm's behavior** than the data scientists. As it is unlikely that they could gather this lead by our dashboard, there must be an unobserved variable, which needs further investigation. Besides most of the models only achieve a mediocre rating, the LIME module stands out being the only one scoring above 6. We suspect the reason is that this is the only module that gives direct insight into the algorithm instead of indirectly examining the prediction results. Therefore, it is rather unusual that the users also consider LIME to be the best module for its capabilities to **detect biases**. The globally operating scatter plot is only second. Apparently, there is no tendency such as

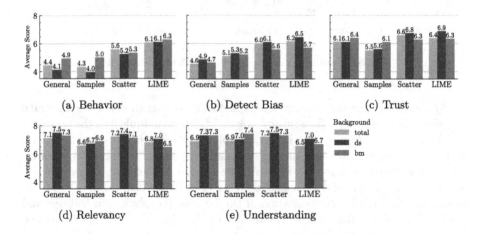

Fig. 6. Overall improvements through XAI dashboards

global or local methods are superior to detect biases. Detecting biases scores a little higher than the aspect about understanding the behavior, but still, there is room for improvement. Users with a biomedical background seem to prefer the modules 1 and 2 when it comes to judging the **trustworthiness** of an algorithm. For the other modules, it is the opposite. Again, module 3 (Morphology, Scatter Plots) and module 4 (LIME) on average score highest for all user groups. On top, biomedical participants believe that the algorithm training and validation module helps them most in assessing trustworthiness.

5   Discussion and Conclusion

The aim of the dashboard is to provide users with a tool for interpretation, explanation and to judge an algorithm's trustworthiness. For this purpose, with respect to the diversity of the target groups, it proved beneficial to combine individual modules into a so-called XAI dashboard, as the overall dashboard was rated higher in all aspects than the single modules. When asking participants, if they would use the presented tools, the average participant is very positive with 7 out of 8 points.

In the qualitative interviews conducted in advance, respondents indicated that data scientists were more interested in more technical approaches such as the LIME module, while biomedical scientists would be more interested in morphological features and cell samples. Although this is confirmed to some extent, interestingly, the scatter plot of morphological features also emerged as the favorite module of data scientists. Thus, it can be concluded that while there are differences, there is certainly overlap in the relevance of the dashboard for both user groups simultaneously.

However, the users' assessment of understanding the algorithm's behavior and their ability to detect biases is only mediocre and the ultimate goal of making black box models transparent can only be approximated. Moreover, we observed that the domain knowledge might cause the users to be more skeptical versus technology they are familiar with. On the other hand, the explanations and interactivity might convey an exaggerated sense of security and lead users to overestimate their own trust, as can be seen in high ratings of trustworthiness despite the lack of understanding and detecting biases. Here, the responsibility lies with the developers. They must be careful not to mislead the users and stay as transparent as possible. Also legislation and academia require that the system's level of accuracy and its limitations are communicated [4,5].

All in all, there is a high demand for explainability and the ability to understand the decision making process is a crucial prerequisite for the deployment of ML. The shown model-agnostic, surrogate XAI modules, be they local or global, are considered suitable for this purpose by the different user groups. Nevertheless, there are still many other techniques (especially model-specific ones), which could provide even deeper insight. Tracing their impact in the proposed aspects will help to establish high potential technologies like quantitative phase imaging combined with ML in the biomedical domain. From our perspective XAI approaches will be indispensable for interdisciplinary research and clinical decision support systems.

Acknowledgment. The authors would like to especially honor the contributions H. Meier for the implementation of the user interface prototype and the execution of the studies. We also thank of J. Groll for the training and providing the underlying machine learning models.

References

1. Alberts, B.: Molecular biology of the cell. WW Norton & Company (2017)
2. Barcia, J.J.: The giemsa stain: its history and applications. Int. J. Surg. Pathol. **15**(3), 292–296 (2007)
3. Doshi-Velez, F., Kim, B.: Towards a rigorous science of interpretable machine learning. arXiv:1702.08608 (2017)
4. Gilpin, L.H. et al.: Explaining explanations: an overview of interpretability of machine learning. In: 5th International Conference on Data Science and Advanced Analytics, pp. 80–89 (2018)
5. High-Level Expert Group on Artificial Intelligence: Ethics Guidelines for Trustworthy AI. European Commission (2019)
6. Holzinger, A., Langs, G., Denk, H., Zatloukal, K., Müller, H.: Causability and explainability of artificial intelligence in medicine. Wiley Interdisciplinary Reviews: Data Mining and Knowledge Discovery 9(4) (2019)
7. Horton, S., et al.: The Top 25 laboratory tests by volume and revenue in five different countries. Am. J. Clin. Pathol. **151**(5), 446–451 (2018)
8. Klenk, C., Heim, D., Ugele, M., Hayden, O.: Impact of sample preparation on holographic imaging of leukocytes. Opt. Eng. **59**(10), 102403 (2019)
9. Krizhevsky, A. et al.: ImageNet classification with deep convolutional neural networks. In: Advances in Neural Information Processing Systems, vol. 25, pp. 1097–1105 (2012)
10. Lam, V.K., et al.: Machine Learning with Optical Phase Signatures for Phenotypic Profiling of Cell Lines. Cytometry A **95**(7), 757–768 (2019)
11. LeCun, Y., et al.: Deep learning. Nature **521**(7553), 436–444 (2015)
12. Murdoch, W.J., et al.: Definitions, methods, and applications in interpretable machine learning. Proc. Natl. Acad. Sci. **116**(44), 22071–22080 (2019)
13. Nguyen, T.H., et al.: Automatic Gleason grading of prostate cancer using quantitative phase imaging and machine learning. J. Biomed. Opt. **22**(3), 036015 (2017)
14. Nguyen, T.L., et al.: Quantitative Phase Imaging: Recent Advances and Expanding Potential in Biomedicine. Am. Chem. Soc. **16**(8), 11516–11544 (2022)
15. Paidi, S.K., et al.: Raman and quantitative phase imaging allow morpho-molecular recognition of malignancy and stages of B-cell acute lymphoblastic leukemia. Biosens. Bioelectron. **190**, 113403 (2021)
16. Park, Y., Depeursinge, C., Popescu, G.: Quantitative phase imaging in biomedicine. Nat. Photonics **12**(10), 578–589 (2018)
17. Ribeiro, M.T., Singh, S., Guestrin, C.: "Why should i trust you?": explaining the predictions of any classifier. In: Proceedings of the 22nd International Conference on Knowledge Discovery and Data Mining SIGKDD, pp. 1135–1144 (2016)

18. Rudin, C.: Stop explaining black box machine learning models for high stakes decisions and use interpretable models instead. Nature Mach. Intell. **1**, 206–215 (2019)
19. Ugele, M. et al.: Label-Free High-Throughput Leukemia Detection by Holographic Microscopy. Advanced Science 5(12) (2018)
20. Vellido, A.: The importance of interpretability and visualization in machine learning for applications in medicine and health care. Neural Comput. Appl. **32**(24), 18069–18083 (2020)
21. Zhang, Q.S., Zhu, S.C.: Visual interpretability for deep learning: a survey. Front. Inf. Technol. Electron. Eng. **19**(1), 27–39 (2018)

Federated Learning to Improve Counterfactual Explanations for Sepsis Treatment Prediction

Christoph Düsing[✉] and Philipp Cimiano

CITEC, Bielefeld University, Bielefeld, Germany
{cduesing,cimiano}@techfak.uni-bielefeld.de

Abstract. In recent years, we have witnessed both artificial intelligence obtaining remarkable results in clinical decision support systems (CDSSs) and explainable artificial intelligence (XAI) improving the interpretability of these models. In turn, this fosters the adoption by medical personnel and improves trustworthiness of CDSSs. Among others, counterfactual explanations prove to be one such XAI technique particularly suitable for the healthcare domain due to its ease of interpretation, even for less technically proficient staff. However, the generation of high-quality counterfactuals relies on generative models for guidance. Unfortunately, training such models requires a huge amount of data that is beyond the means of ordinary hospitals. In this paper, we therefore propose to use federated learning to allow multiple hospitals to jointly train such generative models while maintaining full data privacy. We demonstrate the superiority of our approach compared to locally generated counterfactuals on a CDSS for sepsis treatment prescription using various metrics. Moreover, we prove that generative models for counterfactual generation that are trained using federated learning in a suitable environment perform only marginally worse compared to centrally trained ones while offering the benefit of data privacy preservation.

Keywords: Counterfactual explanations · Federated learning · Generative models · Sepsis treatment

1 Introduction

Due to their high utility and adaptability, clinical decision support systems (CDSSs) that build upon recent advancements in machine learning (ML) prevail in assisting healthcare providers with adequate patient treatment [21]. Recent efforts toward increasing the interpretability of these systems exploit explainable artificial intelligence (XAI) techniques in order to increase their reliability and trustworthiness [16]. Ultimately, this allows to make sense of the decision-making of the underlying ML models and also to identify potential misconceptions during the process. Thus, XAI is necessary to reliably use ML for CDSSs [21].

Consequently, XAI is subject to a vast variety of research within the healthcare domain and is successfully deployed in various applications including patient survival prediction [24] and patient assessment [7]. Among the wide range of

© The Author(s), under exclusive license to Springer Nature Switzerland AG 2023
J. M. Juarez et al. (Eds.): AIME 2023, LNAI 13897, pp. 86–96, 2023.
https://doi.org/10.1007/978-3-031-34344-5_11

XAI approaches, counterfactual explanations seem particularly suitable for the healthcare domain in general (e.g., [7,24]) and CDSSs specifically [7]. This is due to their more human-like approach of explaining the model outcome, allowing for straight-forward and fast interpretation [23]. In order to compute counterfactual explanations, typical approaches rely on generative models during counterfactual generation [16], as they are used to compute more realistic examples than conventional approaches [5]. Unfortunately, training such models requires a huge amount of electronic health records (EHRs), exceeding the resources of most hospitals and healthcare providers (see Sect. 3.1 for our estimates). Hence, ordinary hospitals cannot unleash the full potential of counterfactual explanation on their own due to limited availability of data. To overcome this issue and to avoid having to share sensitive data among peers to jointly train such generative models, federated learning (FL) can be used [6]. FL allows a cohort of hospitals to jointly train a ML model while maintaining data privacy [15]. Thus, FL is suitable to address the problem of high-quality counterfactual generation with limited local data. However, FL has not been evaluated regarding its potential to increase the quality of clinical counterfactual explanations yet.

To this end, we investigate the quality of FL-generated counterfactual explanations and demonstrate the superiority of counterfactuals generated using jointly trained autoencoders (AEs) over those locally generated. More precisely, the contributions of our paper are the following: (1) we propose a FL-based solution to overcome the issue of counterfactual generation for CDSSs of hospitals with limited data, (2) we provide in-depth analysis of various counterfactual generation approaches for CDSSs, (3) we measure the effects of data imbalance among hospitals on counterfactual quality, and (4) we provide insights into the clinical application of our approach on a CDSS for sepsis treatment. Thus, our approach enables ordinary hospitals to benefit from counterfactual explanations by overcoming the issue of training a generative model with limited local data. In turn, our results are of great value for researchers and practitioners who seek to enhance the quality of explainable CDSSs.

2   Related Work

XAI finds widespread application within the healthcare domain due to providing interpretations of the inner working of black-box CDSSs relying on ML models [25]. Here, interpretability is critical to reliably utilize ML predictions, as it allows clinicians to understand the decision made and to identify possible errors [21]. While most such XAI techniques highlight input features most relevant to the explanandum, i.e., instance and prediction to be explained (e.g., [2]), counterfactual explanations follow a more human-like approach of explaining [23]. More precisely, counterfactual generating approaches identify changes to the original input that would change the prediction of the underlying ML model and thus explain the prediction by providing a hypothetical, yet similar counterexample [7]. In turn, they have been successfully deployed in various CDSSs in the past [7,24]. However, most counterfactual generation approaches require generative models in order to guide the computation of high-quality counterfactuals, i.e., examples that are close to their initial inputs and a-priori feature distributions [5]. Usually, these models are either AEs [5,14] or generative adversarial networks (GANs), as applied by Mertes et al. [16]. Unfortunately, we find average-sized hospitals holding an insufficient amount of sepsis-related EHRs to train such models (for example, training a GAN for pneumonia detection required a dataset five times larger than the number of sepsis EHRs available at average US hospitals [16]).

To overcome this issue, FL can be applied to jointly train an AE or GAN without having to share data. Therefore, a central communication server orchestrates several rounds of training, each consisting of a local training performed by each hospital and a subsequent aggregation of model updates [15]. Ultimately, a federated model is achieved that incorporates knowledge derived from each hospital's EHRs while privacy is maintained [26]. Thus, FL seems to be a perfect fit for cross-hospital collaborations and was successfully deployed in applications like diabetes detection [13]. However, data imbalance among hospitals has detrimental effects on FL performance, such that the amount of performance decrease depends on the domain of application [12]. To the best of our knowledge, no prior research investigated the potential of FL to improve counterfactual explanations.

3   Methodology

Our proposed approach to improve the quality of counterfactual explanations for ordinary hospitals relies on FL to jointly train a generative model (e.g., an AE) that can be used during counterfactual generation. The rationale behind our proposed approach is: (1) ordinary hospitals hold a moderate amount of EHRs of sepsis patients. While this amount allows them to train ML models such as decision trees assisting with treatment prescription, it is insufficient to train large deep learning (DL) models such as AEs. However, the availability of such DL models is crucial for the generation of counterfactuals of adequate quality [5]. In turn, most hospitals are unable to compute high-quality counterfactual explanations for CDSSs on their own; (2) in order to facilitate counterfactual explanations, hospitals can collaboratively train an AE which is afterwards utilized by all participating hospitals. However, as EHRs contain sensitive information, directly sharing them might not adhere to data privacy regulations. To avoid having to share data directly, FL can be used in order to jointly train an AE and maintain data privacy. Ultimately, our approach allows small- and medium-sized hospitals to partake in the opportunities brought by counterfactual explanations.

In order to investigate the potential of FL for counterfactual generation, we first train a XGBoost classifier [4] to serve as CDSS for sepsis treatment prescription. Afterwards, our work utilizes the Contrastive Explanations Method (CEM) [5] for counterfactual generation. Although CEM can compute counterfactuals without an AE, the authors argue that the availability of an AE further improves their quality [5]. In order to obtain the AEs, we train them in different settings, namely locally, federately, and centrally. Finally, we assess the quality of counterfactuals received from those different settings using a variety of metrics.

3.1 Data Acquisition and Preparation

Our study relies on publicly available data received from the MIMIC-IV repository [8]. Initially, we acquire all relevant patients following the sepsis-3 definition [20]. In collaboration with several ICU clinicians, we identify 56 variables relevant to the assessment of sepsis patients and their respective treatment (see Sect. 4.3 for included variables). Moreover, we identify the five treatments most frequently prescribed for the previously identified patients, namely Vancomycin (2,712 patients), CefePIME (1,285 patients), Piperacillin-Tazobactam (1,088 patients), Meropenem (462 patients), and CefazoLIN (340 patients). Overall, this yields a dataset of 41,098 patients, 5,887 of which were prescribed one of the five treatments. In what follows, we refer to them as labeled and unlabeled patients.

Although estimates of the number of yearly sepsis cases in the US vary, the highest estimate is that about 1.7 million adults develop sepsis per year [17]. Moreover, more than 6,000 hospitals are registered in the US today [1]. Hence, an upper bound for the average number of yearly hospitalizations due to sepsis is 300 per hospital. Considering the availability of EHRs from a 20-year-period and the steady increase in yearly sepsis cases [11], we argue that the number of our labeled patients is a fair estimate of sepsis-related EHRs available to average US hospitals. Lastly, in order to obtain a realistic environment for the application of FL, we split the unlabeled EHRs among nine simulated hospitals.

3.2 Sepsis Treatment Prediction

The CDSS for sepsis treatment that we want to explain using counterfactual explanations is a XGBoost classifier [4] trained using all labeled patient data. It is worth mentioning that applying DL is not an option, due to the above mentioned shortage of sufficient data. Furthermore, achieving state-of-the-art performance on the treatment prescription task is out of scope for this work.

Unfortunately, as previously described, the distribution of the top-5 treatments is very imbalanced, which might harm the classifier's performance. To avoid overestimating the majority classes, we apply oversampling using SMOTE [3] prior to training the classifier. A prestudy revealed that this improves performance in terms of accuracy by 9.35%. Finally, we evaluate the classifier that serves as CDSS using 5-fold cross-validation on the labeled data. Here, the classifiers achieve a mean accuracy of 0.64 and a macro-averaged F1-Score of 0.63.

3.3 Counterfactual Explanation Generation

In this work, we rely on CEM [5] for the generation of counterfactual examples. We chose CEM over alternative approaches like CFRL [19] and CFProto [14] due to its higher overall quality of explanations. Additionally, CEM is not limited to AE-guided counterfactual generation, but is also applicable without a generative model [5]. Thus, we can compare the quality of counterfactuals from these two settings. CEM can be used to compute both pertinent positives and negatives, where the latter is the computation of counterfactual explanations used in this work [5]. More precisely, we evaluate CEM in the following four settings:

- **CEM.** It refers to the application of CEM for the explanation of treatment predictions without an AE. It serves as baseline to compare with.
- **CEM_LAE.** Here, we train an AE on the local data of a single hospital only. This serves to demonstrate the quality of counterfactuals achieved by hospitals limited to their own data only.
- **CEM_FL.** In this setting, we apply FL to jointly train an AE that is afterwards used to guide the counterfactual generation. More precisely, we use FedAvg [15] as FL strategy. We train the AE among all simulated hospitals that hold unlabeled data and the hospital holding labeled data.
- **CEM_AE.** Here, we train an AE among all hospitals at a single, central side. While this would be inapplicable to real-world applications due to data privacy violations, it serves as an upper bound of counterfactuals' quality.

All AEs consist of five hidden layers and have input and output dimensionality of 56. The hidden layers are of size 50, 40, 30, 40, and 50 with ReLU-activation and dropout of 0.2. Each AE is trained for 250 epochs or rounds, respectively.

3.4 Counterfactual Explanation Evaluation

There are several metrics available to evaluate the quality of counterfactual generating approaches. According to Verma et al. [22], there are eight metrics to apply upon a set of counterfactual explanations. However, the actual selection of suitable metrics highly depends on domain and application [22]. For the problem at hand, we decide to implement the following metrics:

- **Validity.** Measuring the fraction of valid counterfactuals, i.e., counterfactuals that actually change the model prediction, is one of the most common metrics in counterfactual evaluation [22], including the healthcare domain [24]. For Validity, higher values are considered better.
- **Proximity.** Proximity quantifies the mean distance of counterfactual examples from their respective original input [22]. It is widely accepted that counterfactuals of low Proximity are preferred [16]. Following Verma et al. [22], we measure Proximity as mean L1 norm of feature-wise distance between original feature vector and counterfactual.
- **Sparsity.** Explanations are more comprehensible if kept short [24]. This is especially true for the healthcare domain, as short explanations are faster to understand when time is limited [24]. Thus we measure Sparsity as fraction of features changed for the counterfactual. Lower values are better.
- **Diversity.** To measure Diversity of the generating approach, we again compute the delta of original features and counterfactuals per explanation. Afterwards, we compute the mean pairwise cosine distance between all deltas. Diversity is considered an indicator for the adaptability regarding the instance to be explained [22]. For Diversity, higher values are better.
- **Closeness.** Lastly, we use the local outlier factor [9] to measure the fraction of counterfactuals falsely considered non-outliers among 5,000 real patients. Thus, Closeness measures the realisticness of counterfactuals [22], where higher values are better again.

Table 1. Evaluation metrics for counterfactual explanation approaches.

	Validity (higher is better)	Proximity (lower is better)	Sparsity (lower is better)	Diversity (higher is better)	Closeness (higher is better)
CEM	**1.0000** ±0.00	6.0239 ±0.83	0.7471 ±0.02	0.9332 ±0.01	0.7433 ±0.05
CEM_LAE	**1.0000** ±0.00	1.7937 ±0.26	0.7486 ±0.02	0.4456 ±0.06	0.7215 ±0.03
CEM_FL	**1.0000** ±0.00	0.5452 ±0.03	0.7279 ±0.01	0.9523 ±0.02	**0.8812** ±0.02
CEM_AE	**1.0000** ±0.00	**0.5289** ±0.02	**0.7268** ±0.00	**0.9647** ±0.01	0.8801 ±0.03

4 Experiments

4.1 Counterfactual Explanation Quality Comparison

First, we apply the four approaches to explain the CDSS output for 500 patients. Afterwards, we evaluate the explanations using previously outlined metrics. These experiments serve to assess the initial hypothesis that counterfactuals generated using jointly trained AEs are of higher quality than locally generated ones. Also, we want to prove the suitability of FL for this task.

Table 1 summarizes the results of five consecutive runs. It shows that both approaches relying on jointly trained AEs, namely CEM_FL and CEM_AE, perform significantly better than those generated locally. The only exception is Validity, as counterfactuals from all approaches are valid. This indicates that hospitals with (below-)average amounts of EHRs suffer from being unable to compute high-quality counterfactuals. Other than that we find CEM_LAE to outperform CEM in most metrics, which confirms previous findings regarding the benefit of AE availability [5]. In terms of FL suitability for the task, CEM_FL performs only marginally worse than CEM_AE. However, we argue that maintaining data privacy during training justifies such a marginal decrease in performance.

4.2 Effects of Data Imbalance on Counterfactual Quality

FL is known to suffer from data imbalance [26]. Hence, we want to determine the impact of data imbalance on counterfactual quality. Therefore, we enforce different magnitudes of data imbalance among simulated hospitals using a dirichlet distribution [12] before applying CEM_FL. Here, the parameter α controls the degree of imbalance, where smaller α correspond to higher imbalance.

Figure 1 shows the effects of data imbalance on the quality of counterfactuals. In fact, with the exception of Validity, higher imbalance correlates with decreasing quality. Both Diversity and Proximity show high robustness to data imbalance, where CEM_FL outperforms the local baselines in terms of Proximity even for the unrealistically imbalanced setting of $\alpha = 0.1$. For the other three metrics: Diversity, Closeness, and Sparsity, CEM_FL performs only worse than local baselines in the most extreme settings of data imbalance. Moreover, we find that CEM_FL performs similar to CEM_AE in settings of low and medium imbalance for all metrics except Closeness, as CEM_FL's Closeness starts to worsen significantly even in settings of medium imbalance.

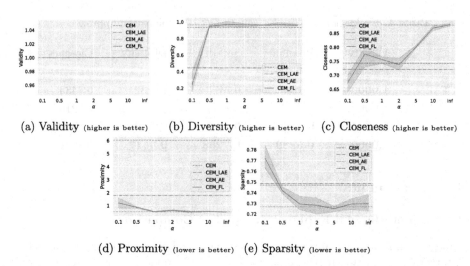

(a) Validity (higher is better)   (b) Diversity (higher is better)   (c) Closeness (higher is better)

(d) Proximity (lower is better)   (e) Sparsity (lower is better)

Fig. 1. Impact of data imbalance among hospitals on counterfactual explanation quality. For CEM_FL, we report mean, min, and max of five consecutive runs.

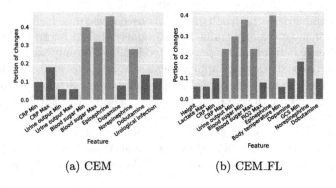

(a) CEM   (b) CEM_FL

Fig. 2. Fraction of counterfactuals in which the respective feature deviates from the original state. The threshold for reporting is set to 0.05.

From these findings, we can conclude that FL offers great potential to hospitals for counterfactual generation in most settings, except those of severe data imbalance ($\alpha = 0.1$). In fact, CEM_FL even performs similar to CEM_AE, although it maintains data privacy during training. Hence, FL is suitable to increase the quality of counterfactual explanations for hospitals with limited data.

4.3   Clinical Application

Evaluating counterfactuals for CDSSs requires to gather profound knowledge of both counterfactual examples and their respective class predictions [16]. Therefore, we compare counterfactuals generated using CEM and CEM_FL to provide insights into possible implications during clinical application.

From previous findings we know that CEM_FL outperforms CEM in terms of Sparsity. However, sparse counterfactuals are not better per se [10]. Instead, the context of applications greatly affects the perceived quality [10]. To study explanation quality, prior literature suggests to monitor features frequently changed by the generating approach [18]. Figure 2 shows that although CEM has higher Sparsity, i.e., more features change per counterfactual, the number of features exceeding the threshold of 0.05 is lower than for CEM_FL. This means that CEM is limited to a narrow selection of features that are permuted. CEM_FL on the other hand changes a broader spectrum of features. We argue that this is evidence for the superiority of CEM_FL as (1) clinicians value variance in feature permutation as being repeatedly presented similar counterfactuals harms attention towards them, and (2) higher variance in permutation indicates a higher specificity regarding the patient at hand.

Taking a closer look at the most frequently altered features reveals that both approaches consider *epinephrine*, and *norepinephrine* most relevant. Features such as *CRP*, *urine output*, *dopamine*, and *dobutamine* are frequently altered, too. These finding are mostly in line with criteria for the Sequential Organ Failure Assessment (SOFA) score [20]. For CEM_FL, *CRP* and *urine output* are even among those most frequently changed features. Moreover, it considers *FiO2* and *GCS*, both relevant to SOFA, too [20]. Ultimately, CEM_FL covers a broader and more relevant spectrum of features.

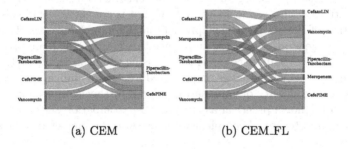

(a) CEM   (b) CEM_FL

Fig. 3. Transition from the original sepsis treatment prediction (left) to the counterfactual's prediction (right) for (a) CEM and (b) CEM_FL.

Lastly, Fig. 3 depicts the change in treatment prediction from the original input to the counterfactual prediction. The results prove a higher diversity in output classes for CEM_FL. More precisely, counterfactuals computed using CEM are all predicted as one of the three majority classes. CEM_FL however computes counterfactual examples of all five classes and shows an overall higher diversity of prediction transitions. Again, this substantiate our previous claim of specificness and quality of CEM_FL-generated counterfactuals.

In order to illustrate how our approach would work in practice, we discuss the behaviour of CEM and CEM_FL explaining the CDSS output for the case of an exemplary sepsis patient. For the patient with the recommendation of Vancomycin, CEM_FL suggests that a decrease in *bilirubin* by 0.34 mg/dl and an increase in *epinephrine* of 0.64 µg/kg/min would change the recommendation to CefePIME, thus proposing that the high *bilirubin* and the low *epinephrine* are the causes for its decision. CEM on the other hand offers a similar explanation but also requires *norepinephrine* to be increased significantly by 1.74 µg/kg/min. We provide additional examples and a rudimentary CDSS here.

5   Conclusion, Limitations, and Future Work

In this paper, we present an approach to overcome the issue of high-quality counterfactual explanation generation for CDSSs for sepsis treatment. In particular, we propose to incorporate FL to jointly train a generative model that guides counterfactual generation and improves the quality and, in turn, value of counterfactual explanations. This complements existing research that usually assumes a sufficient amount of data to be available. Our results indicate that most hospitals can greatly benefit from FL participation and that CEM_FL yields high-quality counterfactuals in most settings. Compared to a centrally trained AE, as applied in CEM_AE, using FL achieves similar results but offers the benefit of data privacy preservation. Regarding the effect of data imbalance, we show that FL improves clinical counterfactual generation in all but highly imbalanced settings. Regarding potential limitations, the classifier that serves as CDSS has some potential for improvement. And although it is not the scope of this work to improve state-of-the-art treatment prediction, this might have an effect on the counterfactual generation process and hence the quality of explanations.

In our study, we evaluate counterfactuals using widely accepted metrics. However, we believe that user studies on preferred aspects of clinical counterfactuals and the perceived quality of explanations would further complement our work. Also, our work ignores practical aspects of FL participation such as increase in training time and technical complexity that need to be managed by hospitals.

Acknowledgement. This research was partially funded by the German Federal Ministry of Health as part of the KINBIOTICS project.

References

1. American Hospital Association, et al.: Fast facts on us hospitals (2022)
2. Caicedo-Torres, W., Gutierrez, J.: Iseeu: visually interpretable deep learning for mortality prediction inside the icu. J. Biomed. Inf. **98** (2019)
3. Chawla, N.V., Bowyer, K.W., Hall, L.O., Kegelmeyer, W.P.: Smote: synthetic minority over-sampling technique. J. Artif. Intell. Res. **16** (2002)
4. Chen, T., Guestrin, C.: Xgboost: a scalable tree boosting system. In: ACM International Conference on Knowledge Discovery and Data Mining, pp. 785–794 (2016)

5. Dhurandhar, A., Chen, P.Y., Luss, R., Tu, C.C., Ting, P., Shanmugam, K., Das, P.: Explanations based on the missing: Towards contrastive explanations with pertinent negatives. Advances in neural information processing systems 31 (2018)

6. Fang, M.L., Dhami, D.S., Kersting, K.: Dp-ctgan: differentially private medical data generation using ctgans. In: International Conference on AI in Medicine, pp. 178–188. Springer, Cham (2022). https://doi.org/10.1007/978-3-031-09342-5_17

7. Jia, Y., McDermid, J., Habli, I.: Enhancing the value of counterfactual explanations for deep learning. In: Tucker, A., Henriques Abreu, P., Cardoso, J., Pereira Rodrigues, P., Riaño, D. (eds.) AIME 2021. LNCS (LNAI), vol. 12721, pp. 389–394. Springer, Cham (2021). https://doi.org/10.1007/978-3-030-77211-6_46

8. Johnson, A., Bulgarelli, L., Pollard, T., Horng, S., Celi, L.A., Mark, R.: Mimic-iv (2022). https://doi.org/10.13026/rrgf-xw32

9. Kanamori, K., Takagi, T., Kobayashi, K., Arimura, H.: Dace: Distribution-aware counterfactual explanation by mixed-integer linear optimization. In: IJCAI (2020)

10. Keane, M.T., Kenny, E.M., Delaney, E., Smyth, B.: If only we had better counterfactual explanations: five key deficits to rectify in the evaluation of counterfactual xai techniques. arXiv preprint arXiv:2103.01035 (2021)

11. Lagu, T., Rothberg, M.B., Shieh, M.S., Pekow, P.S., Steingrub, J.S., Lindenauer, P.K.: Hospitalizations, costs, and outcomes of severe sepsis in the united states 2003 to 2007. Crit. Care Med. **40**(3), 754–761 (2012)

12. Li, Q., Diao, Y., Chen, Q., He, B.: Federated learning on non-iid data silos: an experimental study. In: Proceedings of IEEE 38th ICDE, pp. 965–978 (2022)

13. Lincy, M., Kowshalya, A.M.: Early detection of type-2 diabetes using federated learning. IJSRST (2020)

14. Van Looveren, A., Klaise, J.: Interpretable counterfactual explanations guided by prototypes. In: Oliver, N., Pérez-Cruz, F., Kramer, S., Read, J., Lozano, J.A. (eds.) ECML PKDD 2021. LNCS (LNAI), vol. 12976, pp. 650–665. Springer, Cham (2021). https://doi.org/10.1007/978-3-030-86520-7_40

15. McMahan, B., Moore, E., Ramage, D., Hampson, S., Arcas, B.A.: Communication-efficient learning of deep networks from decentralized data. In: Artificial Intelligence and Statistics, pp. 1273–1282. PMLR (2017)

16. Mertes, S., Huber, T., Weitz, K., Heimerl, A., André, E.: Ganterfactual-counterfactual explanations for medical non-experts using generative adversarial learning. Frontiers in artificial intelligence 5 (2022)

17. National Institute of General Medical Sciences: Sepsis (2022)

18. Nguyen, T.M., Quinn, T.P., Nguyen, T., Tran, T.: Counterfactual explanation with multi-agent reinforcement learning for drug target prediction. arXiv preprint arXiv:2103.12983 (2021)

19. Samoilescu, R.F., Van Looveren, A., Klaise, J.: Model-agnostic and scalable counterfactual explanations via reinforcement learning. arXiv preprint arXiv:2106.02597 (2021)

20. Singer, M., Deutschman, C.S., Seymour, C.W., Shankar-Hari, M., Annane, D., Bauer, M., Bellomo, R., Bernard, G.R., Chiche, J.D., Coopersmith, C.M., et al.: The third international consensus definitions for sepsis and septic shock (sepsis-3). JAMA **315**(8), 801–810 (2016)

21. Tonekaboni, S., Joshi, S., McCradden, M.D., Goldenberg, A.: What clinicians want: contextualizing explainable machine learning for clinical end use. In: Machine Learning for Healthcare Conference, pp. 359–380. PMLR (2019)

22. Verma, S., Dickerson, J., Hines, K.: Counterfactual explanations for machine learning: a review. arXiv preprint arXiv:2010.10596 (2020)

23. Wachter, S., Mittelstadt, B., Russell, C.: Counterfactual explanations without opening the black box: automated decisions and the gdpr. Harv. JL & Tech. (2017)
24. Wang, Z., Samsten, I., Papapetrou, P.: Counterfactual Explanations for Survival Prediction of Cardiovascular ICU Patients. In: Tucker, A., Henriques Abreu, P., Cardoso, J., Pereira Rodrigues, P., Riaño, D. (eds.) AIME 2021. LNCS (LNAI), vol. 12721, pp. 338–348. Springer, Cham (2021). https://doi.org/10.1007/978-3-030-77211-6_38
25. Woensel, W.V., et al.: Explainable clinical decision support: towards patient-facing explanations for education and long-term behavior change. In: International Conference on AI in Medicine, pp. 57–62. Springer, Cham (2022). https://doi.org/10.1007/978-3-031-09342-5_6
26. Xu, J., Glicksberg, B.S., Su, C., Walker, P., Bian, J., Wang, F.: Federated learning for healthcare informatics. J. Healthcare Inf. Res. (2021)

Explainable AI for Medical Event Prediction for Heart Failure Patients

Weronika Wrazen[1]([✉]), Kordian Gontarska[1,2], Felix Grzelka[1],
and Andreas Polze[1]

[1] Hasso Plattner Institute, University of Potsdam, Potsdam, Germany
weronika.wrazen@hpi.de
[2] Technische Universität Berlin, Berlin, Germany

Abstract. The past decade has witnessed significant progress in deploying AI in the medical field. However, most AI models are considered black-boxes, making predictions neither understandable nor interpretable by humans. This limitation is especially significant when they contradict clinicians' expectations based on medical knowledge. This can lead to a lack of trust in the model. In this work, we propose a pipeline to explain AI models. We used a previously devised Neural Network model to present our approach. It predicts the daily risk for patients with heart failure and is a part of a Decision Support System. In our pipeline, we deployed DeepSHAP algorithm to receive global and local explanations. With a global explanation, we defined the most important features in the model and their influence on the prediction. With local explanation, we analyzed individual observations and explained why a specific prediction was made. To validate the clinical relevance of our results, we consulted them with medical experts and made a literature review. Moreover, we described how the proposed pipeline can be integrated into Decision Support Systems. With the above tools, medical personnel can analyze the root of decisions and have insights into how medical parameters should be changed to improve the patient's health state.

Keywords: XAI · Machine Learning · Heart Failure · Healthcare

1 Introduction

Nowadays, it is observed that people are increasingly affected by various diseases and would benefit from frequent or even constant medical care. However, that growth is not proportional to available clinical personnel. Some patients live far from a medical center or ignore early symptoms of a disease without reporting them to a doctor. As a result, they cannot be provided with appropriate care and treatment in time. Introducing a Clinical Decision Support System (CDSS) as help for both medical staff and patients can be a solution to this issue. More and more often, Artificial Intelligence (AI) solutions are assembled into CDSS as healthcare assistance. It can be used to e.g. detect cancer, or predict health failure risk based on vital signals [4,19]. However, most AI solutions are considered *black-boxes*. A human does not know exactly why the model received

© The Author(s), under exclusive license to Springer Nature Switzerland AG 2023
J. M. Juarez et al. (Eds.): AIME 2023, LNAI 13897, pp. 97–107, 2023.
https://doi.org/10.1007/978-3-031-34344-5_12

a specific outcome. As a result, they do not understand a prediction. Sometimes, the decisions of an AI may contradict the clinician's expectations based on state-of-the-art medical knowledge. It leads to a lack of trust in the model. As Amann et al. point out: there is a medical, legal, ethical, and societal necessity to involve Explainable AI (XAI) [2]. They mention XAI can trace model bias and solve disagreements between AI and medical experts.

Cardiovascular diseases are the main cause of non-communicable disease mortality in the world, hence it is crucial to recognize patients in critical condition as soon as possible [21]. This paper is a part of the Telemed5000 project and follows our previous works on a CDSS for heart failure patients. The first work investigated the effectiveness of telemedical interventions [8]. The second one aimed to scale the Telemedical Center (TMC) capacity up to 5000 patients. In that work, we developed a Deep Neural Network for predicting the daily per-patient risk for one of the following medical events: intervention of the Telemedical Center, rehospitalization, or death [7]. Both projects were run in collaboration with the Charité-Universitätsmedizin Berlin.

The objective of the paper is to propose a pipeline for explaining the medical AI models in a transparent and understandable way. It combines XAI methodology with medical expertise. Consequently, medical personnel can check the consistency of predictions with state-of-the-art knowledge. It is especially important when the model is used in healthcare to assess patients' health. End-users have to trust it and be able to recognize any disagreement. The second aim of the pipeline is to give developers information on how the model can be improved. Additionally, we describe an exemplary implementation of the pipeline into CDSS to simplify the analysis. We use the above-mentioned Deep Neural Network to be explained as an example.

2   Related Work

XAI is being increasingly used to improve the transparency and performance of models in medicine. Moreno-Sanchez used SHapley Additive exPlanations (SHAP) and feature permutation to explain XGBoost model that predicts the survival of patients with heart failure [16]. Dave et al. also employed different XAI techniques to explain XGBoost model for detecting the presence of heart diseases [6]. They involved Local Interpretable Model-Agnostic Explanations (LIME) and SHAP. Jia et al. involved Counterfactual Explanations to define what changes in the original input shall be made to receive the opposite outcome [10]. They predicted readiness for weaning from mechanical support. They used the MIMIC-III data set to develop a Convolutional Neural Network. However, in the above-mentioned research, results were not consulted with a medical expert or medical literature regarding their clinical relevance. Pawar et al. proposed a theoretical approach to explain medical models in conjunction with clinical knowledge to increase model transparency and to enable result tracking [17]. However, they did not present any exemplary explanation or a real use case. Wang et al. presented a theory-driven framework for designing human-centered XAI based on philosophy and psychology [23]. They took into consideration how people reason and understand systems.

3   Materials and Methods

3.1   Database and Machine Learning Model

We use a proprietary source database from The Telemedical Interventional Management in Heart Failure II (TIM-HF 2) clinical trial. It consists of daily self-measurements of 746 patients with heart failure conducted for one year [11]. The dataset was split into 3 sets: train, validation, and test. Each patient was assigned exactly to one set in a way to preserve the distribution of samples and events per patient across all sets. The predictor takes two classes, 1 (a medical event) and 0 (no medical event). The number of medical events in the dataset is 3,837, which constitutes approximately 2% of the database.

Features included in the model can be divided into two groups: dynamic and static. Dynamic ones consist of daily transmission of body weight, systolic and diastolic blood pressure, pulse rate, analysis of the heart rhythm conducted with an ECG monitor, and a self-rated health state (1-excellent, 5-poor). Additionally, we include the difference in fluid status expressed as weight: between the event day and 1, 3, and 8 days before. Static features were defined at the beginning of the clinical trial and were constant throughout the research. We include symptoms of heart failure (exertional dyspnoea, dyspnea at rest, peripheral edema, cervical vein congestion, NYHA class, and pulmonary rales), baseline health state assessed by a patient at the beginning of a clinical trial (0-poor, 100-excellent), social factors (living alone, smoker, self-catering, complaints, anxiety, breaks during the day, being depressive, having a lower appetite, and memory difficulties), and comorbidities of heart failure (diabetes, and coronary heart disease). Categorical features were one-hot encoded in the preprocessing step. In total, we have 53 features [7].

3.2   Global and Local Explanations

In this work, we use DeepSHAP algorithm for global and local explanations [14]. *Global explanation* analyzes the general behavior of the model. It describes what effect each feature has on the model's decision. It allows us to find the most important features. *Local explanation* focuses on one observation at a time. It defines why the specific decision was made and can be a suggestion on how to change the values of features to receive other predictions [12].

DeepSHAP is an estimation of Shapley values, called *SHAP value*. It describes what impact the individual feature has on the prediction in a separate observation. Positive SHAP value results in increasing the prediction score, whereas negative in decreasing [14]. SHAP values ϕ are estimated separately for each observation i and feature j and represent *local importance*. To estimate *global feature importance* I_j, we average the absolute Shapley values for each feature across all observations based on (1).

$$I_j = \frac{I}{n} \sum_{i=1}^{n} |\phi_j^{(i)}| \tag{1}$$

Additionally, we use *Most Relevant First (MoRF)* curve to quantitatively assess the fidelity of XAI algorithms [18]. It proceeds from progressive removing features from the most important one to the least and monitoring a decrease in the model performance. The higher fidelity, the higher decrease in the performance at the beginning of the curve. In our model, the performance is estimated with Area under the ROC curve (AUROC), ergo MoRF curve shall converge to 0.5, which is the value of the model random guess. To simulate the removal of continuous features, we randomly sampled new features' values from a normal distribution of the original features. Binary features were permuted. For categorical features, we permuted an original feature and then create new one-hot encoding. The last has a limitation that we do not remove one binary feature at a time (for example ex-smoker in smoking status), but a whole categorical one (ex-smoker, no-smoker, smoker). Nevertheless, that approach prevents creation of unrealistic data. We also plot two MoRF curves for randomly removed features as a baseline. Each simulation was conducted 10 times to estimate variability.

3.3   Model Explanation Pipeline

Our proposed pipeline consists of the following steps.

1. Global explanation
   (a) Estimation of global importance I_j for each feature j.
   (b) Creating the MoRF curve for XAI algorithm and comparing it with baseline curves. It gives medical personnel visualization of XAI reliability.
   (c) Defining the most and the least important features based on MoRF curve and the global importance I_j.
   (d) Plotting the summary plot with cumulative SHAP values for selected features for defining their effect on the prediction. At this point, a medical practitioner gets a general overview of the model.
   (e) Analysis of dependence plots of SHAP values versus feature values to discover non-linearity, and to define high-risk or baseline ranges (with no influence on the prediction). This analysis allows medical personnel to check consistency with state-of-the-art knowledge and receive new insight into how the risk changes based on the features' value.
   (f) Optional: Creating the MoRF curve for clustered features, based on defined requirements. In our case, clustering is done based on devices used for daily measurements. It allows us to define what contribution has each group of features. It not only enables comparing results with state-of-the-art knowledge but also indicates the least and the most important devices. It can be helpful, for example when the funds are limited and there is a need to discard some devices.
2. Local explanation

(a) Analysis of SHAP values for an individual observation to define what contribution has each feature with the defined value. It enables medical personnel to understand why a specific prediction was made.

(b) Analysis of the dependence plot for a specific feature to receive information on how its value shall be changed to reduce the risk. It can be a suggestion of what diagnosis and therapy could be introduced to improve a patient's health state, assuming that there is a causal effect.

4   Results and Discussion

Global Explanation. Figure 1 presents MoRF curves for DeepSHAP and random explanations, which are considered baseline and MoRF curves for three devices: weight scale, blood pressure (BP) monitor, and ECG monitor. It is observed that the DeepSHAP curve quickly decreases and outperforms Random trials. It also indicates there are 10 most important features. Discarding them leads to a decrease in the AUROC from 0.840 (full model) to 0.550. After discarding the 20 most important features AUROC converges to 0.50 and can be treated as a random guess. The weight scale has the highest contribution to the prediction and removing it leads to decreasing AUROC to 0.731 (-13.0%). The second most important device is the BP monitor and removing it results in AUROC equals 0.827 (-1.5%). The least important device is the ECG monitor and the AUROC obtained after removing ECG-related features equals 0.833 (-0.8%).

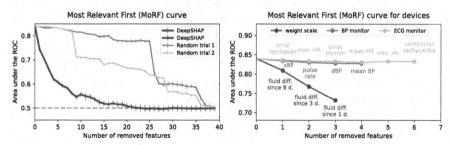

Fig. 1. MoRF curves. On the left: for checking the fidelity of DeepSHAP in comparison to random explanations (random trials). On the right: for each device used for daily measurement conducted by patients at home.

Figure 2 presents a summary plot for the 10 most important features. The *self-assessed health state on the event day* has the highest contribution to the prediction and indicates that the worse a patient feels the higher the risk of a medical event. The further most important features are related to *fluid status* and indicate patients with fluid overload are also more subjected to a higher risk. The opposite relationship is observed for patients whose *baseline health state* was higher at the beginning of the clinical trial. Moreover, both high (red points) and low (blue points) values of *blood pressure systolic (sBP)* and *fluid difference since 8 days ago* increase the risk of a medical event. It suggests that the relationship between those features and the prediction is non-linear. This nonlinearity is proven with dependence plots. To check the medical relevance of

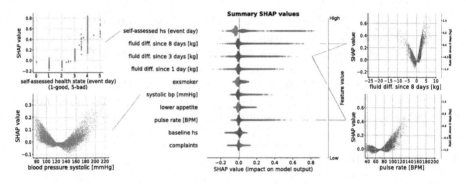

Fig. 2. In the middle: The summary plot represents a global explanation. It collects SHAP values estimated for each observation and feature separately. A high SHAP value indicates the feature's value in the observation increases the prediction's output score. On the left: Dependence plot of SHAP values versus feature values across the database. On the right: Dependence plots with interaction visualization. It informs if the influence of one feature is dependent on the value of the second feature. (Color figure online)

our outcomes we analyzed them with cardiovascular literature and 2 independent medics (Table 1). The 10 most important features indicated by the model were found to be consistent with the medical knowledge except *smoking status*. According to the medical expertise *smoking* shall lead to higher risk than *exsmoking*. Both medical experts agreed there is a relationship between *pulse rate (PR)* and *fluid status* but we received two different causal interpretations. In the first one, high *PR* causes *fluid overload*, whereas in the second one, the *fluid underload* increases *PR*. To correctly interpret the interaction we shall more precisely analyze what was the cause and effect in each observation. XAI analysis indicates that ECG parameters do not have a high contribution to the prediction. The most important parameter of ECG is atrial fibrillation (AFib). It is the 13th most important feature. This is not consistent with medical knowledge. Stegmann et al. proved that patients with heart failure and AFib have a higher risk of hospitalization or death than those with a normal heart rhythm, ergo it is important to monitor ECG [20]. Ventricular tachycardia (VT) is the 45th most important feature and by analyzing Fig. 1 we can assume its contribution is infinitesimal. It is also not consistent with medical knowledge. After consultation with a doctor, the possible reason why AFib and VT occurred to be not so important is the fact it is difficult to capture the moment of their occurrence during a 2-min ECG measurement, especially VT.

Local Explanation. Figure 3 presents the SHAP force plot for a correctly classified observation as high risk. It can be observed that the highest contribution to increased risk has *sBP* with the value of 192 mmHg, and *fluid difference since 3 days* equals 1.1 kg. Analysis of dependence plots suggests that those values are too high and there is a need to decrease them. *Self-assessed health state* with a grade 2 decreases the risk. All above is consistent with medical knowledge presented in Table 1, where the existence of *fluid overload* and *SBP* over 140 mmHg lead to higher risk.

Table 1. Comparison of XAI outcomes with literature and medical expertise, obtained through consultation with doctors. sBP - systolic blood pressure, PR - pulse rate.

Model explanation	Literature	Medical expertise
Better self-assessed health state on the event day results in lower risk	Mortality risk for patients whose self-assessed health state was excellent/very good was lower than for good/fair/poor. The relationship is proportional [3].	The way how patients feel is valuable information. Deterioration of about 2, or grade 4 or 5 results in higher risk.
Fluid overload results in higher risk. The optimal fluid diff. over days is 0 kg.	Fluid overload is a major clinical problem leading to hospitalizations of patients with heart failure [5]. It remains a primary issue for patients with chronic heart failure [15].	If the heart does not work correctly, kidneys do not work properly, leading to fluid overload. Weight gain>1 kg (1 d), >2 kg (3 d), or >2.5 kg (8 d) is high risk.
Fluid underload in 8 days results in higher risk	Side effect of some medicines, like diuretics [13].	Med 1: Patients loose more fluid, than intake. Med 2: It is related to dehydration, that worsens prognosis.
Being ex-smoker results in higher risk.	Smoking is one of the major risk factors for people with heart disease. The risk almost vanishes after 5–10 y. of abstention [13, 21].	Med 1: It is always better to quit smoking but if they smoked long enough they could develop coronary heart disease. Med 2: The parameter is problematic while many lie about quitting. Also, it is important to analyse when they stopped, and how long they smoked.
The baseline for sBP is 118 - 135 mmHg. Lower and higher sBP results in higher risk.	Hypertension (sBP > 140) is one of the major risk factors [13, 21]. A mean sBP for hospitalized patients was 144± 32.6 mmHg [1].	The ideal sBP is 125 mmHg. sBP < 90 and sBP >140 shall be considered high risk.
Having lower appetite results in higher risk.	1) Lower appetite can result from fluid accumulation in the digestive system, which receives less blood. 2) Nutrition is important in heart disease prevention. Especially consumption of cereal fiber, fruit, vegetables, and polyunsaturated fatty acids [13, 21].	It is a symptom of advanced heart failure, especially in coronary heart disease. If blood circulation is damaged, due to heart failure patients have lower appetite.
The baseline for PR is 65–75 BPM. Higher PR is more risky than lower, especially over ∼100 BPM.	Moderate baseline PR is 65–80. Both increase and decrease are associated with higher risk [22, 24].	PR < 50 and PR > 100 shall be considered high risk. Increased PR can lead to cardiac arrhythmia and stroke.
Interaction between PR and fluid status. By PR>70, fluid underload leads to a higher risk than fluid overload.	Fluid underload (dehydration) leads to decrease in circulating blood. As a result, the heart tries to compensate it by beating faster, increasing heart rate [9].	Med 1: With increased PR and by heart damage there is not enough circulation to bring water out, what leads to fluid overload. Med 2: By dehydration, PR increases to compensate the demands of peripheral tissues.
Higher baseline heath status results in lower risk.	It was self assessed by patients at the beginning of the clinical trial, ergo interpretation is identical as for self-assessed heath state.	
Complaining more results in lower risk.	No information was found.	Complaining shall be acknowledged as a related to health state. It shall not be ignored. It is an indicator that something with patient's health is wrong.

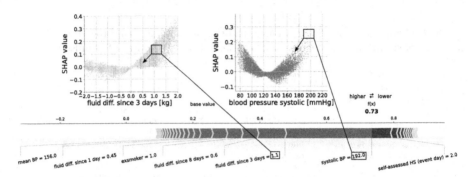

Fig. 3. The force plot (lower) indicates the direction and magnitude of the SHAP value for a separate feature. Features marked as red increase the prediction score, whereas marked as blue decrease. Dependence plots (upper) present the change needed to decrease the score. BP - blood pressure, HS - health state. (Color figure online)

4.1 Possible Implementation in CDSS

The proposed pipeline can be integrated into a CDSS to make the interface more ergonomic for medical personnel. First of all, we can highlight the most important features, so the practitioners can easily focus on them at the beginning. The other solution is to add an option of sorting based on features' importance or alphabetic order making the analysis more convenient. Secondly, we can enable to display of a dependence plot for a selected feature with a marked area where the value for an observation is located. It can give practitioners a quick overview on how the parameter shall be changed. We can also collect new SHAP values of daily incoming data and update dependence plots for a general population or only for one specific patient. New observations can be marked with different colors whats gives the possibility to obtain information about existing trends.

5   Conclusion and Future Work

The main aim of the paper was to introduce a pipeline, which transparently and understandably can explain a medical AI model. To achieve that we involved DeepSHAP algorithm. We used a Neural Network model for predicting the daily risk for patients with heart failure to visualize our approach. The proposed pipeline gave us a broader view of the model's behavior. We distinguished the model's most significant features and their effect on the prediction. We determined also no-risk and increased-risk features' ranges. With analysis of individual SHAP values, we obtained information on why a specific decision was made by the model. With the assistance of dependence plots, we could further receive suggestions on how features need to be changed to improve a patient's health state. It can help medical personnel in e.g. deciding if a specific medicine shall be changed to reduce blood pressure or fluid status. Consequently, they can shorten the time of the diagnosis and obtain a broader view of the patient's state.

Our results were compared with the literature and discussed with medical experts. Most features' interpretation agreed with the medical expertise. However, we also obtained some disagreement, for example, the model wrongly indicates that being an ex-smoker leads to a higher risk than being a smoker. After consultation, we got to know that for that feature it is also important to analyze how long people smoked, when they stopped, or maybe they lied about quitting smoking. The other disagreement concerns the low importance of ECG parameters, especially atrial fibrillation, and ventricular tachycardia. It can be explained by the fact that we took into consideration only records from one day, whereas those abnormal heart rhythms could be difficult to recognize during 2 min records.

Taking the above into consideration, we conclude that the main aim of the paper was achieved. The second aim of the paper regarding obtaining information on how to improve the model was also achieved. We shall analyze smoking status in more details, and also consider introducing time series inputs to observe how the occurrence of atrial fibrillation and ventricular tachycardia change within days. This would give us insight, into whether patients have those irregularities often or if is it just one observation. In this paper, we were not able to describe every 53 features but we think this work can be an instruction about how to create XAI documentation for a model, so medical personnel can check the importance and medical interpretation of every feature. We believe the proposed pipeline can become a best practice during developing medical models and a standard approach.

Acknowledgment. We thank Prof. Dr. med. Friedrich Köhler and his team for the access to the database and valuable feedback regarding our evaluation. This research has been supported by the Federal Ministry for Economic Affairs and Energy of Germany as part of the program *Smart Data* (01MD19014C).

References

1. Adams, K.F., Fonarow, G.C., Emerman, C.L., et al.: Characteristics and outcomes of patients hospitalized for heart failure in the united states: Rationale, design, and preliminary observations from the first 100, 000 cases in the acute decompensated heart failure national registry (ADHERE). Am. Heart J. **149**(2), 209–216 (2005). https://doi.org/10.1016/j.ahj.2004.08.005
2. Amann, J., Blasimme, A., Vayena, E., et al.: Explainability for artificial intelligence in healthcare. BMC Med. Inform. Decis. Mak. **20**, 310 (2020)
3. Bundgaard, J.S., Thune, J.J., Torp-Pedersen, C., et al.: Self-reported health status and the associated risk of mortality in heart failure: the DANISH trial. J. Psychosom. Res. **137**, 110220 (2020). https://doi.org/10.1016/j.jpsychores.2020.110220
4. Chen, Y., Qi, B.: Representation learning in intraoperative vital signs for heart failure risk prediction. BMC Med. Inform. Decis. Mak. **19**, 260 (2019)
5. Costanzo, M.R., Guglin, M.E., Saltzberg, M.T., et al.: Ultrafiltration versus intravenous diuretics for patients hospitalized for acute decompensated heart failure. J. Am. Coll. Cardiol. **49**(6), 675–683 (2007). https://doi.org/10.1016/j.jacc.2006.07.073

6. Dave, D., Naik, H., Singhal, S., Patel, P.: Explainable AI meets healthcare: a study on heart disease dataset. CoRR abs/2011.03195 (2020)

7. Gontarska, K., Wrazen, W., Beilharz, J., Schmid, R., Thamsen, L., Polze, A.: Predicting medical interventions from vital parameters: towards a decision support system for remote patient monitoring. In: Tucker, A., Henriques Abreu, P., Cardoso, J., Pereira Rodrigues, P., Riaño, D. (eds.) AIME 2021. LNCS (LNAI), vol. 12721, pp. 293–297. Springer, Cham (2021). https://doi.org/10.1007/978-3-030-77211-6_33

8. Heinze, T., Wierschke, R., Schacht, A., von Löwis, M.: A hybrid artificial intelligence system for assistance in remote monitoring of heart patients. In: Corchado, E., Kurzyński, M., Woźniak, M. (eds.) HAIS 2011. LNCS (LNAI), vol. 6679, pp. 413–420. Springer, Cham (2011). https://doi.org/10.1007/978-3-642-21222-2_50

9. Jenkins, S.: Sports science handbook: V. 1. Multi Science Publishing (2005)

10. Jia, Y., McDermid, J., Habli, I.: Enhancing the value of counterfactual explanations for deep learning. In: Tucker, A., Henriques Abreu, P., Cardoso, J., Pereira Rodrigues, P., Riaño, D. (eds.) AIME 2021. LNCS (LNAI), vol. 12721, pp. 389–394. Springer, Cham (2021). https://doi.org/10.1007/978-3-030-77211-6_46

11. Koehler, F., Koehler, K., Deckwart, O., et al.: Efficacy of telemedical interventional management in patients with heart failure (tim-hf2): a randomised, controlled, parallel-group, unmasked trial. Lancet (2018)

12. Kopitar, L., Cilar, L., Kocbek, P., Stiglic, G.: Local vs. global interpretability of machine learning models in type 2 diabetes mellitus screening. In: Marcos, M., Juarez, J.M., Lenz, R., Nalepa, G.J., Nowaczyk, S., Peleg, M., Stefanowski, J., Stiglic, G. (eds.) KR4HC/TEAAM -2019. LNCS (LNAI), vol. 11979, pp. 108–119. Springer, Cham (2019). https://doi.org/10.1007/978-3-030-37446-4_9

13. Kumar, P.: Kumar & clark's medical management and therapeutics. W B Saunders (2011)

14. Lundberg, S.M., Lee, S.I.: A unified approach to interpreting model predictions. In: Adv Neural Inf Process Systs 30, pp. 4765–4774. Curran Associates, Inc. (2017)

15. Miller, W.L.: Fluid volume overload and congestion in heart failure. Circ. Heart Fail 9(8) (2016). https://doi.org/10.1161/circheartfailure.115.002922

16. Moreno-Sanchez, P.A.: Development of an explainable prediction model of heart failure survival by using ensemble trees. In: 2020 IEEE International Conference on Big Data (Big Data), pp. 4902–4910 (2020)

17. Pawar, U., O'Shea, D., Rea, S., O'Reilly, R.: Explainable AI in healthcare. In: 2020 Int. Conf. Cyber Situational Aware. Data Anal. Assess. CyberSA, pp. 1–2 (2020)

18. Samek, W., Binder, A., Montavon, G., et al.: Evaluating the visualization of what a deep neural network has learned. IEEE Trans. Neural Netw. Learn. Syst. 28(11), 2660–2673 (2017). https://doi.org/10.1109/TNNLS.2016.2599820

19. Shen, L., Margolies, L.R., Rothstein, J.H., et al.: Deep learning to improve breast cancer detection on screening mammography. Sci. Rep. 9, 12495 (2019)

20. Stegmann, T., Koehler, K., Wachter, R., et al.: Heart failure patients with atrial fibrillation benefit from remote patient management: insights from the TIM-HF2 trial. ESC Heart Fail. 7(5), 2516–2526 (2020). https://doi.org/10.1002/ehf2.12819

21. Thom, T., Haase, N., Rosamond, W., et al.: Heart disease and stroke statistics—2006 update. Circulation 113(6) (2006). https://doi.org/10.1161/circulationaha.105.171600

22. Tian, J., Yuan, Y., Shen, M., et al.: Association of resting heart rate and its change with incident cardiovascular events in the middle-aged and older Chinese. Sci. Rep. 9 (2019). https://doi.org/10.1038/s41598-019-43045-5

23. Wang, D., Yang, Q., Abdul, A., Lim, B.Y.: Designing theory-driven user-centric explainable AI, p. 1–15. ACM (2019)
24. Zhang, D., Wang, W., Li, F.: Association between resting heart rate and coronary artery disease, stroke, sudden death and noncardiovascular diseases: a meta-analysis. CMAJ **188**(15), E384–E392 (2016). https://doi.org/10.1503/cmaj.160050

Adversarial Robustness and Feature Impact Analysis for Driver Drowsiness Detection

João Vitorino[1]([✉]) [iD], Lourenço Rodrigues[2] [iD], Eva Maia[1] [iD], Isabel Praça[1] [iD], and André Lourenço[2] [iD]

[1] Research Group On Intelligent Engineering and Computing for Advanced Innovation and Development (GECAD), School of Engineering, Polytechnic of Porto (ISEP/IPP), 4249-015 Porto, Portugal
{jpmvo,egm,icp}@isep.ipp.pt
[2] CardioID Technologies, 1959-007 Lisboa, Portugal
{lar,arl}@cardio-id.com

Abstract. Drowsy driving is a major cause of road accidents, but drivers are dismissive of the impact that fatigue can have on their reaction times. To detect drowsiness before any impairment occurs, a promising strategy is using Machine Learning (ML) to monitor Heart Rate Variability (HRV) signals. This work presents multiple experiments with different HRV time windows and ML models, a feature impact analysis using Shapley Additive Explanations (SHAP), and an adversarial robustness analysis to assess their reliability when processing faulty input data and perturbed HRV signals. The most reliable model was Extreme Gradient Boosting (XGB) and the optimal time window had between 120 and 150 s. Furthermore, the 18 most impactful features were selected and new smaller models were trained, achieving a performance as good as the initial ones. Despite the susceptibility of all models to adversarial attacks, adversarial training enabled them to preserve significantly higher results, so it can be a valuable approach to provide a more robust driver drowsiness detection.

Keywords: adversarial robustness · explainability · machine learning · heart rate variability · driver drowsiness detection

1 Introduction

The European Road Safety Observatory estimates that 15 to 20% of all road crashes are related to drowsy driving [1]. Therefore, it is essential to monitor drivers and provide timely warnings of excessive fatigue, improving accountability and leading to protective measures. Most of the currently available monitoring mechanisms rely on lane positioning and steering wheel dynamics, but these only detect changes in driving behavior, so a warning is only provided after a driver is impaired [2].

To detect anomalous alterations before any behavioral impairment occurs, a solution can be to infer drowsiness levels directly from a driver's Heart Rate Variability (HRV) signals, which reflect the alertness state of an individual [3]. Due to the rise of consumer electronics like smart watches with Photoplethysmography sensors, as well as

J. M. Juarez et al. (Eds.): AIME 2023, LNAI 13897, pp. 108–113, 2023.
https://doi.org/10.1007/978-3-031-34344-5_13

off-the-person Electrocardiogram (ECG) recording devices like the CardioWheel [4], it is becoming easier to seamlessly gather physiological data from drivers.

Research efforts have been made on the usage of different features and time windows. In the SleepEYE project [5], subjects drove in a Swedish highway while recording ECG and registering Karolinska Sleepiness Scale (KSS) levels. Machine Learning (ML) models achieved over 65% accuracy, but attention was raised for the impact that class imbalance and subject-independent approaches can have [6]. There has been work on subject-dependent approaches and models agnostic to the sensor acquiring the signal, but it was observed that drivers would tend to wrongly estimate their KSS levels, which resulted in unrealistic progressions and numerous misclassifications [7].

Since HRV monitoring mechanisms are prone to motion and lack of contact artifacts, robustness against faulty input data, either produced by corrupted data readings or by malicious attackers, is a very desirable property. This work presents multiple experiments with HRV time windows and ML models, followed by a feature impact and adversarial robustness analysis to assess their reliability.

2 Analysis Methodology

This section describes the utilized dataset, ML models, and the performed analysis. To perform drowsiness detection with ML, it was essential to use a labelled dataset with reliable samples of awake and drowsy signals. The DROZY [8] multimodal dataset was used because it contains both HRV signals and Psychomotor Vigilance Task (PVT) reaction times of individuals that performed three sessions: (i) fully awake, (ii) moderately tired, and (iii) sleep deprived. The PVT values were used in a methodical data labeling process where the first session of each subject was used as a baseline to identify anomalous reaction times in the other two, according to the Z-Score metric. A total of 31 HRV features were computed, and the dataset was split into training and holdout sets with 70% and 30% of the samples. Table 1 summarizes the computed features.

Table 1. Full utilized features and their domains.

Time		Frequency		Nonlinear
HR Mean	NN20	VLF Peak Frequency	HF Peak Frequency	DFA alpha 1
HR Std	pNN20	VLF Absolute Power	HF Absolute Power	SD1
HR Min	NN50	VLF Relative Power	HF Relative Power	SD2
HR Max	pNN50	VLF Logarithmic Power	HF Logarithmic Power	SD2/SD1 Ratio
SDSD	SDNN	LF Peak Frequency	Total Power	
RMSSD		LF Absolute Power	LF Normalized Power	
		LF Relative Power	HF Normalized Power	
		LF Logarithmic Power	LF/HF Ratio	

To ensure an unbiased analysis of the adversarial robustness of the considered ML algorithms, the methodology introduced in [9] was replicated for drowsiness detection. Two training approaches were utilized: regular and adversarial training. In the former, the models were created with the original training sets, whereas in the latter, the training data was augmented with realistic adversarial examples with the Adaptive Perturbation Pattern Method (A2PM) [10]. This data augmentation created a single perturbation in a copy of each drowsy sample, accounting for the feature correlations. Afterwards, model-specific adversarial sets were generated by full adversarial evasion attacks that created as many data perturbations as necessary in the holdout sets until every drowsy sample was misclassified as awake or 30 attack iterations were performed.

Distinct models were created using four ML algorithms: Support Vector Machine (SVM) [11], K-Nearest Neighbors (KNN) [12], Extreme Gradient Boosting (XGB) [13], and Light Gradient Boosting Machine (LGBM) [14]. They were fine-tuned with a grid search over well-established hyperparameters and a 5-fold cross-validation was performed with the F1-Score metric, which is adequate for imbalanced data [15]. Shapley Additive Explanations (SHAP) [16] was used to explain their predictions and the features with highest impact were selected to create new smaller models.

3   Results and Discussion

This section presents the obtained results and the performed analysis. Multiple experiments were performed to assess the viability of different window sizes and ML algorithms. The highest F1-Score, 81.58%, was reached by the SVM model that processed windows of 120 s: SVM 120 (see Fig. 1). Overall, the highest results were obtained with windows of 120 and 150 s, which suggests that this may be the optimal range for the classification of awake and drowsy samples.

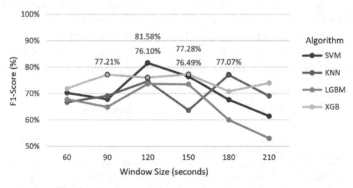

Fig. 1. Window and algorithm experimental results.

The six models with F1-Scores above 75% were chosen for the subsequent feature impact analysis with SHAP. The feature that was particularly relevant to the classification in all six models was the SD2/SD1 ratio, of the nonlinear domain. Nonetheless, most time domain features reached high Shapley values. Regarding the frequency-domain, the relative powers were generally more impactful than the absolute and normalized powers. This suggests that training with relative powers, in percentages of the total power of a sample, can lead to a better generalization and transferability to individuals with different HRV signals, such as healthy athletes and smokers.

The most relevant features were selected according to the feature impact rankings, resulting in 18 numerical features. The new models created with fewer features reached the same F1-Scores on the original holdout set as the initial models with 31 features. This indicates that the selected features provide sufficient information for the classification of awake and drowsy samples. Table 2 summarizes the selected features.

Table 2. Selected features and their domains.

Time			Frequency	Nonlinear
HR Mean	NN20	NN50	VLF Logarithmic Power	DFA alpha 1
HR Min	pNN20	pNN50	LF Relative Power	SD2
HR Max	SDSD	SDNN	HF Relative Power	SD2/SD1 Ratio
			LF/HF Ratio	

Adversarial attacks performed with A2PM were used to analyze the robustness of the smaller models. Even though all six models created with regular training had high F1-Scores on the original holdout set, numerous misclassifications occurred. Most models exhibited significant performance declines across all windows sizes, with only XGB 90 being able to keep an F1-Score above 60% after the attack (see Fig. 2). In contrast, the models created with adversarial training kept significantly higher scores. XGB 150 stood out for preserving an F1-Score of 79.19% throughout the entire attack, which highlighted the adversarially robust generalization of XGB.

Regarding the adversarially trained KNN 180, even though several misclassifications occurred, it reached an F1-Score of 83.33% on the original holdout set, a value even higher than the 81.58% previously obtained by SVM 120. Therefore, besides successfully improving robustness against adversarial examples, the utilized adversarial training approach can also improve their generalization to regular samples.

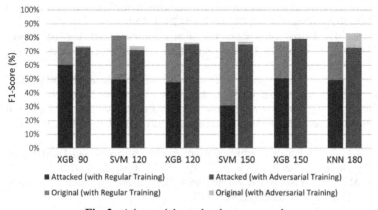

Fig. 2. Adversarial attack robustness results.

4 Conclusions

This work addressed the use of ML for driver drowsiness detection from feature impact and adversarial robustness perspectives. Despite the good detection results in regular samples, the adversarial attacks caused significant performance declines in most models, especially in SVM. Nonetheless, XGB stood out for being the least susceptible and preserving high F1-Scores with adversarial training, which highlights its robust generalization to faulty input data. Therefore, ML models can significantly benefit from adversarial training with realistic examples to provide a more robust driver drowsiness detection. This is a pertinent research topic to be further explored in the future.

Acknowledgments. This work was done and funded in the scope of the European Union's Horizon 2020 research and innovation program, under project VALU3S (grant agreement no. 876852). This work has also received funding from UIDP/00760/2020. A publicly available dataset was utilized in this work. The data can be found at: https://hdl.handle.net/2268/191620.

References

1. European Commission, Road safety thematic report – Fatigue, (2021). https://road-safety.transport.ec.europa.eu/statistics-and-analysis/data-and-analysis/thematic-reports_en
2. Oliveira, L., et al.: Driver drowsiness detection: a comparison between intrusive and non-intrusive signal acquisition methods, In: 7th Eur. Work. Visu. Info. Process., pp. 1–6 (2018)
3. Chowdhury, A., et al.: Sensor Applications and Physiological Features in Drivers' Drowsiness Detection: A Review. IEEE Sens. J. (2018). https://doi.org/10.1109/JSEN.2018.2807245
4. Lourenço, A., et al.: CardioWheel: ECG Biometrics on the Steering Wheel, In: Machine Learning and Knowledge Discovery in Databases, pp. 267–270 (2015)
5. Fors, C., et al.: Camera-based sleepiness detection: final report of the project Sleep-EYE, (2011) https://www.academia.edu/34322032/Camera_based_sleepiness_detection_Final_report_of_the_project_SleepEYE
6. Silveira, C.S., et al.: Importance of subject-dependent classification and imbalanced distributions in driver sleepiness detection in realistic conditions. IET Intell. Transp. Syst. **13**(2), 347–355 (2019). https://doi.org/10.1049/iet-its.2018.5284
7. Rodrigues, L.: Driver Drowsiness Detection with Peripheral Cardiac Signals, Instituto Superior Técnico, (2021) https://fenix.tecnico.ulisboa.pt/cursos/mebiom/dissertacao/1128253548922289
8. Massoz, Q., et al.: The ULg multimodality drowsiness database (called DROZY) and examples of use. In: IEEE Winter Conf. on Appl. of Computer Vision, pp. 1–7 (2016)
9. Vitorino, J., et al.: Towards Adversarial Realism and Robust Learning for IoT Intrusion Detection and Classification. Ann. Telecommun. (2023). https://doi.org/10.1007/s12243-023-00953-y
10. Vitorino, J., et al.: Adaptative Perturbation Patterns: Realistic Adversarial Learning for Robust Intrusion Detection, Future Internet, vol. 14(4) (2022) https://doi.org/10.3390/fi14040108
11. Hearst, M.A., et al.: Support vector machines. IEEE Intell. Syst. their Appl. **13**(4), 18–28 (1998). https://doi.org/10.1109/5254.708428
12. Cover, T., Hart, P.: Nearest neighbor pattern classification. IEEE Trans. Inf. Theory **13**(1), 21–27 (1967). https://doi.org/10.1109/TIT.1967.1053964

13. Chen, T., Guestrin, C.: XGBoost: a scalable tree boosting system In: Proceedings of the ACM Int. Conf. on Knowl. Discov. and Data Min., (2016) https://doi.org/10.1145/2939672.293 9785

14. Ke, G., et al.: LightGBM: a highly efficient gradient boosting decision tree, in Advan. in Neural Info. Process. Sys., 2017, pp. 3147–3155 (2017)

15. Vitorino, J., et al.: A Comparative Analysis of Machine Learning Techniques for IoT Intrusion Detection. Found. Pract. of Sec. (2022). https://doi.org/10.1007/978-3-031-08147-7_13

16. Lundberg, S. M., Lee, S.-I.: A Unified Approach to Interpreting Model Predictions, in Advan. in Neural Info. Process. Sys., pp. 4765–4774 (2017)

Computational Evaluation of Model-Agnostic Explainable AI Using Local Feature Importance in Healthcare

Seda Polat Erdeniz[1,3]([⊠]), Michael Schrempf[1,2], Diether Kramer[1],
Peter P. Rainer[2], Alexander Felfernig[3], Trang Tran[3], Tamim Burgstaller[3],
and Sebastian Lubos[3]

[1] Die Steiermärkische Krankenanstaltengesellschaft m. b. H. (KAGes), Graz, Austria
`seda.polaterdeniz@kages.at`
[2] Medical University Graz, Graz, Austria
[3] Graz University of Technology, Graz, Austria
`http://www.kages.at/`, `https://www.medunigraz.at/`,
`http://www.ase.ist.tugraz.at/`

Abstract. Explainable artificial intelligence (XAI) is essential for enabling clinical users to get informed decision support from AI and comply with evidence-based medical practice. In the XAI field, effective evaluation methods are still being developed. The straightforward way is to evaluate via user feedback. However, this needs big efforts (applying on high number of users and test cases) and can still include various biases inside. A computational evaluation of explanation methods is also not easy since there is not yet a standard output of XAI models and the unsupervised learning behavior of XAI models. In this paper, we propose a computational evaluation method for XAI models which generate *local feature importance* as explanations. We use the output of XAI model (local feature importances) as features and the output of the prediction problem (labels) again as labels. We evaluate the method based a real-world tabular electronic health records dataset. At the end, we answer the research question: "How can we computationally evaluate XAI Models for a specific prediction model and dataset?".

Keywords: machine learning · explainable AI · healthcare

1 Introduction

Healthcare professionals usually need to be convinced by the explanations of AI-based prediction results in Clinical Decision Support Systems (CDSS) [6,9]. Particularly, in cases where the CDSS results in predictions that are not in the line with a clinician's expectations, explainability allows verification whether the parameters taken into account by the CDSS make sense from a clinical point-of-view. In other words, explainability enables the resolution of disagreement between an AI system and healthcare professionals. In such cases, the CDSS

J. M. Juarez et al. (Eds.): AIME 2023, LNAI 13897, pp. 114–119, 2023.
https://doi.org/10.1007/978-3-031-34344-5_14

needs to employ XAI techniques in order to provide reasons why certain predictions have been made by the AI model.

Since the XAI methods do not have standard outputs, there is also no standard way to evaluate their outputs. In order to evaluate XAI models, two approaches can be considered: user evaluation [6] or computational evaluation [10]. Both options have challenges. User evaluations need big efforts (applying on high number of users and test cases) and can still include various biases inside. Computational evaluation is also not straightforward, since it is an unsupervised learning, we can not apply a performance evaluation for supervised learning models. Some researchers applied several methods such as Clustering, Trojaning, and Removing important features. Clustering [3] provides an idea about the similarity between the prediction-explanation pairs, but it does not give us a direct information about the explanation quality. Trojaning [4] is indeed very intuitive for image datasets, but it can not be applied so easily on datasets where the relation between the input and output is not direct. Removing important features [1] cannot be efficiently applied to evaluate the local explanations since the important features of one local explanation are not the same for another. Consequently, these methods can not solve our needs.

Therefore, we proposed an XAI evaluation method by involving supervised learning models trained on the local feature importance values. It is converting the unsupervised nature of XAI models into a supervised model for evaluation. We used local feature importances (a part of the output of XAI models) to build feature datasets and used predicted prognosis probabilities and observed prognoses as labels. As the prediction problem, we used the dataset of MACE risk prediction [8] at the public healthcare provider Steiermärkische Krankenanstaltengesellschaft m.b.H. (KAGes). At the end, we evaluated two famous XAI models "LIME" and "SHAP" based on this real-world prognosis prediction problem. The proposed evaluation approach in this paper can be applied to evaluate any other XAI model with local features importances and it can be used to improve the prediction model.

2   Method

Due to the drawbacks of the available computational evaluation methods mentioned, in this research we propose an XAI evaluation approach which has two evaluation functions as shown in Algorithm 1.

Prerequisites of running the XAI Evaluator are:

- Data (D, L): a labeled binary classification dataset (e.g., features: laboratory values of patients; label: having a cardiovascular attack or not)
- Prediction Model Results (P): classification results of a classification model (e.g., Random Forest Classifier)
- Explanation Model Results (FI): local feature importances generated by an explanation model (e.g., LIME) for results of prediction model (P)

After the prerequisites are met, we can employ our two XAI Evaluator functions separately. *"Evaluation Model - Classifier (EMC)"* function takes the observations (D), corresponding classes (L) and local feature importances generated by an XAI model (FI). This function provides a classification performance (in terms of AUROC) of a classification model which is trained on the weighted dataset (observation values multiplied with local feature importances) and the expected classes (L). The other function *"Evaluation Model - Regressor (EMR)"* takes the observations (D), predicted probabilities generated by a prediction model (P) and local feature importances generated by an XAI model (FI). This function provides the error rate (in terms of RMSE) of a regression model which is trained on the weighted dataset (observation values multiplied with local feature importances) and the predicted probabilities (P).

Algorithm 1: XAI Evaluator

---

1 function EMC (D, L, FI);
   Input : D (observations), L (classes) and FI (local feature imp.)
   Output: $AUROC$
2 $DFI \leftarrow D \times FI$;
3 $trainX, testX, trainY, testY \leftarrow train\_test\_split(D,L)$;
4 $RF\_EMC\_model \leftarrow train\_classifier(\text{trainX, trainY})$;
5 $Pred\_classes \leftarrow RF\_EMC\_model.predict(\text{testX})$;
6 return AUROC$(Pred\_classes, testY)$;
7 function EMR (D, P, FI);
   Input : D (observations), P (predicted proba.) and FI (local feature imp.)
   Output: $RMSE$
8 $DFI \leftarrow D \times FI$;
9 $trainX, testX, trainY, testY \leftarrow train\_test\_split(D,P)$;
10 $RF\_EMC\_model \leftarrow train\_regressor(\text{trainX, trainY})$;
11 $Pred\_proba \leftarrow RF\_EMC\_model.predict(\text{testX})$;
12 return RMSE$(Pred\_proba, testY)$;

---

3   Results

In our experiments, we applied the proposed approach on MACE Prognosis Prediction Dataset with 2k real-world observations at the public healthcare provider Steiermärkische Krankenanstaltengesellschaft m.b.H. (KAGes) [8].

In order to build the labeled binary-classification dataset, EHR data of KAGes was used. Due to the coverage of more than 90% of all hospital beds in Styria, KAGes has access to approximately two million longitudinal patient histories. For every inpatient stay in a KAGes hospital, the following structured information is available: demographic data (e.g. age, gender), visited departments, diagnoses (coded with ICD-10 format), laboratory data (coded in LOINC format), health care procedures (coded in Austrian procedure codes format) and nursing assessment (e.g. visual impairment and orientation).

As the prediction model, we trained a Random Forest binary classifier [2] as a supervised learning model since it gives the best classification performance among the compared methods. As the explanation model, we used two model agnostic XAI frameworks LIME [7] and SHAP [5] which are both calculating local feature importance and have been widely recognized as the state-of-the-art in XAI. We implemented a LIME XAI model using LIME Tabular Data Explainer. We set the number of explanations as 10 which provides the most important 10 features for each observation since more than 10 explanations was taking a very long amount of execution time with our dataset. Therefore, it provided the most important 10 features with their importance values and the remaining features are assigned with feature importance 0. We implemented a SHAP XAI model as a *TreeExplainer* option since it is used to explain an RF model. Then, SHAP values were calculated using the *shap_values* function.

For *XAI Evaluator*, a supervised learning model should be used which can be trained as a classifier or a regressor. Therefore, we used again *Random Forest* as in *Prediction Model* for both Evaluation Models: "Classifier" and "Regressor". As *Evaluation Model "Classifier"*, we implemented a Random Forest Classifier model. As *Evaluation Model "Regressor"*, we implemented a Random Forest Regressor model. Besides, we developed three versions of the local feature importance dataset which is used by *Evaluation Model* for training and test.

We used RMSE to evaluate "Evaluation Model - Regressor (EMR)" performance and AUROC to evaluate "Evaluation Model - Classifier (EMC)". With these results, we aimed to answer the research question for classification problems in healthcare: "How can we select an XAI Model to explain predictions of a prediction model?".

To measure the regression fit between the prediction model and the explanation model, we proposed using the local feature importance values produced by XAI models to compare their performances. For this purpose, we employed *"Evaluation Model - Regressor" (EMR)* to compare the regression error of XAI models. In this approach, we used a supervised learning model to approach again the output of the prediction model. We observed that *EMR on LIME* (RMSE:0.16) gives slightly better results than *EMR on SHAP* (RMSE:0.19). This means, LIME model is more compatible with our prediction model based on this particular dataset, so it can be preferred as the explanation model.

We also compared the classification fit between the prediction model and the explanation model. For this purpose, we employed *"Evaluation Model - Classifier" (EMC)* to compare the classification performance of XAI models. We observed that using *EMC on LIME* give us a better classification performance (AUROC:0.87) than *Prediction Model (PM)* (AUROC:0.74) and *EMC on SHAP* (AUROC:0.77) in terms of AUROC. This also supports the results of EMR and we say that LIME model is more compatible with our prediction model based on this particular dataset, so it can be preferred as the explanation model.

4 Conclusion

Model agnostic XAI models do not have a standard performance evaluation method yet. On one hand, they can be evaluated via user feedback which can have many biases and are not scalable. On the other hand, several computational evaluation methods by recent researches can not be effectively applied for explaining local explanations. In this paper we proposed a computational evaluation method using local feature importance values. We have tested the proposed model on LIME and SHAP explanations with a Random Forest prediction model based on a real-world healthcare data[1]. As observed, applying the proposed approach on the explanations with local feature importance values, it is possible to computationally compare the performance of XAI models using local feature importance in order to select one of these models as the explanation model for a specific prediction model and dataset[2].

Acknowledgement. This study received approval from the Ethics Committee of the Medical University of Graz (approval no. 30-146 ex 17/18) and was supported by a research grant from ERA PerMed within the project called "PreCareML". The source codes of this study are publicly available in PreCareML/XAI Repository.

References

1. Alvarez Melis, D., Jaakkola, T.: Towards robust interpretability with self-explaining neural networks. Advances in Neural Information Processing Systems 31 (2018)
2. Breiman, L.: Random forests. Mach. Learn. **45**(1), 5–32 (2001)
3. Gramegna, A., Giudici, P.: SHAP and LIME: an evaluation of discriminative power in credit risk. Frontiers in Artificial Intelligence, p. 140 (2021)
4. Lin, Y.S., Lee, W.C., Celik, Z.B.: What do you see? evaluation of explainable artificial intelligence (xai) interpretability through neural backdoors. arXiv preprint arXiv:2009.10639 (2020)
5. Lundberg, S.M., Lee, S.I.: A unified approach to interpreting model predictions. Advances in Neural Information Processing Systems 30 (2017)
6. Polat Erdeniz, S., et al.: Explaining machine learning predictions of decision support systems in healthcare. In: Current Directions in Biomedical Engineering, vol. 8, pp. 117–120. De Gruyter (2022)
7. Ribeiro, M.T., Singh, S., Guestrin, C.: "why should i trust you" explaining the predictions of any classifier. In: Proceedings of the 22nd ACM SIGKDD International Conference on Knowledge Discovery and Data Mining, pp. 1135–1144 (2016)
8. Schrempf, M., Kramer, D., Jauk, S., Veeranki, S.P., Leodolter, W., Rainer, P.P.: Machine learning based risk prediction for major adverse cardiovascular events. In: dHealth, pp. 136–143 (2021)

---

[1] https://erapermed.isciii.es/.

[2] https://github.com/precareml/xai.

9. Tran, T.N.T., Felfernig, A., Trattner, C., Holzinger, A.: Recommender systems in the healthcare domain: state-of-the-art and research issues. J. Intell. Inf. Syst. **57**, 171–201 (2021)
10. van der Waa, J., Nieuwburg, E., Cremers, A., Neerincx, M.: Evaluating XAI: a comparison of rule-based and example-based explanations. Artif. Intell. **291**, 103404 (2021)

Batch Integrated Gradients: Explanations for Temporal Electronic Health Records

Jamie Duell[1]([⊠])[ID], Xiuyi Fan[2][ID], Hsuan Fu[3][ID], and Monika Seisenberger[1][ID]

[1] School of Mathematics and Computer Science, Swansea University, Wales, UK
{853435,m.seisenberger}@swansea.ac.uk

[2] Nanyang Technological University, Singapore, Singapore
xyfan@ntu.edu.sg

[3] Department of Finance, Insurance and Real Estate, Université Laval, Quebec City, Canada
hsuan.fu@fsa.ulaval.ca

Abstract. eXplainable Artifical Intelligence (XAI) is integral for the usability of black-box models in high-risk domains. Many problems in such domains are concerned with analysing temporal data. Namely, we must consider a sequence of instances that occur in time, and explain why the prediction transitions from one time point to the next. Currently, XAI techniques do not leverage the temporal nature of data and instead treat each instance independently. Therefore, we introduce a new approach advancing the *Integrated Gradients* method developed in the literature, namely the *Batch-Integrated Gradients* (Batch-IG) technique that (1) produces explanations over a temporal batch for instance-to-instance state transitions and (2) takes into account features that change over time. In Electronic Health Records (EHRs), we see patient records can be stored in temporal sequences. Thus, we demonstrate Batch-Integrated Gradients in producing explanations over a temporal sequence that satisfy proposed properties corresponding to XAI for EHR data.

1 Introduction

Due to modern computation capabilities, black-box models in recent Artificial Intelligence (AI) literature often take the form of Deep Neural Networks (DNNs), the complexity of such methods enable an increase in accuracy. At the same time, the increase in accuracy is usually associated with a decrease in model interpretability. Explainable Artificial Intelligence (XAI) is a common approach for increasing the transparency of black-box AI methods, XAI encompasses an increase in importance as desire to use accurate predictive models is inherited within high-risk domains. The need is emphasised by the GDPR's "*right to an explanation*" [7]. Feature-attribution is a commonly used method for XAI, where the aim is to determine how each feature influences the prediction of an instance, the landmark papers for this are introduced by the authors in [4,5].

Despite recent success in XAI methods in developing feature attribution explainers such as SHapley Additive exPlanations (SHAP) [4] and Local Interpretable Model-Agnostic Explanations (LIME) [5], the temporal nature of data is often neglected when

---

Jamie Duell is supported by the UKRI AIMLAC CDT, funded by grant EP/S023992/1.

J. M. Juarez et al. (Eds.): AIME 2023, LNAI 13897, pp. 120–124, 2023.
https://doi.org/10.1007/978-3-031-34344-5_15

developing XAI methods for tabular data [8]. Whilst there does exist the application of existing XAI methods to temporal data (e.g., [6, 10]), to our knowledge, there is no local explanation method is designed to focus on the temporal nature of the data and the associated change in prediction across instances. Hereinafter, we propose the adaptation of the Integrated Gradients (IG) method [9] to adhere to temporal data. Following this, we propose properties inherent from IG that conform to ideal properties for temporal data. Comparing with existing approaches to XAI, our method deviates away from instance based attribution and instead determines attribution with respect to the change in prediction value (probability) for regression (classification).

The occurrence of temporal data can be seen often in healthcare [2]. Healthcare data is often stored in the form of Electronic Health Records (EHR) and an explanation is necessary when providing black-box predictions. Therefore, we consider an EHR case to demonstrate the proposed approach to produce explanations.

Consider a breast cancer patient in the Simulacrum dataset[1]:

Age: 78, Sex: Female, Site: 0, Morph: 8500, Weight: 84, Height: 1.67, **Dose Administration: 600 → 300 → 450 → 50**, Chemo Radiation: No, Regimen Outcome Description: N/A, **Admin Route: Subcutaneous → Subcutaneous → Intravenous → Oral**, Regimen Time Delay: No, Regimen Stopped Early: No, **Cycle Number: 1 → 3 → 3 → 5**, Cancer Plan: 2, Ethnicity: J, Behaviour: Malignant, Grade: G3, CReg Code: L1201, T Best: 2, N Best: 0, M Best: 0, Laterality: Left, CNS: Y1, ACE: 9, Performance: 1, Clinical Trial: Yes

There are four records of this patient, representing the sequences of treatments the patient has received. From these records, we observe that three features of this patient has gone through the following changes: drug dose administration (from **600** to **50**), cycle number (from **1** to **5**) and drug administration route (from **subcutaneous** to **oral**), while other features have remain unchanged. In the context of XAI, we pose the question:

How does each of these changing features affect the patient's survival?

Answering such a question is critical for medical decision making [1]. Yet, existing feature attribution algorithms in XAI [3] cannot directly answer this question, as they treat each record as an independent instance and do not consider temporal changes. In other words, state-of-the-art XAI explainers such as the SHAP and LIME would consider the above as four separate patients and provide attribution values to all features, instead of only analyzing the changing ones.

2   Method

We introduce Batch-IG as an extension to Integrated Gradients (IG) [9] for temporal explainability over batches of time-based data, such that we explain a collection

---

[1] https://simulacrum.healthdatainsight.org.uk/ - The Simulacrum is a synthetic dataset developed by Health Data Insight CiC derived from anonymous cancer data provided by the National Cancer Registration and Analysis Service, which is part of Public Health England.

of sequential instances and determine the attribution with respect to each point within the time batch. We represent a time batch as a matrix $\chi \in \mathbb{R}^{N \times J}$. A time batch contains N time points, where each time point is a vector, such that we have $\chi = \langle \mathbf{x}_1, \ldots, \mathbf{x}_t, \ldots, \mathbf{x}_N \rangle$.

Analysing the behaviour of the black-box model between time points helps to determine the behaviour of the model with respect to data of a temporal nature, so we analyse where between time points that a feature had the greatest change in importance. The insight provided into the change in partial derivatives with respect to time intervals could lead to deeper insight to which point the feature had greater importance when altering in prediction.

Consider a neural network $f : \mathbb{R}^{1 \times J} \rightarrow \mathbb{R}$, the the accumulation of gradients between points are given by a fraction of the difference denoted by $\alpha \times (\mathbf{x}_{t+1} - \mathbf{x}_t)$, where $\alpha \in [0, 1]$, integrating over α from 0 to 1 accumulates partial derivatives over a straight line path from the prior time point to the current time point with respect to the jth feature dimension. We introduce Batch-IG, with an iterative function using the prior time-step t as the baseline, and following step $t + 1$ as the target, therefore we have

$$\text{Batch-IG}(\chi) := \sum_{t=1}^{N-1} \sum_{j=1}^{J} \left((x_{t+1}^j - x_t^j) \times \int_{\alpha=0}^{1} \frac{\partial f(\mathbf{x}_t + \alpha \times (\mathbf{x}_{t+1} - \mathbf{x}_t))}{\partial x^j} \Delta\alpha \right).$$

This can then be computed through the Riemann approximation method, namely:

$$\text{Batch-IG}^{\mathcal{R}}(\chi) := \sum_{t=1}^{N-1} \sum_{j=1}^{J} \left((x_{t+1}^j - x_t^j) \times \sum_{k=1}^{M} \frac{\partial f(\mathbf{x}_t + \frac{k}{M} \times (\mathbf{x}_{t+1} - \mathbf{x}_t))}{\partial x^j} \times \frac{1}{M} \right)$$

where $(\mathbf{x}_t + \frac{k}{M} \times (\mathbf{x}_{t+1} - \mathbf{x}_t))$ takes an initial point in time for an instance $\mathbf{x}_t$, and integrates with respect to $x_t^j \in \mathbf{x}_t, \forall j$ over $\frac{k}{M}$ steps, where $\frac{k}{M} \in [0, 1]$ and $0 < k \leq M$ to approximate the path integral between $\mathbf{x}_{t+1}$ and $\mathbf{x}_t$ such that, sub-intervals have equal lengths between both points. Therefore, a larger value of M will allow for a more accurate approximation of Batch-IG between points $\mathbf{x}_{t+1}$ and $\mathbf{x}_t$.

3   Experiment

From the Simulacrum dataset, we isolate a cohort of patients with the ICD-10 code "C50" *Malignant neoplasm of breast*. We group patients by their patient unique identifiers, and within these groups we order the patients by cycle number to maintain temporally organised patient data as a means for generating explanation examples. We want to obtain explanations of the form, *"given features that change during the course of patient treatment, how do the changes effect the survival prediction probability?"*. Standard feature-attribution methods such as SHapley Additive exPlanations (SHAP) [4] and Local Interpretable Model-Agnostic Explanations (LIME) [5] gives explanations of the form, *"given an instance, which features attributed towards the prediction probability?"*, yet, this is potentially problematic, as when we observe temporal data groups of data belonging to the same patient should not be viewed independently. For example, let us consider a temporal batch of instances of the same patient where the alterations at $\{t_0, t_1, t_2, t_3\}$ then we have the patient instance state transitions:

Dose Administration: 600 → 300 → 450 → 50,
Admin Route: Subcutaneous → Subcutaneous → Intravenous → Oral,
Cycle Number: 1 → 3 → 3 → 5.

The following predictions given at each time interval:

t0 = 94.49%, t1 = 95.87%, t2 = 95.92%, t3=93.82% towards the class ≥ 6 Months survival.

Therefore, upon generating explanation with respect to the introduced patient cycle, we see that the dose administration feature is the only attributed feature under all 3 transitions, such that t0 : 600 → t1 : 300 → t2 : 450 → t3 : 50, whereas the cycle number between the time intervals $t1 \rightarrow t2$ does not change, the attribution given for the cycle number feature is only evident in Figs. 1a and c. The explanation highlights that the cycle number in the earlier cycles attribute towards a probability of longer survival, whereas in the final recorded cycle, the later cycle numbers attribute towards a shorter survival. Similarly, drug administration has positive influence over longer survival earlier in the cycle and negative influence in the final recorded cycle (Fig. 1).

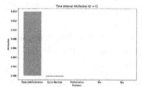 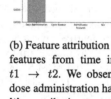 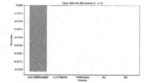

(a) Feature attribution for the features from time interval $t0 \rightarrow t1$. We observe the dose administration had **positive** attribution towards the class ≥ 6 Months and the cycle number transition from 1 → 3 had **negative** attribution towards the ≥ 6 Months class.

(b) Feature attribution for the features from time interval $t1 \rightarrow t2$. We observe the dose administration had **positive** attribution towards the class ≥ 6 Months. This time interval was observed during the drug cycle 3, where there exists only change to the drug administration.

(c) Feature attribution for the features from time interval $t2 \rightarrow t3$. We observe the dose administration had **negative** attribution towards the class ≥ 6 Months and the cycle number transition from 3 → 5 also had **negative** attribution towards the ≥ 6 Months class.

Fig. 1. The attribution of the controllable features F^d for a breast cancer patient through a set of 3 recorded cycles, with two instances under the same cycle, such that Cycle Number = {1,3,3,5}, such that the set of controllable features are given by F^d = {"Dose Administration", "Cycle Number", "Admin Route"}. Observing each sub-figure, we determine that the most influential feature in altering predictions between time points is given by the adjustment to the drug dose administration for the patient.

4   Conclusion

In this paper, we identify a gap in current literature surrounding XAI for temporal data. To combat this, we introduce Batch-IG as a modification to the IG framework to consider time and both dynamic and static features. Futhermore, we propose properties that should be satisfied when considering temporal applications of XAI. Upon introducing the given properties, we provide a qualitative comparison between Batch-IG, SHAP, DeepSHAP and LIME. Similarly, we provide a quantitative comparison on a controlled example comparing the same methods. From this, we determine that for temporal EHR the proposed Batch-IG method adheres to all introduced properties and provides the true attribution from the controlled example.

Limitations of the proposed approach are the requirement of knowledge (e.g. the temporal data needs to be linked via an identifier) with respect to temporal sequences within the data. The model specificity of such approach limits the ML models that can be applied in order to use the Batch-IG framework. Similarly, as with other current methods, Batch-IG also assumes independence between features.

References

1. Amann, J., Blasimme, A., Vayena, E., Frey, D., Madai, V.I.: Explainability for artificial intelligence in healthcare: a multidisciplinary perspective. BMC Med. Inf. Decis. Making **20**(1), 310 (2020)
2. Batal, I., Valizadegan, H., Cooper, G.F., Hauskrecht, M.: A temporal pattern mining approach for classifying electronic health record data. ACM Trans. Intell. Syst. Technol. **4**(4), 2508044 (2013). https://doi.org/10.1145/2508037.2508044
3. Loh, H.W., Ooi, C.P., Seoni, S., Barua, P.D., Molinari, F., Acharya, U.R.: Application of explainable artificial intelligence for healthcare: A systematic review of the last decade (2011–2022). Comput. Methods Programs Biomed. **226**, 107161 (2022)
4. Lundberg, S.M., Lee, S.: A unified approach to interpreting model predictions. In: Advances in NeurIPS 30: Annual Conference on NeurIPS, pp. 4765–4774 (2017)
5. Ribeiro, M.T., Singh, S., Guestrin, C.: "Why should I trust you?": Explaining the predictions of any classifier. In: Proceedings of the 22nd ACM SIGKDD International Conference on Knowledge Discovery and Data Mining, pp. 1135–1144. KDD 2016, Association for Computing Machinery, New York, NY, USA (2016)
6. Schlegel, U., Arnout, H., El-Assady, M., Oelke, D., Keim, D.A.: Towards a rigorous evaluation of XAI methods on time series. In: 2019 IEEE/CVF International Conference on Computer Vision Workshop (ICCVW), pp. 4197–4201 (2019). https://doi.org/10.1109/ICCVW.2019.00516
7. Selbst, A.D., Powles, J.: Meaningful information and the right to explanation. Int. Data Priv. Law **7**(4), 233–242 (2017)
8. Simic, I., Sabol, V., Veas, E.E.: XAI methods for neural time series classification: a brief review. CoRR abs/2108.08009 (2021). https://arxiv.org/abs/2108.08009
9. Sundararajan, M., Taly, A., Yan, Q.: Axiomatic attribution for deep networks. In: ICML2017: Proceedings of the 34th International Conference on Machine Learning, pp. 3319–3328. JMLR.org (2017)
10. Veerappa, M., Anneken, M., Burkart, N., Huber, M.F.: Validation of XAI explanations for multivariate time series classification in the maritime domain. J. Comput. Sci. **58**, 101539 (2022). https://doi.org/10.1016/j.jocs.2021.101539

Improving Stroke Trace Classification Explainability Through Counterexamples

Giorgio Leonardi, Stefania Montani[✉], and Manuel Striani

Laboratorio Integrato di Intelligenza Artificiale e Informatica in Medicina DAIRI - Azienda Ospedaliera SS. Antonio e Biagio e Cesare Arrigo, Alessandria - e DISIT -, Università del Piemonte Orientale, Alessandria, Italy
stefania.montani@uniupo.it

Abstract. Deep learning process trace classification is proving powerful in several application domains, including medical ones; however, classification results are typically not explainable, an issue which is particularly relevant in medicine.

In our recent work we tackled this problem, by proposing *trace saliency maps*, a novel tool able to highlight what trace activities are particularly significant for the classification task. A trace saliency map is built by generating artificial perturbations of the trace at hand that are classified in the same class as the original one, called *examples*.

In this paper, we investigate the role of *counterexamples* (i.e., artificial perturbations that are classified in a different class with respect to the original trace) in refining trace saliency map information, thus improving explainability. We test the approach in the domain of stroke.

1 Introduction

Medical process traces, i.e., the sequences of activities implemented to care each single patient within a hospital organization, represent a rich source of information, useful to identify bottlenecks and issues in patient management, and to improve the quality of patient care.

In particular, trace classification [2], which consists in exploiting the logged activity sequences to classify traces on the basis of some categorical or numerical performance properties, can be adopted to verify whether single traces meet some expected criteria, or to make predictions about the future of a running trace (such as, e.g., the remaining time and the needed resources to complete the work). It can thus support a better planning of the required human and instrumental resources, and an assessment of the quality of the provided service, by identifying non-compliances with respect to the expected performance.

State-of-the-art approaches to trace classification resort to deep learning techniques, that are proving very powerful also in medical applications. However deep learning suffers from lack of *explainability* [3]: deep learning architectures are able to stack multiple layers of operations, in order to create a hierarchy of increasingly more abstract *latent* features [4], but the meaning of the latent

J. M. Juarez et al. (Eds.): AIME 2023, LNAI 13897, pp. 125–129, 2023.
https://doi.org/10.1007/978-3-031-34344-5_16

features and their correlation to the original input data are typically difficult to understand. A deep learning classification tool, therefore, does not provide an explanation for its classification results, that are based on latent features. This issue is obviously particularly critical in medicine.

In our previous work [5], we addressed this problem by introducing *trace saliency maps*, a novel tool able to graphically highlight what parts of the traces (i.e., what activities in what positions) are particularly significant for the classification task. A trace saliency map is built by generating artificial perturbations of the trace at hand that are classified in the same class as the original one, called **examples**. The idea is that the artificial traces that share the same class as the original one, also share elements (i.e., activities and their position in the sequence) that are relevant for the classification process. These elements are highlighted in the trace saliency map.

In this paper, we investigate the role of **counterexamples** (i.e., artificial perturbations that are classified in a different class with respect to the original trace) in refining trace saliency map information, by building a **countermap** able to further improve explainability.

We also provide some tests in the domain of stroke management quality assessment.

2   Trace Classification Explainability and the Role of Counterexamples

In our architecture, every process trace is first converted in a numeric vector, where each activity is encoded as an integer; then the trace is classified resorting to a Convolutiona Neural Network (CNN) [1]. In parallel, it is also provided as an input to an Adversarial AutoEncoder (AAE) [6], which generates a neighbourhood of artificial traces, as perturbations of the original one. AAE are in fact probabilistic autoencoders that aim at generating new random items highly similar to the training data. The AAE architecture includes an encoder, a decoder and a discriminator. The discriminator identifies if a randomly generated latent instance can be considered valid or not, i.e., if it can be confused with the instances coming from the true prior distribution. Technical details about the CNN and the AAE we exploited are outside the scope of this paper, and can be found in [5].

In the approach, artificial traces are generated in the latent feature space; after generating an artificial trace, the discriminator checks its validity; the valid artificial trace is then decoded into the original feature space. Finally, the artificial trace is classified by the CNN architecture.

Artificial traces classified in the same class as the original one are called *examples*, and are adopted to built the trace saliency map, as described in [5].

Figure 1 (left) shows the visualization of a trace saliency map, as provided by our tool. Activities in the trace that were conserved in the examples, and are thus important for classification, are in green. Activities that frequently changed in the set of the examples, and are therefore unimportant, are in red. Activities whose importance is somewhere in the middle are in yellow.

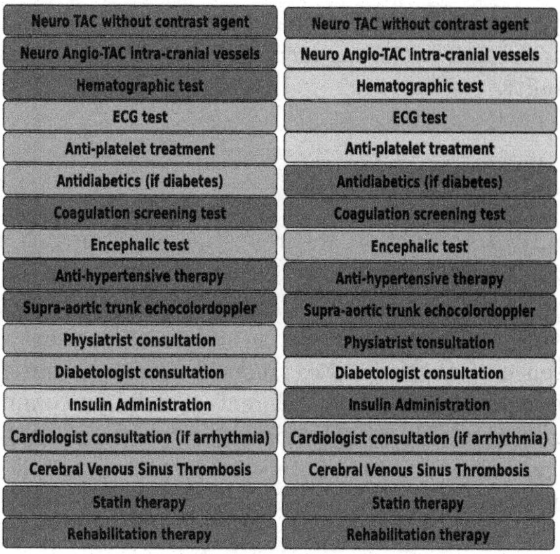

Fig. 1. A trace saliency map (left) and the corresponding countermap (right) (Color figure online)

In the following, we describe how we use the artificial traces classified in the opposite class with respect to the original one.

Given the counterexample i_j ($j \in [1, n]$) in the neighbourhood of a real trace x, we adopt it to build a **countermap**, as follows: (1) for each encoded activity ac_k in trace x, we calculate the average of all the corresponding encodings of the counterexamples, obtaining the average values μ_k with $k \in [1, m]$; (2) we then calculate the absolute value of the difference between ac_k and the average value μ_k. The countermap of x is the sequence of these differences ($|ac_k - \mu_k|$ with $k \in [1, m]$), activity by activity, from left to right.

Since we exploited counterexamples, activities associated to a high difference in the countermap are those that provide the greater contribution to classification: as a matter of fact, these activities changed with respect to the original trace, and this change led to a change in the classification output as well. In our interface, they will thus be highlighted in green. On the other hand, activities associated to a small difference in the countermap are not very important, because classification changed even if they were conserved in the artificial traces: they will be highlighted in red. Activities associated to a difference value somehow in the middle will be highlighted in yellow. Countermaps can be exploited to enforce or better characterize the output of trace saliency maps. Indeed, if an activity is in green both in the trace saliency map and in the countermap, it is certainly important for classification, while if it is in red in both maps, it is certainly not relevant. Therefore, identical colors can enforce the role of a specific activity in determining the classification output. On the other hand, when the colors of an activity do not match in the two maps, the countermap can be used as an instrument to refine the importance of such an activity in determining the classification output, thus better characterizing the situation and improving the overall explainability.

3   Results

As anticipated in the Introduction, we are testing our approach in the domain of stroke management quality assessment. In this field, it is necessary to distinguish between simpler patients and patients affected by complex/rare conditions. This distinction has to be made referring to clinical data and patient's characteristics, such as the presence of co-morbidities. Indeed, heterogeneous patients (simple vs. complex) are supposed to follow different guideline recommendations, thus originating different traces; in complex patient traces additional procedures and more frequent diagnostic/monitoring steps are expected, to follow the evolution over time of their critical situation. (Binary) classification of the logged traces can thus be adopted, to verify if the actual performance of a hospital is satisfactory, or if a better organization of activities and human resources is needed.

In our experiments, first we classified all traces resorting to the CNN black box architecture mentioned in Sect. 2, reaching an accuracy of 82% (detailed classification results are however outside the scope of this paper, and can be found in [5]). We then generated the trace saliency map and the countermap for each trace, and calculated the *error* percentage, defined as follows:

$$error = \frac{|discord|}{|total|} * 100$$

where $|discord|$ is the number of activities that are highlighted in green in a trace saliency map and in red in the corresponding cuntermap, or viceversa, while $|total|$ is the number of activities in all the traces in the log.

In our experimental work, the *error* percentage reached a value of 33.83% on simple patients, and a value of 26.88% on complex ones, demonstrating that the countermap information reinforces the trace saliency map information in over two thirds of the cases.

In addition to these quantitative measure, we also conducted a qualitative analysis of the map/countermap pairs, verifying that, in most cases, the findings they highlight are coherent with medical knowledge. In the following, we discuss in detail one case. Figure 1 compares a trace saliency map (left), obtained on a complex patient trace, to the corresponding countermap (right), activity by activity. As the figure shows, a set of activities have proved relevant for classification: indeed, they appear in green in the map (since they were conserved in the examples), and they appear in green in the countermap as well (since they were not conserved in the counterexamples - thus determining the misclassification). In particular, an ECG test in the emergency phase is atypical, and suggests that the patient is experiencing serious cardiovascular problems. Indeed, s/he is later sent to cardiologist consultation, and treated for cerebral venous sinus thrombosis. This result is coherent with medical knowledge: such a co-morbid patient is certainly a complex one, as these activities are normally absent in simpler situations. On the other hand, insulin administration, which appears in yellow in the trace saliency map, is highlighted in red in the countermap. The yellow labelling identifies insulin administration as a possibly important activity for classification (even though not a very important one, not being green). This uncertain interpretation is clarified by the countermap, that excludes insulin administration from the list of the relevant activities: indeed, insulin therapy is normally absent in simpler patients, but in this case the patient at hand is better characterized by the activities that manage her/his serious cardiovascular complications, and insulin is not the key information to determine the classification output. In this way, the countermap not only enforces the information provided by the saliency map, when colours are fully coherent, but also refines and clarifies the saliency map information, resolving possible ambiguities. The countermap information is thus able to further characterize the patient at hand's specific situation, providing an even more explainable classification output to the end user.

4   Conclusions

In this paper, we have proposed to combine the use of trace saliency maps (that we introduced in [5]) and countermaps, in order to improve the explainability of deep learning process trace classification. Specifically, our experiments in the domain of stroke management show that countermaps can enforce and refine the information provided by saliency maps, thus better characterizing the subset of activities that justify the classification output. In the future, we wish to test the approach resorting to different deep learning classifiers - a step which will require a reduced effort, since our definition of saliency maps and countermaps does not depend on the specific black box architecture that has been selected for classification. Moreover, we plan to conduct further experiments in the hospital setting, collecting a feedback on the utility of the tool and also on the usability of its interface.

References

1. Alom, M.Z., et al.: A state-of-the-art survey on deep learning theory and architectures. Electronics **8**(3), 292 (2019)
2. Breuker, D., Matzner, M., Delfmann, P., Becker, J.: Comprehensible predictive models for business processes. MIS Q. **40**, 1009–1034 (2016)
3. Guidotti, R., Monreale, A., Ruggieri, S., Turini, F., Giannotti, F., Pedreschi, D.: A survey of methods for explaining black box models. ACM Comput. Surv. **51**(5), 1–4 (2019)
4. LeCun, Y., Bengio, Y., Hinton, G.E.: Deep learning. Nature **521**(7553), 436–444 (2015)
5. Leonardi, G., Montani, S., Striani, M.: Explainable process trace classification: An application to stroke. J. Biomed. Informatics **126**, 103981 (2022)
6. Makhzani, A., Shlens, J., Jaitly, N., Goodfellow, I., Frey, B.: Adversarial autoencoders (2016)

Spatial Knowledge Transfer with Deep Adaptation Network for Predicting Hospital Readmission

Ameen Abdel Hai[1]([⊠]), Mark G. Weiner[2], Alice Livshits[3], Jeremiah R. Brown[4], Anuradha Paranjape[3], Zoran Obradovic[1], and Daniel J. Rubin[3]

[1] Computer and Information Sciences, Temple University, Philadelphia, PA, USA
`{aabdelhai,zoran.obradovic}@temple.edu`
[2] Weill Cornell Medicine, New York, NY, USA
[3] Lewis Katz School of Medicine, Temple University, Philadelphia, PA, USA
`daniel.rubin@tuhs.temple.edu`
[4] Department of Epidemiology and Biomedical Data Science, Geisel School of Medicine at Dartmouth, Hanover, NH, USA

Abstract. A hospital readmission risk prediction model based on electronic health record (EHR) data can be an important tool for identifying high-risk patients in need of additional support. Performant readmission models based on deep learning approaches require large, high-quality training datasets to perform optimally. Utilizing EHR data from a source hospital system to enhance prediction on a target hospital using traditional approaches might bias the dataset if distributions of the source and target data are different. There is a lack of an end-to-end readmission model that can capture cross-domain knowledge. Herein, we propose an early readmission risk temporal deep adaptation network, ERR-TDAN, for cross-domain spatial knowledge transfer. ERR-TDAN transforms source and target data to a common embedding space while capturing temporal dependencies of the sequential EHR data. Domain adaptation is then applied on a domain-specific fully connected linear layer. The model is optimized by a loss function that combines distribution discrepancy loss to match the mean embeddings of the two distributions and the task loss to optimize predicting readmission at the target hospital. In a use case of patients with diabetes, a model developed using target data of 37,091 patients from an urban academic hospital was enhanced by transferring knowledge from high-quality source data of 20,471 patients from a rural academic hospital. The proposed method yielded a 5% increase in F1-score compared to baselines. ERR-TDAN may be an effective way to increase a readmission risk model's performance when data from multiple sites are available.

Keywords: Transfer Learning · Readmission Prediction · Electronic Health Records data · Machine Learning · Deep Learning

J. M. Juarez et al. (Eds.): AIME 2023, LNAI 13897, pp. 130–139, 2023.
https://doi.org/10.1007/978-3-031-34344-5_17

1 Introduction

Hospital readmission is an undesirable outcome and a driver of high financial costs. Approximately, 20% of Medicare discharges had readmission within 30-days, corresponding to $20+ billion in hospital costs [1]. Identifying patients with higher risk of readmission would enable the targeting of interventions to those at greatest need, optimizing the cost-benefit ratio.

Previously, we published a risk deep learning (DL) model based on electronic health records (EHR) data collected from an urban academic hospital that predicts the risk of unplanned, 30-day readmission among patients with diabetes. We used a sequential model, long short-term memory (LSTM). Performance was adequate (F-1 Score 0.80), and results showed that this LSTM model can capture temporal dependencies of the EHR data [2].

Performant readmission models based on DL techniques require large, high-quality training data to perform optimally. Utilizing EHR data from a source hospital system to enhance prediction on a target hospital using traditional approaches enlarge dataset bias which might deteriorate performance due to distributional difference of the source and target datasets, resulting in statistically unbounded risk for the target tasks [3]. Traditional approaches are designed for a specific data type, and not capable of generalizing to other temporal data.

Transfer learning approaches have been explored for hospital readmission with the objective to improve learning at the target population by exploiting information from a related source population. In [4, 5], classical transfer learning was employed to address data scarcity using a relevant source dataset. In [6], classical transfer learning techniques were explored as to what extent can transfer learning benefit learning on target tasks by fine-tuning pre-trained models in the healthcare domain. However, there is still a need for an end-to-end model to perform cross-domain spatial knowledge transfer and predictive learning in a unified learning framework while capturing temporal dependencies for hospital readmissions.

In this paper, we propose an early readmission risk temporal deep adaptation network, ERR-TDAN, to perform cross-domain spatial knowledge transfer from EHR data of different sites and perform predictive learning. Deep Adaption Network (DAN) utilizes deep convolutional neural network (CNN) and generalizes it to the domain adaptation setting through learning transferable latent features between source and target domains for computer vision tasks [3, 7]. Motivated by the success of DAN in numerous transfer learning tasks in computer vision, we employed the idea of learning transferable features of temporal data by matching the source and target domain distributions in the latent feature space. We tailored it for hospital readmission using EHR data and optimized for the target task.

The aims of this study were as follows: 1) To develop a hospital readmission framework using EHR data that transfers knowledge between a rural academic hospital and an urban academic hospital to enhance predictions on the urban academic hospital. 2) To study the optimal amount of retrospective EHR data needed for future predictions. 3) To study the duration of optimal performance. Experiments conducted show that ERR-TDAN can enhance hospital readmission prediction.

2   Deep Adaptation Network

Domain adaptation is a form of transfer learning commonly used in computer vision to address the problem of learning using data from two related domains but under different distributions [3, 8]. Domain adaptation can help improve the performance of a model by learning transferable features to minimize the gap between the source and target domains in an isomorphic latent feature space. DAN generalizes deep CNN for computer vision applications to utilize domain adaptation techniques to learn transferable feature representation in the latent embedding space [7]. Motivated by the success of DAN in various computer vision tasks [9–11], we utilized the idea of DAN for transferring cross-domain spatial knowledge tailored for predicting hospital readmission on EHR data and optimized to enhance predictions on the target, rather than generalizing on both domains. A direct comparison to DAN is not applicable since DAN is modified for computer vision tasks using CNN. CNNs capture spatial correlations and are unable to capture temporal correlations of EHR data [3]. Thus, we employed the idea of DAN and tailored it for hospital readmission on EHR data to capture temporal dependencies using LSTM layers, establish cross-domain knowledge transfer, and optimized it for the target task using a customized loss function.

3   The Proposed ERR-TDAN Framework

An early readmission risk framework based on temporal deep adaptation network was developed to enhance prediction on the target data collected from Temple University Hospital System (TUHS) by establishing spatial knowledge transfer from a source data with higher quality features collected from Penn State University Hospital System (PSUHS). The model was developed using data as defined by the National Patient-Centered Clinical Research Network (PCORnet) Common Data Model (CDM) [12].

We applied a hospital readmission LSTM model that we previously published using EHR data collected from TUHS [2]. When trained on TUHS data and tested on the following year TUHS data this model F-1 score was 0.80. We trained and tested the same method on EHR data collected from PSUHS, where performance was better (F1-score 0.91). The 11% increase in F-1 score was achieved since EHR data from PSUHS contained fewer missing data, denser features, and less erroneous data. However, training and evaluating the same method on data from both domains affected the performance (F-1 score 0.79) since the model struggled to generalize and converge due to training data drawn from different distributions. To address this limitation, we employed the idea of DAN, tailored for hospital readmission on EHR data that captures temporal correlations and enhances target prediction through learning transferable features via domain-specific fully connected linear layers to explicitly reduce the domain discrepancy. DAN generalizes on both domains for computer vision tasks, whereas in our study we tailored this technique for hospital readmission using EHR data and optimized on the target task, instead of generalizing on both domains. To accomplish this, the hidden embeddings of the domain-specific layers are embedded to a reproducing kernel Hilbert space through maximum mean discrepancy (MMD), to match the mean embeddings of two domain distributions. The model was optimized via a customized loss function.

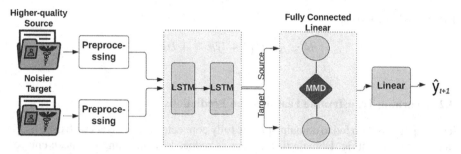

Fig. 1. The proposed method framework, ERR-TDAN. It comprises of three main processes. 1) LSTM layers learn hidden representation of the input of both source and target domains. 2) Deep adaption network structure with fully connected layers is constructed to match the mean embeddings of different domains drawn from different distributions. 3) The matched embeddings are then passed to a fully connected layer with sigmoid function for binary classifications. The model is optimized through a customized loss function that penalizes on domain discrepancy of both source and target, and task loss to optimize for the target task.

Figure 1 presents the proposed framework, ERR-TDAN which consists of the following main processes: 1) LSTM's input was data from both source and target to learn hidden representation to map source and target data to a common embedding while capturing temporal dependencies of the EHR data. 2) To match the embedding distributions of the source and target domains, deep adaptation network scenario is established through fully connected linear layers constructed to match the mean embeddings of different domain distributions. The hidden representation is embedded through a reproducing kernel Hilbert space to transfer knowledge and bridge the gap between two distributions via MMD to reduce domain discrepancy. 3) The matched embeddings are then passed to a fully connected layer with a sigmoid function to classify if a patient is likely to be readmitted or not. In backpropagation, we optimize the model on the target domain using a customized loss function that combines the domain discrepancy loss and binary cross entropy loss. The following sections illustrate the framework in more detail.

3.1 Representation Learning of Temporal EHR Data with LSTM

Initially, we utilized LSTM with two recurrent layers to form a stacked LSTM to learn hidden data representation embedded to a common latent feature space of the temporal EHR data of the source and target domains. LSTM, a sequential model capable of capturing temporal correlations, is commonly used for sequential tasks and is proven to be effective for hospital readmission using EHR data [2, 13]. LSTM takes as an input a 3-dimensional tensor of stacked source and target data. LSTM is structured based on basic neural network, but neurons of the same layer are connected, enabling a neuron to learn from adjacent layers, in addition to learning from outputs of the previous layers and the input data. Hence, neurons include two sources of inputs, the recent past and the present. A dropout of 0.1 was applied between the first and second LSTM layers. To add nonlinearity, we utilized ReLU activation function on the output of LSTM (embeddings), formulated as follows:

$$b^t = ReLU\left(b + Wh^{t-1} + Ux^t\right) \tag{1}$$

3.2 Learning Transferable Features and Predictions

The output b^t is then fed to domain-specific fully connected linear layers with deep adaptation network setting. Domain discrepancy is reduced by matching the mean embeddings of the source and target distributions. Hidden representation of the linear layers embedded through a reproducing kernel Hilbert space to bridge the gap between two distributions and transfer knowledge via MMD. MMD measures the distance of the source and target distributions in the embedding space. MMD distance measure was originally used to determine whether two samples are drawn from the same distribution and measures how distant the samples are [14]. In this study, we utilized MMD to learn transferable features between source and target domains to enhance prediction on the target. MMD was utilized as one of the two components of the loss function to minimize the domain discrepancy. The loss function is explained in more detail in the next section. MMD is defined as:

$$MMD_{loss}\left(D^S, D^T\right) = \|\frac{1}{n}\sum_{i=1}^{n}\phi\left(d_i^S\right) - \frac{1}{m}\sum_{j=1}^{m}\phi\left(d_j^T\right)\|_H^2, \tag{2}$$

where D^S and D^T denote source and target data respectively, ϕ denotes the Gaussian kernel function, H denotes the Hilbert space, and n and m denote the number of observations of the source and target sets, respectively. The temporal embeddings of the LSTM are then fed into fully connected layers with MMD loss to measure the distance between two distributions and reduce domain discrepancy.

Prediction. The matched embeddings are then fed into a linear layer with output of 1 with sigmoid activation function for predictions $\hat{y}$ [2].

3.3 Model Optimization via a Customized Loss Function

We tailored the loss function for hospital readmission on the target domain by combining Binary Cross Entropy (BCE) loss to measure the error of reconstruction, applied on the target task only, and MMD loss applied on both source and target to reduce domain discrepancy. Since the aim of this study is to enhance prediction on the target domain using higher-quality source data, we reduced the weight of the MMD loss via the penalty parameter γ and optimized the loss on the target domain. Loss function used in the proposed ERR-TDAN model is defined as follows:

$$BCE_{loss} = (x, y) = L = \{l_1, \ldots, l_N\}^T, l_n$$
$$= -w_n[y_n \cdot logx_n + (1 - y_n) \cdot log(1 - x_n)],$$
$$TOTAL_{loss} = \frac{1}{L^N}\sum_{t=1}^{L^N}\left(BCE_{loss}(x, y) + \gamma MMD_{loss}(d^S, d^T)\right), \tag{3}$$

where x and y are the predictions and ground truth for a given batch respectively. L denotes loss. N is the batch size, w is a rescaling weight given to the loss of each batch element, γ is the penalty parameter of domain discrepancy. To optimize for the target task, we determined empirically that 0.5 value of γ is appropriate.

4 Data

We collected data from an urban academic hospital, TUHS, and a rural academic hospital, PSUHS, between July 1st, 2010, and December 31st, 2020. We extracted data on encounters, demographics, diagnosis, laboratory tests, medication orders, procedures, and vital signs. in the cohort of patients with diabetes was defined as previously described [2]. Data preprocessing, handling of missingness of data, different number of recordings per encounter, learning embeddings to reduce dimensionality, address sparse feature vectors, and data representation were performed as presented in [2]. Additional features were aggregated to assist with learning temporal dependencies, including duration of stay in days, and number of days since the prior encounter.

We obtained a total of 1,421,992 encounters corresponding to 20,471 patients for PSUHS, and a total of 3,023,267 encounters corresponding to 37,091 patients for TUHS. The class distributions were as follows. TUHS: 28,107 for the negative class (no readmission), and 8,984 for the positive class (readmitted within 30-days); PSUHS: 18,775 for the negative class and 1,696 for the positive class.

The characteristics of the samples from the two sites were different. For instance, 4.9% of patients were Hispanic at PSUHS, whereas TUHS contained large Hispanic population of 22%. Other differences included race and tobacco use. The numbers of unique ICD-9 and ICD-10 codes, and vital recordings at PSUHS were larger than that at TUHS.

Patient encounters were sequentially ordered by admission date and represented in a 3-dimensional tensor for the LSTM model, where each patient's data is represented as a 2-dimensional matrix in which features of each encounter are represented in a 1-dimensional array while a second dimension represents different hospitalizations of that patient. The third dimension is used to encode hospitalization information of different patients.

5 Experimental Setup and Results

We hypothesize that it is feasible to enhance readmission predictions on target data of TUHS using a source data from a relevant domain under different distribution. In this section, we conduct extensive experiments to evaluate the performance of the proposed model, ERR-TDAN and compare it to baselines. F-1 score, precision, recall (sensitivity), specificity, and accuracy were used to evaluate the model's performance [15]. We randomly selected different patients for training and testing. Experiments were iterated 10 times; results were presented based on the mean and two-sided 95% confidence interval (CI). Moreover, we address the following research questions to evaluate optimal performance of the model.

5.1 Can We Enhance Readmission Risk Prediction for a Target Hospital by Utilizing Data from Another Hospital?

We randomly split TUHS and PSUHS data to 70% training, 10% validation, and 20% testing. Then, we concatenated training data of both domains, and fed to the ERR-TDAN. We tested the model on TUHS using 7,418 patients, of whom 1,557 had a readmission.

Table 1 presents a comparative analysis to evaluate the proposed method, ERR-TDAN compared to alternative baselines. Table 1 shows that ERR-TDAN yielded a 5% increase in F1-score when compared to a model we previously published for hospital readmission on EHR data collected from TUHS, and 3% increase using a generalized version of ERR-TDAN (G-ERR-TDAN) of the domain adaptation framework with MMD loss without optimizing on the target task. G-ERR-TDAN results provide evidence that optimizing on the target task enhances target's predictions is superior to generalizing on both domains.

Table 1. Performance of the proposed method, ERR-TDAN and three alternatives tested on the target domain (TUHS) enhanced by a related source data (PSUHS). The Average F1, Recall/Sensitivity, Specificity, and accuracy and their corresponding two-sides 95% confidence interval (CI) on 10 experiments on training and testing patients' data selected completely at random.

Model	Train	F1-score	Recall	Specificity	Accuracy
[2]	TUHS	0.80 ± 0.003	0.81 ± 0.002	0.94 ± 0.010	0.81 ± 0.002
LSTM	TUHS + PSUH	0.79 ± 0.007	0.81 ± 0.006	0.95 ± 0.008	0.81 ± 0.005
G-ERR-TDAN	TUHS + PSUH	0.82 ± 0.001	0.81 ± 0.001	0.92 ± 0.002	0.81 ± 0.001
ERR-TDAN	TUHS + PSUH	**0.85** ± 0.002	0.84 ± 0.002	0.91 ± 0.003	0.84 ± 0.001

5.2 What is the Retrospective Optimal Amount of EHR Data Needed for Future Predictions?

We conducted extensive experiments to find the optimal amounts of patient's historical data needed for the model to perform optimally. Our objective was to determine a size of training data so that further enlargements do not improve predictions of hospitalization risk. The model was trained on varying t and tested on $t + x$, where t denotes a period in the past and $t + x$ denotes a period in the future. For a fair comparison, $t + x$ was a fixed test dataset of 2020, and trained on varying training sets of t, , including 6 months (July-December of 2019), 1 year (2019), 2 years (2018–2019), 3 years (2017–2019), 4 years (2016–2019), and 5 years (2015–2019) look-back time. For instance, training on 2019, and testing on 2020 (1 year look-back) to test if learning on 1 year of historical EHR data from the past is sufficient to perform optimally.

Figure 2 (left) shows that 1 year of historical data are optimal to predict readmission since it yielded the highest F1-score with least amounts of data required.

5.3 How Often Do We Need to Retrain the Model to Achieve Optimal Performance?

Concept and covariate shifts are one of the major reasons model performances degrade overtime. Monitoring data drift helps avoid performance degradation. Thus, we conducted experiments to study the lifetime of the proposed model. Based on the optimal look-back time of question 2, we trained the model on EHR data collected in 2015 and tested it with 1, 2, 3, 4, and 5 future gaps. For instance, training on data collected in 2015 and testing in 2020 to experiment if the model's performance would degrade after 5 years. We iterated this over various models trained on 1 year of data collected in 2015, 2016, and 2017 and tested for readmissions on future instances.

Figure 2 (right) shows that F1-score decreased over time due to data drift. Performance was relatively stable when tested on 1 and 2 years in the future. There was a significant decrease in F1-score when used to predict readmissions with 3 years gap between training and testing. On overage, F1-score degraded 0.6% when used 3 years later, 3.5% when used 4 years later, and 7% when used 5 years later. Therefore, to maintain optimal performance of hospital readmission models on EHR data, retraining the model every 3 years may avoid model degradation and maintain optimal performance.

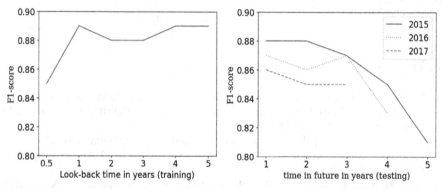

Fig. 2. (left) Presents the retrospective optimal amount of EHR data needed for future predictions. Results show that 1 year of historical data are sufficient to predict hospital readmission from the leading year. (right) Presents the lifetime of the model to maintain and achieve optimal performance. Three different models were developed on data collected from 2015, 2016, and 2017 to predict future instances with 1 to 5 years gap. Results show that to maintain optimal performance, the proposed framework may benefit from retraining every three years.

6 Discussion and Conclusion

We examined the hypothesis that it is feasible to enhance hospital readmission risk predictions on EHR data using data collected from a related source domain. ERR-TDAN model trained on joint TUHS, and PSUHS data yielded a 5% increase in F1-score when compared to an LSTM model trained on TUHS only, 6% increase in F1-score when compared to LSTM model trained on both TUHS and PUSH, and 3% increase when compared to a generalized version of ERR-TDAN (G-ERR-TDAN) aimed to generalize on both domains. Furthermore, conducted experiments showed that one year of historical data is sufficient to predict readmission. We studied the lifetime of the model to avoid performance degradation due to data drift over time. Experiments suggest that retraining the ERR-TDAN framework every three years avoids performance degradation.

We propose a framework, ERR-TDAN that establishes spatial knowledge transfer based on a temporal deep adaptation network tailored for hospital readmission on EHR data and optimized for the target task. ERR-TDAN can enhance readmission predictions of the target task using higher quality data from a related source domain under different distributions by matching the mean embeddings to reduce domain discrepancy. This is the first end-to-end transfer learning framework based on domain adaptation for hospital readmission. A deployment challenge for the proposed framework is that it requires training data from both source and target domains which might be difficult to obtain. In a planned follow up study we will evaluate applicability of the proposed method for prospective applications. In addition, we will compare the proposed hospital readmission method to alternatives aimed to learn from integrated data with explanatory variables of various quality.

Acknowledgments. This research was supported by the National Health Institute (NIH) under grant number R01DK122073.

References

1. McIlvennan, C.K., Eapen, Z.J., Allen, L.A.: Hospital readmissions reduction program. Circulation **131**(20), 1796–803 (2015). https://doi.org/10.1161/CIRCULATIONAHA.114.010270. PMID: 25986448; PMCID: PMC4439931
2. Hai, A.A., et al.: Deep learning vs traditional models for predicting hospital readmission among patients with diabetes. In: Proceedings of AMIA 2022 Annual symposium, Washington, D.C. (2022)
3. Long, M., Cao, Y., Wang, J., Jordan, M.: Learning transferable features with deep adaptation networks. In: Proceedings of the 32nd International Conference on Machine Learning. Proceedings of Machine Learning Research, vol. 37, pp. 97–105 (2015)
4. Helm, J.E., Alaeddini, A., Stauffer, J.M., Bretthauer, K.M., Skolarus, T.A.: Reducing hospital readmissions by integrating empirical prediction with resource optimization. Prod. Oper. Manag. **25**(2), 233–257 (2016)
5. Desautels, T., et al.: Prediction of early unplanned intensive care unit readmission in a UK tertiary care hospital: a cross-sectional machine learning approach. BMJ Open **7**(9), e017199 (2017)
6. Gupta, P., Malhotra, P., Narwariya, J., Vig, L., Shroff, G.: Transfer learning for clinical time series analysis using deep neural networks. J. Healthc. Inform. Res. **4**(2), 112–137 (2019). https://doi.org/10.1007/s41666-019-00062-3
7. Long, M., Cao, Y., Wang, J., Jordan, M.: Learning transferable features with deep adaptation networks. In: Proceedings of 32nd International Conference on Machine Learning, pp. 97–105 (2015)
8. Wang, M., Deng, W.: Deep visual domain adaptation: a survey. Neurocomputing **312**, 135–153 (2018)

9. Deng, W., Zheng, L., Ye, Q., Kang, G., Yang, Y., Jiao, J.: Image-image domain adaptation with preserved self-similarity and domain- dissimilarity for person re-identification. In: Proceedings of IEEE/CVF Conference on Computer Vision and Pattern Recognition, pp. 994–1003, June 2018

10. Chen, Y., Li, W., Sakaridis, C., Dai, D., Van Gool, L.: Domain adaptive faster R-CNN for object detection in the wild. In: Proceedings of IEEE/CVF Conference on Computer Vision and Pattern Recognition, pp. 994–1003, June 2018

11. Hosseini-Asl, E., Keynton, R., El-Baz, A.: Alzheimer's disease diagnostics by adaptation of 3D convolutional network. In: Proceedings of IEEE International Conference on Image Processing (ICIP), pp. 126–130, September 2016

12. Elixhauser A SC PL. Clinical classifications software (ccs): Agency for healthcare research and quality (2014). http://www.hcup-us.ahrq.gov/toolssoftware/ccs/ccs.jsp. Accessed 27 Dec 2021

13. Hochreiter, S., Schmidhuber, J.: Long short-term memory. Neural Comput. **9**(8), 1735–1780 (1997)

14. Gretton, A., Borgwardt, K.M., Rasch, M.J., Schölkopf, B., Smola, A.: A kernel two-sample test. J. Mach. Learn. Res. **13**, 723–773 (2012)

15. Powers DMW. Evaluation: From precision, recall and f-measure to roc., informedness, markedness & correlation. J. Mach. Learn. Technol. **2**, 37–63 (2011)

Dealing with Data Scarcity in Rare Diseases: Dynamic Bayesian Networks and Transfer Learning to Develop Prognostic Models of Amyotrophic Lateral Sclerosis

Enrico Longato[1]($\boxtimes$) ⓘ, Erica Tavazzi[1] ⓘ, Adriano Chió[2] ⓘ, Gabriele Mora[2] ⓘ, Giovanni Sparacino[1] ⓘ, and Barbara Di Camillo[1,3] ⓘ

[1] Department of Information Engineering, University of Padova, Padova, Italy
enrico.longato@dei.unipd.it
[2] Neuroscience Department "Rita Levi Montalcini", University of Torino, Torino, Italy
[3] Department of Comparative Biomedicine and Food Science, University of Padova, Padova, Italy

Abstract. The extremely low prevalence of rare diseases exacerbates many of the typical challenges to prognostic model development, resulting, at the same time, in low data availability and difficulties in procuring additional data due to, e.g., privacy concerns over the risk of patient reidentification. Yet, developing prognostic models with possibly limited in-house data is often of interest for many applications (e.g., prototyping, hypothesis confirmation, exploratory analyses).

Several options exist beyond simply training a model with the available data: data from a larger database might be acquired; or, lacking that, to sidestep limitations to data sharing, one might resort to simulators based, e.g., on dynamic Bayesian networks (DBNs). Additionally, transfer learning techniques might be applied to integrate external and in-house data sources.

Here, we compare the effectiveness of these strategies in developing a predictive model of 3-year mortality in amyotrophic lateral sclerosis (ALS, a rare neurodegenerative disease with <0.01% prevalence) using the in-house dataset of a single ALS clinic in Milan, Italy (N = 116). We test several combinations of direct and transfer-learning-mediated development based on additional real data from the Italian PARALS register (N = 568). We also train two DBNs, one for each dataset, and use them to simulate large numbers of virtual subjects whose variables are linked by the same probabilistic relationships as in the real data.

We show that, compared to a baseline model developed on the smaller dataset (AUROC = 0.633), the largest performance increase was obtained using data simulated using a DBN trained on the larger PARALS register (AUROC = 0.734).

Keywords: Amyotrophic Lateral Sclerosis · Dynamic Bayesian Networks · Transfer Learning

J. M. Juarez et al. (Eds.): AIME 2023, LNAI 13897, pp. 140–150, 2023.
https://doi.org/10.1007/978-3-031-34344-5_18

1 Introduction

In the field of rare diseases, many of the typical challenges to prognostic model development are compounded by the unavoidable, hard limit set by their extremely low prevalence. For instance, as, by definition, only very few people are affected by a given rare disease in a given geographical area, and they all tend to converge to the same clinical centres of excellence for treatment, privacy concerns related to data reidentification are magnified. In turn, this might reflect into a reduced ability or willingness on the part of the data owners to share their data and create multicentre databases of sufficient sample size for robust model development. Despite these limitations, developing prognostic models with (usually limited) in-house data is often of interest for many applications, such as conducting exploratory data analyses, prototyping new ideas, or simply verifying that relationships seen during clinical trials or reported in the literature apply to the subpopulation of interest represented in the in-house dataset.

A priori, there might be several reasonable alternatives for developing such models beyond using the available, in-house data as-is. A first alternative would be acquiring additional data, in ways that range from agreements with larger clinical centres (where a critical mass of data would be presumably available) to making use of open data. Oftentimes, however, this is an unworkable solution as, on the one hand, as previously mentioned in the context of privacy, there are many considerations that might prevent direct data sharing; on the other, open data related to a given rare disease might not exist, or they might be incompatible with the in-house data (e.g., in terms on input variables). A second alternative would be resorting to simulated data [15] in place of real patient data, which may have the added benefit of limiting privacy-related concerns. In practice, this might mean developing an in-house simulator to augment [11] the in-house data, or relying on an external simulator, based on presumably higher-quality data, which may be directly accessible for augmentation or only via a repository of pre-simulated data. Many effective simulators of this kind are based on dynamic Bayesian networks [5], a class of probabilistic graphical models able to capture the time-dependent conditional probability relationships that exist within a dataset so that they can be used to generate arbitrary amounts of in-silico patients data starting from realistic initial conditions [10,12,14]. Furthermore, regardless of how one has acquired additional data, be it directly or via simulation, there is also a question of whether and how these should be integrated with the in-house data. One possibility, in this sense, would be to use transfer learning (TL) as a knowledge sharing approach [16] between a source domain characterised by greater data availability and a target domain represented by the in-house data.

While the feasibility of these options is mainly determined by what is accessible in terms of external data, simulation models, or internal know-how, the relative performance of each approach remains an open question. Hence, in this work, we conduct a systematic analysis of each possible strategy based on a real-life case study, namely the prediction of 3-year mortality in amyotrophic lateral sclerosis (ALS), a fatal neurodegenerative disease with an extremely low prevalence of less than 1 in 10,000 individuals, based on an in-house dataset from a single tertiary ALS clinic (the ALS centre of the Istituti Clinici Scientifici Maugeri IRCCS in Milan, Italy), possibly enriched by data extracted from a population-based ALS register, namely the Piemonte and Valle d'Aosta ALS (PARALS) register, either directly or via transfer learning, with or without simulation via DBNs. In doing so, we aim at providing a preliminary set of evidence that might inform future model development efforts with small-sample data.

2   Study Population and Data Preprocessing

In this work, we analysed the demographic and clinical data collected during the routine screening visits of two real-world cohorts of ALS patients: the first dataset, named MIMA (short for Milan - Maugeri) in the following, included the data of 175 ALS patients treated in the ALS centre of the Istituti Clinici Scientifici Maugeri IRCCS in Milan, Italy, and represented our in-house database; the second, much larger dataset, consisting of 843 subjects, was extracted from the PARALS register [4], a population-based register of two Italian regions.

Both data sources included static information collected at diagnosis and longitudinal information acquired during each patient visit. The variables collected at diagnosis were: sex, age at onset, site of onset (spinal or bulbar), diagnostic delay, body-mass index (BMI) both premorbid and at diagnosis, forced vital capacity (FVC) at diagnosis, the result of a genetic test on the main ALS-related genes (namely C9orf72, FUS, SOD1, and TARDBP), ALS familiality, and presence of frontotemporal dementia (FTD). Longitudinal information comprised the use of non-invasive ventilation (NIV), the administration of percutaneous endoscopic gastrostomy (PEG), and the values of a functional scale, namely the Milano-Torino staging (MiToS) system [2], made up of four binary variables reflecting the presence or absence of impairment in the walking/self-care, swallowing, communicating, and breathing domains. Additionally, the date of death or tracheostomy (a standard composite outcome in ALS clinical trials, considered equivalent to, and henceforth referred to as, death [13]) was recorded.

These data were collected in the context of the *CompALS* project, an Italian-Israeli collaboration. The study was approved by the ethical committees of the coordinating and participating centres. Written informed consent to participate in the study was obtained from all the patients or their legal representatives, and databases were anonymised according to the Italian privacy protection legislation.

2.1   Data Preprocessing

Let the *baseline* be the first visit recorded for each subject (not necessarily coinciding with the date of diagnosis). We cleaned both datasets by first removing the subjects with a death date preceding their baseline, as well as discarding the visits performed after the tracheostomy intervention. Then, to ensure that our methodologies could be based on data with sufficiently uniform longitudinal information, we filtered out the subjects with no visits after the first 9 months of visits from baseline. Finally, in order to better define the passage of time as the disease progresses, we introduced two additional variables for each visit, specifically the time since onset (time since onset, TSO), and the time between the current visit and the next visit (time between visits, TBV), with the latter only serving as a support variable for the training of the DBNs (see Sect. 3.1).

After the preprocessing, the MIMA dataset consisted of 116 subjects, for a total of 639 visits, and the PARALS dataset consisted of 568 subjects, for a total of 6431 visits. See Table 1, in columns 3 and 6, for a full description of the MIMA and PARALS datasets, respectively.

Finally, to allow the development and the evaluation of the predictive models according to pipeline described in Sect. 3, we split the MIMA dataset into a training (70% of the data, 82 subjects) and a test sets (30%, 34 subjects).

Table 1. Demographic and clinical characteristics of the real MIMA and PARALS study populations after pre-processing (in grey) and of the respective cohorts simulated with DBNs. The continuous and categorical variables are reported as means±SD, and frequencies (proportions), respectively. Times are expressed in months/years from the subject's baseline visit.

Variable	Levels	MIMA real (n=116)	MIMA simulated with the MIMA DBN (n=8200)	MIMA simulated with the PARALS DBN (n=8200)	PARALS real (n=568)	PARALS simulated with the PARALS DBN (n=56800)
N visits total		639	64740	79347	6431	616679
N visits per patient		5.51 ± 3.33	7.90 ± 3.36	9.68 ± 6.40	11.32 ± 6.80	10.86 ± 6.90
Sex	Female	60 (51.7%)	4300 (52.4%)	4300 (52.4%)	270 (47.5%)	27000 (47.5%)
Familiality	Yes	2 (1.7%)	100 (1.2%)	100 (1.2%)	54 (9.5%)	5500 (9.7%)
	<NA>	0 (0.0%)	0 (0.0%)	0 (0.0%)	2 (0.4%)	0 (0.0%)
Genetics	Mutated	7 (6.0%)	700 (8.5%)	600 (7.3%)	62 (10.9%)	6400 (11.3%)
	<NA>	22 (19.0%)	0 (0.0%)	0 (0.0%)	26 (4.6%)	0 (0.0%)
FTD	Yes	2 (1.7%)	0 (0.0%)	300 (3.7%)	53 (9.3%)	7100 (12.5%)
	<NA>	5 (4.3%)	0 (0.0%)	0 (0.0%)	198 (34.9%)	0 (0.0%)
Onset site	Bulbar	26 (22.4%)	1500 (18.3%)	1500 (18.3%)	160 (28.2%)	16000 (28.2%)
Age at onset (years)		59.11 ± 9.24	59.74 ± 8.94	59.74 ± 8.94	62.33 ± 11.09	62.33 ± 11.08
Diagnostic delay (months)		12.32 ± 10.97	12.04 ± 10.46	12.04 ± 10.46	12.26 ± 10.67	12.26 ± 10.66
Time between visits (months)		4.76 ± 6.83	5.83 ± 7.77	3.28 ± 2.91	3.21 ± 2.93	3.03 ± 2.61
Time since onset (months)		31.59 ± 27.46	46.94 ± 34.11	47.35 ± 37.23	39.04 ± 34.75	40.47 ± 36.71
BMI premorbid (kg/m^2)		25.85 ± 0.97	24.52 ± 4.55	24.67 ± 4.67	26.10 ± 4.30	26.07 ± 4.26
BMI at diagnosis (kg/m^2)		24.24 ± 4.28	24.40 ± 4.58	24.55 ± 4.70	25.13 ± 4.39	25.13 ± 4.36
FVC at diagnosis (%)		88.19 ± 21.13	88.40 ± 20.78	87.73 ± 19.91	93.82 ± 21.86	93.80 ± 21.83
NIV	Administered	67 (57.8%)	7270 (88.7%)	3974 (48.5%)	221 (38.9%)	27005 (47.5%)
PEG	Administered	38 (32.8%)	3431 (41.8%)	3040 (37.1%)	176 (31.0%)	22534 (39.7%)
MiToS walking/self-care	Impaired	93 (80.2%)	7940 (96.8%)	5590 (68.2%)	436 (76.8%)	36300 (63.9%)
MiToS swallowing	Impaired	41 (35.3%)	3753 (45.8%)	2330 (28.4%)	187 (32.9%)	17954 (31.6%)
MiToS communication	Impaired	34 (29.3%)	3025 (36.9%)	1975 (24.1%)	148 (26.1%)	15022 (26.4%)
MiToS breathing	Impaired	66 (56.9%)	6706 (81.8%)	3912 (47.7%)	248 (43.7%)	26427 (46.5%)
Survival	Censored	41 (35.3%)	0 (0.0%)	78 (1.0%)	151 (26.6%)	849 (1.5%)
	Dead	75 (64.7%)	8200 (100.0%)	8122 (99.0%)	417 (73.4%)	55951 (98.5%)
Time to death or censoring (months)		57.48 ± 37.63	69.10 ± 37.32	53.93 ± 33.75	52.59 ± 35.78	48.34 ± 35.80

3   Methods

With the aim of systematically evaluating the performance impact of leveraging external data (real or simulated), either directly of integrating them with the in-house data via transfer learning, we devised an experimental pipeline leading to the development of a battery of prognostic models of 3-year mortality. Model development, as described in Sect. 3.2, was based on a standard sequence of: i) cross-validation for hyperparameter tuning, ii) retraining on the whole training set with the optimal hyperparameters, and iii) predicting 3-year mortality on a consistent test set (an unseen subset of the MIMA dataset). We examined the following combinations of training data (MIMA or PARALS, real or simulated as per Sect. 3.1) and TL application.

1. A baseline model developed with the MIMA in-house data.
2. Two models (one with and one without TL) developed with the simulated version of the MIMA data obtained via a DBN trained on the MIMA data.
3. Two models (one with and one without TL) developed with the simulated version of the MIMA data obtained via a DBN trained on the PARALS database.
4. Two models (one with and one without TL) developed with the PARALS dataset.
5. Two models (one with and one without TL) developed with the simulated version of the PARALS dataset obtained via a DBN trained on the PARALS database itself.

3.1   Data Simulation via DBNs

For simulating large amounts of in silico data starting from the available ones, we developed two DBNs. Extending Bayesian Networks, DBNs are probabilistic graphical models that represent a set of static and time-dependent variables and their conditional dependencies via a directed acyclic graph. [5]. After being trained, a DBN can simulate the evolution over time of a population of subjects based on the probability relationships it encoded during the learning phase, starting from an initial condition, such as a set of covariates collected at a baseline visit. This generates an in silico population that reflects the trends seen in the real training population, thus effectively oversampling the original data consistently with their multivariate distribution.

Here, we used the *bnstruct* R package [8] to learn two distinct DBNs on the MIMA training set and on the PARALS dataset, respectively. In what follows, we present the procedure we performed on each dataset. First, for the purpose of learning the network on a more balanced occurrence of the death outcome, we recoded the survival information at each visit as a binary variable answering the question "Will the subject die within the next 9 months?". Second, we imputed the missing data by chained equations using the *mice* R package [1] with default parameters. Third, since *bnstruct* implements discrete-space/discrete-time DBNs, which encode probabilistic relationships among discrete variables over a discrete number of time steps, we discretised the continuous variables according to their tertiles in the learning dataset. We then imposed some constraints on the DBN structure, by grouping the variables into layers (e.g., demographic variables, variables at diagnosis, dynamic variables, see the caption of Fig. 1) and by imposing or forbidding specific relationships among the variables based on literature knowledge (e.g., MiToS score values must be linked to TSO since they should change as the disease progresses) and common sense (e.g., sex cannot be influenced by clinical values), as performed in [14]. Finally, we inferred the DBN structure using the Max-Min Hill-Climbing algorithm with the Bayesian information criterion as the score function, and computed the CPDs' parameters via maximum a posteriori estimation.

To fit the 5-fold cross-validation scheme needed for the analysis (see Sect. 3.2, we also trained an imputer and a DBN for each of the 5 folds of the MIMA and PARALS datasets, while strictly avoiding data leakage between training and validation portions of the data.

After learning each DBN, we used it to simulate the real data: leveraging the learnt network structures and the probability distributions among the covariates, we sampled the latter to simulate the progression of 100 synthetic repetitions of each real patient starting from their baseline condition, a synthetic visit after the other, and ending after 40 virtual visits or on (simulated) death. Specifically, the DBN learned on the MIMA training set was used to simulate the MIMA training set itself, while the DBN learned on the PARALS dataset was used, in turn, to simulate the MIMA training set and the PARALS dataset itself.

3.2   Model Development Strategy

To promote uniformity between the real and simulated datasets, the input data for the models were those of the latest visit within the first 9 months of observation of each subject. They comprised 11 variables, namely sex, age, familiality, genetics, onset site, FTD, diagnostic delay, TSO, premorbid BMI, BMI at diagnosis, FVC at diagnosis.

As the outcome for our study was 3-year mortality (after the input visit), but the data were originally either collected or simulated in a time-to-event fashion (i.e., with a death indicator and its timestamp), we performed a further round of filtering to exclude all subjects for whom it was impossible to determine whether death had occurred at the 3-year mark (i.e., all the survivors censored before then). This resulted in a total of 57 training and 24 test subjects for MIMA, and 497 training subjects for PARALS. Due to the considerable length of the simulations, no further exclusions were needed for the simulated data.

To simulate a typical exploratory study on small-sample data, the base modelling approach for each scenario was a logistic regression with a single hyperparameter, the strength of its L2 regularisation, whose optimal value we searched along an exponential grid from 10^{-2} to 10^4. The TL method was Frustratingly Easy Domain Augmentation (FEDA) [6], a relatively simple but effective feature augmentation technique. Let the source domain be the additional training data for a given scenario as described at the beginning of Sect. 3 and the target domain be the real MIMA data (i.e., the training portion of the in-house dataset). FEDA consists in creating an input matrix with three times as many columns as there are features in the dataset: the first set of columns refers to the source-specific features, the second set to the target-specific features, and the third set to common features. The data from the source and target domain are then stacked and copied so that the source-specific features are non-zero only for the source data, the target-specific features are non-zero only for the target data, and the common features (equal to the source- or target-specific features) are never equal to zero. The resulting input matrix can, finally, be used with any machine learning method to perform TL. Here, we applied TL to the same logistic regression used for training the models without TL.

We followed a consistent pipeline for model development across all scenarios. Specifically, we performed a 5-fold cross-validation to identify the optimal value of the L2 regularisation strength based on the maximum average cumulative/dynamic area under the receiver-operating characteristic curve (AUROC) [9]. For the simple logistic regressions, the validation portion of the data in each cross-validation iteration was 20% of the training data. For the TL versions of the models, given that they use both additional and in-house data for training, the validation portion was always 20% of the real MIMA training set (the same as in the baseline model). We trained the final version of the model using the optimal hyperparameter on the entire training set, and evaluated it on the test portion of the real MIMA dataset, again via the AUROC.

4   Results

4.1   Data Simulation via DBNs

In Fig. 1 we report the DBN structures built on the MIMA training set and on the PARALS dataset, with nodes representing the variables and directed edges entering a node corresponding to the conditional dependency effect of the child node on its parents. Dynamic variables' self-loops (blue nodes in the diagram) show the dependency of a variable on itself at consecutive time points. By analysing and comparing the two networks, we can observe how the MIMA DBN presents considerably fewer edges than those learnt from the PARALS dataset. In some cases, we can observe that some nodes are not connected to any other, meaning that no direct dependency was found among the variables, possibly because of the extremely reduced cardinality of the MIMA data. For the same reason, some conditional independencies are simplified, e.g., swallowing is completely determined by onset site and TSO in the MIMA DBN, while, in the PARALS DBN, the communicating and PEG nodes are also required. Both networks, however, encoded several known relationships (here found in a completely data-driven manner), such as the relation between the age at onset and the onset site [3] or the link between the breathing ability and the need of NIV.

Table 1 reports the cardinality and the characteristics of the MIMA (column 3) and PARALS (column 6) datasets, and of the in silico ALS populations obtained from the MIMA dataset with the DBN trained on the MIMA dataset itself (column 4) and with the DBN trained on the PARALS data (column 5), as well as the in silico ALS populations obtained on the PARALS dataset with the DBN trained on the PARALS dataset itself (column 7). By comparing the two real populations, we observe how the PARALS dataset presents higher rates of ALS familiality (9.5% vs 1.7%), as well as a higher occurrence of FTD at diagnosis (9.3% vs 1.7%). Conversely, the rate of NIV administration is higher in the MIMA dataset (57.6% vs 38.9%), possibly indicating a different intervention protocol in the two data sources. Expanding the comparison to include also the simulated populations, we note several differences in longitudinal variables as simulated by the two DBNs starting from the initial conditions of the MIMA dataset. For instance, as mean real TBV is shorter in the PARALS than in the MIMA data (3.21 vs. 4.76 months), the mean simulated TBV is also longer (5.83 months) when the MIMA data are simulated using the MIMA DBN than when using the PARALS DBN (3.28 months). Similarly, the generally higher real rates of simulated MiToS impairment in the MIMA dataset directly result in higher rates being simulated by the MIMA DBN. These considerations, together with the previous consistency check on the probabilistic links, support the reliability of the DBNs and of the simulation process.

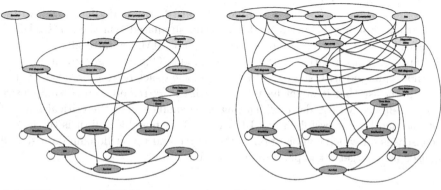

(a) DBN structure (MIMA training set).   (b) DBN structure (PARALS dataset).

Fig. 1. DBN structures built on the real data. The colours of the nodes represent the layers imposed during training: ALS-related demographic variables (yellow); other demographic/premorbid variables (pink); onset/diagnosis variables (orange); diagnostic delay (light blue); TBV (teal); TSO (purple); dynamic variables (blue with self-loops indicating consecutive times); survival (green). (Color figure online)

4.2 Model Performance

Table 2 shows the results of the evaluation of the 11 models. Expectedly, the baseline model, developed only on the very limited in-house MIMA dataset, yielded a mediocre AUROC of 0.633. Simulating the in-house dataset via a DBN also trained on the in-house dataset produced a promising increase of performance up to an AUROC of 0.672. We obtained even better results (0.734 AUROC in both cases, i.e., 16% relative increase) when the DBN used to simulate the data was trained on the larger PARALS dataset. Remarkably, performance was consistent between models developed on a dataset simulated starting from the initial conditions given by the MIMA training set and from the PARALS training set, only the latter of which was used to train the DBN itself. Even more interestingly, the two PARALS-DBN-simulated training sets resulted in better models than the one trained on the real PARALS data (0.702 AUROC). Although it is difficult to definitely claim superiority in light of the limited (but realistic) MIMA test set size, it is apparent that, at least for the purposes of model transfer between the larger PARALS dataset and the in-house MIMA dataset, access to the former was instrumental in considerably increasing performance. Furthermore, it appears that relying on a pre-trained DBN simulator or even on a batch of pre-simulated data is at least not inferior to using the real data from that centre directly. A potential explanation for this fact might be found in the synthesis abilities of DBNs, which might be encapsulating accurate but general enough probabilistic information about the real data to serve a similar function as a regulariser (i.e., promoting generalisation ability and discouraging memorisation of the training examples).

Notably, the TL versions of the models consistently underperformed, to the point that applying TL between the real and simulated (via the MIMA DBN) MIMA data and between the real PARALS and MIMA data resulted in a performance decrease relative to the baseline model. Our experimental framework was not equipped with the tools to untangle the exact reason for this fact, but we hypothesise that it might have to do with the extremely limited sample size of the in-house MIMA data, which might be smaller than the critical mass of data needed for successful TL via FEDA. Another possible culprit is the lack of a consensus on which combination of the source and target domain data should be used as a target for hyperparameter optimisation.

Table 2. Performance evaluation on the MIMA test set across model development scenarios. From left to right, the table reports the training set for the model, the training set for the DBN that produced the simulated data (if applicable), whether logistic regression was used with or without transfer learning, and the AUROC on the test.

Model training data	DBN training data	Method	AUROC
MIMA		Logistic regression	0.633
MIMA (simulated)	MIMA	Logistic regression	0.672
MIMA (simulated)	MIMA	Transfer learning	0.602
MIMA (simulated)	PARALS	Logistic regression	0.734
MIMA (simulated)	PARALS	Transfer learning	0.656
PARALS		Logistic regression	0.703
PARALS		Transfer learning	0.625
PARALS (simulated)	PARALS	Logistic regression	0.734
PARALS (simulated)	PARALS	Transfer learning	0.648

5   Discussion and Conclusions

In the present work, we set up an experimental framework to analyse the performance impact of a wide range of possible strategies to develop more robust prognostic models of ALS mortality starting from the kind of small-sample data that is expected of a typical clinical centre for such a rare disease. We explored a number of scenarios characterised by different levels of (real or simulated) data access, and investigated the possible usefulness of transfer learning in combining the limited in-house data with a larger, additional data sources. Remarkably, our results suggest that the information encoded in a DBN on the joint probability distribution of a sufficiently large database (here, a few hundred examples) might be an excellent starting point for data simulation and subsequent model development for, e.g., exploratory purposes on a small-sample dataset. Moreover, data simulated via a DBN appear to yield similar, or possibly better, results compared to the real data themselves, reinforcing the idea that DBNs might be an effective tool for in silico patient generation.

Future developments to address the main limitations of this work include: testing a wider range of modelling approaches (e.g., non-linear methods), investigating the role of sample size and the goodness of fit of the DBN in the apparent inferiority of TL methods by artificially varying the amount of data available, and acquiring additional data to confer greater statistical validity to these preliminary results via hypothesis testing [7] and validation based on repeated holdout or cross-validation.

Acknowledgements. This research was supported by the University of Padova project C94I19001730001, by the Italian Ministry of Health grant RF-2016-02362405, and by the Italian Ministry of Education, University and Research (PRIN) grant 2017SNW5MB.

References

1. van Buuren, S., Groothuis-Oudshoorn, K.: mice: Multivariate imputation by chained equations in R. J. Stat. Softw. Articles **45**(3), 1–67 (2011)
2. Chiò, A., Hammond, E.R., Mora, G., Bonito, V., Filippini, G.: Development and evaluation of a clinical staging system for amyotrophic lateral sclerosis. J. Neurol. Neurosurg. Psychiatry **86**(1), 38–44 (2015)
3. Chio, A., Logroscino, G., Hardiman, O., et al.: Prognostic factors in ALS: a critical review. Amyotroph. Lateral Scler. **10**(5–6), 310–323 (2009)
4. Chió, A., Mora, G., Moglia, C., Manera, U., Canosa, A., et al.: Secular trends of amyotrophic lateral sclerosis: the Piemonte and Valle d'Aosta register. JAMA Neurol. **74**(9), 1097–1104 (2017)
5. Dagum, P., Galper, A., Horvitz, E.: Dynamic network models for forecasting. In: Uncertainty in Artificial Intelligence, pp. 41–48. Elsevier (1992)
6. Daumé III, H.: Frustratingly easy domain adaptation. arXiv preprint arXiv:0907.1815 (2009)
7. Demšar, J.: Statistical comparisons of classifiers over multiple data sets. J. Mach. Learn. Res. **7**, 1–30 (2006)

8. Franzin, A., Sambo, F., Di Camillo, B.: BNStruct: an R package for Bayesian network structure learning in the presence of missing data. Bioinformatics **33**(8), 1250–1252 (2017)

9. Heagerty, P.J., Zheng, Y.: Survival model predictive accuracy and ROC curves. Biometrics **61**(1), 92–105 (2005)

10. Marini, S., Trifoglio, E., Barbarini, N., et al.: A dynamic Bayesian network model for long-term simulation of clinical complications in type 1 diabetes. J. Biomed. Inform. **57**, 369–376 (2015)

11. Pezoulas, V.C., Grigoriadis, G.I., Gkois, G., Tachos, N.S., et al.: A computational pipeline for data augmentation towards the improvement of disease classification and risk stratification models: a case study in two clinical domains. Comput. Biol. Med. **134**, 104520 (2021)

12. Roversi, C., Tavazzi, E., Vettoretti, M., Di Camillo, B.: A dynamic Bayesian network model for simulating the progression to diabetes onset in the ageing population. In: 2021 IEEE EMBS International Conference on Biomedical and Health Informatics (BHI), pp. 1–4. IEEE (2021)

13. Schmidt, E.P., Drachman, D.B., Wiener, C.M., Clawson, L., Kimball, R., Lechtzin, N.: Pulmonary predictors of survival in amyotrophic lateral sclerosis: use in clinical trial design. Muscle Nerve: Off. J. Am. Assoc. Electrodiagnos. Med. **33**(1), 127–132 (2006)

14. Tavazzi, E., Daberdaku, S., et al.: Predicting functional impairment trajectories in ALS: a probabilistic, multifactorial model of disease progression. J. Neurol. 1–21 (2022)

15. Viceconti, M., Henney, A., Morley-Fletcher, E.: In silico clinical trials: how computer simulation will transform the biomedical industry. Int. J. Clin. Trials **3**(2), 37–46 (2016)

16. Weiss, K., Khoshgoftaar, T.M., Wang, D.D.: A survey of transfer learning. J. Big Data **3**(1), 1–40 (2016). https://doi.org/10.1186/s40537-016-0043-6

Natural Language Processing

A Rule-Free Approach for Cardiological Registry Filling from Italian Clinical Notes with Question Answering Transformers

Tommaso Mario Buonocore[1]([envelope])(iD), Enea Parimbelli[1](iD), Valentina Tibollo[2](iD), Carlo Napolitano[2](iD), Silvia Priori[2](iD), and Riccardo Bellazzi[1](iD)

[1] Department of Electrical, Computer and Biomedical Engineering,
University of Pavia, Pavia, Italy
buonocore.tms@gmail.com, {enea.parimbelli,riccardo.bellazzi}@unipv.it
[2] LISRC, Department of Molecular Medicine
and Department of Molecular Cardiology, ICS Maugeri, Pavia, Italy
{valentina.tibollo,carlo.napolitano,silvia.priori}@icsmaugeri.it

Abstract. The huge volume of textual information generated in hospitals constitutes an essential but underused asset that could be exploited to improve patient care and management. The encoding of raw medical texts into fixed data structures is traditionally addressed with knowledge-based models and complex hand-crafted rules, but the rigidity of this approach poses limitations to the generalizability and transferability of the solutions, in particular for a non-English setting under data scarcity conditions. This paper shows that transformer-based language representation models have the right characteristics to be employed as a more flexible but equally high-performing clinical information retrieval system for this scenario, without relying upon a knowledge-driven component. We demonstrate it pragmatically on the extraction of clinical entities from Italian cardiology reports for patients with inherited arrhythmias, outperforming the previous ontology-based work with our proposed transformer pipeline under the same setting and exploring a new rule-free approach based on question answering to automate cardiological registry filling.

Keywords: Natural Language Processing · Deep Learning · Biomedical Text Mining · Information Extraction · Question Answering

1 Introduction

The digitization of health services and clinical care processes has led healthcare organizations to routinely produce an ever-increasing amount of textual data: reports, nursing notes, discharge letters, and insurance notes are just some of the types of digital documents clinicians must deal with every day [10]. Due to its high information content, this source is fundamental to populating medical

J. M. Juarez et al. (Eds.): AIME 2023, LNAI 13897, pp. 153–162, 2023.
https://doi.org/10.1007/978-3-031-34344-5_19

registries, which play a pivotal role in retrospective studies, clinical trials, and treatment surveillance, improving health outcomes and reducing the costs of health care.

For highly prevalent diseases, manually retrieving information from textual sources can be time-consuming and expensive, requiring domain experts to devote significant effort, which often proves overwhelming and unsustainable in the medium to long term [9]. The automatic encoding of unstructured information in structured registries, however, implies dealing with several technical challenges for information retrieval (IR) systems, with traditional rule-based and dictionary-based NLP approaches still requiring a lot of manual effort by multiple professionals to set up. Due to their rigid design, these methods have been proven difficult to adapt and reuse even in marginally different situations, posing limitations for low-resource scenarios such as non-English speaking local medical institutions working in lower data regimes.

1.1 Research Goal

This work aims at evaluating the potential of a flexible, rule-free approach based on the transformer architecture in the context of clinical text processing and registry curation, considering the implementation of an Information Extraction (IE) system for Italian cardiology reports as a practical clinical case. This pilot will pave the way to wider usage of more transferable and easy-to-generalize NLP techniques to process notes in the clinical practice, assisting medical practitioners in a repetitive and routine part of their working activity, even when the available data are limited. Model and source code are released as part of the publication[1]

1.2 Related Work

There have been many systems developed to process and extract information from English clinical narratives, mainly relying upon either rule-based systems (i.e., regular expressions, sometimes ontology-enhanced [1,7]) or machine learning [6], where clinical IE is approached as a supervised learning problem on annotated texts. The machine-learning approach has grown in popularity after the recent advancements of deep learning models like the transformer, avoiding the time-consuming manual feature engineering required by traditional machine-learning models like support vector machines and conditional random fields while getting closer to human-level performance in several NLP benchmarks.

In the clinical domain, different works explored transformer-based models for clinical information extraction. Yang et al. [12] systematically explored the most used encoder architectures for extracting various types of clinical concepts from English public datasets, while Wei et al. [11] used BERT to identify the type of relations between different clinical entities. Mulyar et al. [4] instead evaluated BERT for clinical information extraction in a multitask learning context, taking into account common clinical tasks like named entity recognition (NER), semantic text similarity (STS), and natural language inference (NLI). To the best of

---

[1] Source code available at https://github.com/detsutut/icsm-cardio-nlp.

our knowledge, only Percha et al. [5] approached the problem of clinical registry curation with a transformer model, proposing a solution based on applying out-of-the-box, pre-trained transformers for NLI.

However, there has been limited research on developing similar systems for non-English languages, particularly for Italian. One possible reason for this is the lack of shared resources for training and evaluating these systems, as well as the smaller coverage of the Italian UMLS Metathesaurus[2] Viani et al. [7] proposed an ontology-driven approach to identify events and their attributes from episodes of care included in Italian medical records, using non-annotated reports belonging to patients with rare arrhythmias and building a domain-specific ontology that includes events and attributes to be extracted, with related regular expressions. In the same context, supervised methods based on recurrent neural networks have been evaluated as well [8]. We are not aware of any research regarding transformer-based IE systems for Italian cardiology reports.

2   Materials and Methods

2.1   Dataset

The corpus used in this work includes 75 Italian cardiology reports belonging to the same number of patients with inherited arrhythmias, such as Brugada Syndrome and Long QT Syndrome. Reports contain family and personal history, information on performed tests, the diagnosis, and conclusions with possible drug prescriptions. The corpus is provided by the Molecular Cardiology Laboratories of the ICS Maugeri hospital in Pavia, Italy, where the information written in the reports is manually entered by domain experts into the Transatlantic Registry of Inherited Arrhythmogenic Diseases (TRIAD). This database collects information about patients' diseases, therapies, Holter and ECG measurements, mutations, implanted devices and events such as cardiac arrests or syncope.

2.2   Transformer-Based Models

Our work revolves around one of the most widely used transformer-based models: BERT (Bidirectional Encoder Representations from Transformers). BERT is a language model used for any kind of sequence classification task that preserves only the encoder part of the original transformer architecture. The BERT model is trained jointly for Masked Language Modelling (MLM) and Next Sentence Prediction (NSP) on BooksCorpus (800M words) and English Wikipedia (2.5B words) [3]. After this preliminary optimization over a large volume of unlabeled, general-purpose data (i.e., pre-training), the model typically undergoes a further training phase that optimizes the model's parameters to perform a specific task on a specific domain (i.e., fine-tuning), usually on a much smaller dataset, like our set of Italian cardiology reports. At the time of writing, reliable pre-trained

---

[2] The metathesaurus statistics for different languages over the years can be found at
    https://www.nlm.nih.gov/research/umls/archive/archive_home.html.

BERT weights are available for several languages (either monolingual or multilingual) and domains, making this approach easily transferable and generalizable to other use cases with minimal effort.

2.3   Experimental Design

To test our hypothesis, we first check whether the transformer architecture constitutes an advancement in the state of the art for cardiology events extraction, then we extend the model as a question-answering (QA) system to address the automatic filling of the arrhythmias registry fields related to the ECG Holter diagnostic test section, taking into account different types of variables.

Identifying Clinical Events with BERT. First, we replicated the work of Viani et al. to identify clinical events in different categories, replacing the previous architecture based on a combination of recurrent neural networks and knowledge bases with the standalone BERT. The problem is formulated as a NER task with relevant events divided into four semantic categories: problems (e.g., syncopal episode, Brugada Syndrome), tests (e.g., electrocardiogram, effort stress test), treatments (mostly drug prescriptions), and occurrences (e.g., admission, medical visit). For this task, we place an additional drop-out layer and a linear layer on top of the BERT model, as shown in Fig. 1, so that every token embedding passes through the same dense network with the same softmax. This way, we can train our model to tag each token of the sequence with one of the categories defining the entities. The dataset is annotated by two annotators using the IOB2 tagging scheme, keeping the widest tag in case of concept overlapping, and split into training, validation and test set following the 70:15:15 ratio. As a baseline, we refer to the same dictionary lookup approach described in Viani et al., which relies on string matching using the Italian version of UMLS and additional sources.

Registry Curation with a QA Approach. While NER is the most common approach for clinical IE, the information collected in a disease registry is very granular, with hundreds of fields (i.e., categories) to be filled in that might also change across different iterations, thus making extractive QA more suitable for this fine-grained IE scenario. In extractive QA, given a context and a question, the model returns an answer extracted directly from the provided context. This minimizes the risk of replying with factoids. In this case, the context is the single clinical note, and the question determines the specific information we need to extract. The answer returned by the QA model is the text span of the context that the model identifies as the answer to the question. Therefore, there will be two different heads on the top of the BERT architecture: one to predict the starting token, and the other to predict the ending token. Everything in between will be returned as the answer. The BERT-based pipeline for registry filling with QA is shown in Fig. 2.

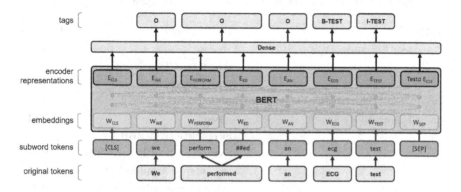

Fig. 1. The architecture of a BERT-based NER model for clinical event identification. Input word-level annotations are broken down into sub-words to be compliant with BERT WordPiece tokenization. The encoder then produces an embedded version of the input tokens, passed to a shared dense classification layer for tagging. Then, output word-level predictions are reconstructed following a majority voting approach.

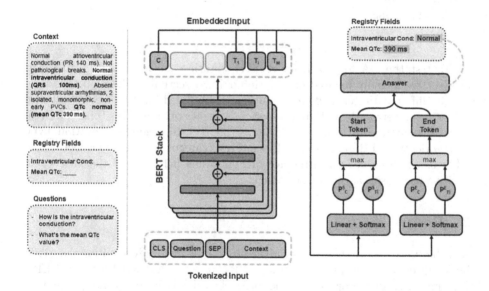

Fig. 2. BERT-based extractive QA pipeline for registry filling. Context and question are tokenized and concatenated in the input sequence, interspersed with the CLS and SEP special tokens. The encoded representation of each input token is then passed to two separate heads to calculate the most probable context span to take as the answer. The CLS special token is used to address the absence of a valid answer in the context (span starting and ending at index 0, i.e., the CLS token), while the SEP token defines the context boundaries in the input sequence (questions are fundamental to generate accurate token embeddings, but answers are extracted from context only).

Table 1. Holter registry field names, values, and relative frequencies.

Field Name	Field Values	Frequency
ST-segment Elevation Type	absent	0.92
	type 2 (saddleback type)	0.05
	type 1 (coved type)	0.00
	both	0.03
Bundle Branch Block Type	absent	0.84
	incomplete right BBB	0.08
	complete right BBB (RBBB)	0.06
	left anterior hemiblock (LAHB)	0.00
	complete left BBB (LBBB)	0.00
	LAHB+RBBB	0.02
Presence of Symptoms	reported	0.02
	not reported	0.98
Presence of Isolated Ventric. Arrhythmias	yes	0.40
	no	0.60
Average QT Interval	N.A	0.05
Average Heart Rate	N.A	0.06

To test the QA approach for registry curation, in accordance with cardiologists, we initially focused on the Holter ECG section of the reports, selecting six different Holter fields to be automatically filled in: Average QT Interval (numeric), Average Heart Rate (numeric), Presence of Symptoms (binary), Presence of Isolated Ventricular Arrhythmias (binary), ST-segment Elevation Type (categorical), and Bundle Branch Block Type (categorical). After inspecting the reports, we discarded the Presence of Symptoms field due to the absence of positive responses, as shown in Table 1.

3   Results

In Table 2, we report the results for the semantic category recognition described by Viani et al., in terms of precision (PR), recall (RE), and F1 score. BERT ITA indicates our proposed methodology based on the fine-tuning of a pre-trained, monolingual, Italian version of BERT. For this purpose, we identified three different candidate pre-trained weights: the first set is obtained from a recent Wikipedia dump and other general-purpose Italian texts[3] the second is the Italian transposition of the BioBERT model, trained on PubMed Abstracts, and the third is an augmented version of the second, trained also on web-crawled Italian data [2]. While the latter gives the best performance, which we report in Table 2, all three checkpoints lead to a performance improvement.

---

[3] Model repository: https://huggingface.co/dbmdz/bert-base-italian-xxl-cased.

Table 2. Overall results for the named entity recognition task.

Method	Precision	Recall	F1
Dictionary Lookup (baseline)	81.1	52.4	63.7
BI-LSTM-CRF-CHAR	86.8	88.9	88.2
BI-LSTM-CRF-CHAR + Dictionary Lookup	88.6	91.7	88.6
BERT ITA	**90.8**	**92.6**	**91.7**

Concerning the QA approach experiment for Holter ECG section registry filling, the results are shown in Table 3 for the same pre-training checkpoint, reporting the size of the test sample, the exact match (EM), and the F1 score for both single registry fields and overall. In the context of question answering, the F1 metric is computed over the individual tokens in the prediction against those in the true answer, and a higher number of shared words lead to a higher F1. The EM metric, instead, is a strict all-or-nothing accuracy metric that compares the answers at the character level, meaning that a single different character is sufficient to consider the answer misclassified. An additional checkpoint candidate based on intermediate training over a set of generic-purpose question-answer pairs in Italian has been considered for this task, but it leads to lower performances compared with the biomedical checkpoints, thus suggesting a higher downstream impact of in-domain data training for this scenario.

[...]

Esame obiettivo: Il paziente si presenta in condizioni di benessere soggettivo. Nega dolore ed allergie. Altezza 170 cm; peso 67 kg.

ECG Holter. Esame eseguito in terapia con Pantoprazolo e Magaldrato (Riopan). ECG registrato con derivazioni precordiali modificate: V1 V2 in II SIC, V3 V4 in III SIC, V5 V6 in IV SIC, in posizione parasternale destra e sinistra. Ritmo sinusale regolare con normali variazioni circadiane della frequenza cardiaca. FC minima 50 bpm, massima 117 bpm, media 80 bpm. Conduzione atrioventricolare ai limiti superiori della norma con tratti notturni di BAV di primo grado (PR massimo 240 ms; PR medio 190 ms, stabile). Blocco di branca destra completo (QRS max 130ms). Non aritmie. Pattern Brugada di tipo 1 in II e III SIC con sopraslivellamento misurato fino a 6.0 mm .7.0 mm, 3.0 mm, 5.0 mm, in V1 e V2, rispettivamente in II 2 III SIC. Tale aspetto si configura pattern Brugada di tipo 1. QTc medio 384 ms.

Conclusioni. Il quadro clinico si presenta stabile. Al controllo odierno non evidenzia di pattern spontaneo per Sindrome di Brugada.

[...]

QTC average	HR average	Isolated VA	ST elevation	Bundle branch block
384 ms	80 bpm	Absent	Absent	Complete right bundle branch block (RBBB)

Fig. 3. Example of annotated Holter ECG section and the correspondent expected output in the TRIAD registry. The expected value for the ST-segment Elevation Type is Absent because medical practitioners report only elevations on the IV derivation in the TRIAD registry.

Table 3. Overall and category-wise results for the QA task with BERT.

TRIAD Fields	Size	F1	EM
ST-segment Elevation Type	4	89.1	45.0
Bundle Branch Block Type	8	100.0	100.0
Presence of Isolated Ventricular Arrhythmias	9	78.1	60.0
Average QT Interval	9	100.0	100.0
Average Heart Rate	10	98.0	98.0
Overall	40	93.5	85

4   Discussion

The results we obtained show a significant performance improvement over the previous state of the art for the semantic category recognition of clinical events examined in this paper, suggesting that simple transformer-based models like BERT can be employed as a rule-free alternative for clinical tasks even without any explicit reference to biomedical knowledge, present instead in all the methods used in the previous work. Moreover, the absence of rigidly codified rules and ad hoc developed dictionaries for the identification of concepts gives the model the ability to reasonably manage typos and abbreviations, very common in clinical practice. The error analysis highlights that the misclassification of the events is often not linked to an intrinsically incorrect evaluation, but to a more conservative attitude of the model, i.e., error type 6 in Table 4. For instance, when evaluating the progressive bradycardization bigram, both the annotator and the NLP model agree in recognizing bradycardization as a problem (PRO), but only the model believes the previous term contributes to the event identification. The results obtained from this preliminary experiment have therefore highlighted not only good accuracy but also excellent generalization capabilities, obtained starting from a very small dataset, which is a fundamental factor for the practical development of models that require preliminary annotation.

When using the same transformer-based strategy for QA on Holter ECG registry fields, results show a strong adherence between the model and the annotators, with a total overlap between the answers for Bundle Branch Block Type and Average QT Interval. Results highlight an almost perfect F1 and EM for continuous variables, and lower - but still greatly positive - values for categorical and binary variables. The ST-segment Elevation Type field shows a considerable mismatch between F1 and EM. This is an expected behavior, as the answers highlighted by the annotators consist of lengthy sentences, increasing the likelihood of the model to start/end the answer too late/early in the sentence, and considering that the test set includes only four annotations for this TRIAD field.

Table 4. NER error type examples. English translation of named entities, in order of appearance: almost suppressed TSH, pregnancy, ematochemical (tests), ACE inhibitors, contractile abnormalities, progressive bradycardization, general checkup.

Named Entities	Tag Sequence		Error type and description
	Actual	Predicted	
TSH pressochè soppresso	O	B-PROB	*Type 1.* Suppressed TSH is arguably
	O	O	a problem that may be considered
	O	O	by cardiologists in this context
Gravidanza	B-OCC	O	*Type 2.* Missed event
Ematochimici	B-TEST	B-PROB	*Type 3.* Event detected, wrong type
ACE inibitori	B-TREAT	B-TREAT	*Type 4.* Event detected, correct type,
	I-TREAT	B-TREAT	but B/I indicators are misplaced
Anomalie contrattili	B-PROB	B-PROB	*Type 5.* Multi-word event only
	I-PROB	O (O, I-PROB)	partially detected
Bradicardizzazione progressiva	B-PROB	B-PROB	*Type 6.* Detected event extends over
	O	I-PROB	the expected boundaries
Controllo generale	B-OCC	B-TEST	*Type 7.* Mix of previous errors
	O	I-TEST	

5   Conclusion

Our work makes use of a practical cardiological use-case to demonstrate how the recent advances in NLP regarding pre-trained language models can help in reducing the gap between AI and clinical practice by making it possible to facilitate the compilation of registers and databases, query clinical documents, and identify clinical concepts. By leveraging the transformer architecture and the pretrain-then-finetune paradigm, in this work we explore a rule-free IR environment suitable with limited data availability and compatible with less-resourced languages like Italian, thanks to the widespread availability of pre-trained weights for different languages and domains. The high adherence shown in Table 4 also highlights how a simple BERT-based extractive QA pipeline can be a valuable solution for filling in both continuous and free-text registry fields. However, a supervised extractive QA approach based on annotations cannot be directly employed for binomial and categorical registry fields, as it is not designed to perform inference over the extracted answer, as shown in Fig. 3 for the ST-segment Elevation Type field. Integrating this inferential step in the context of a rule-free transformer-based system without losing the advantages of extractive QA is a challenging task that we aim to address in a future publication with a multi-task framework based both on extractive QA and multiple-choice QA and implemented through conditional generative transformer models like T5.

References

1. Aronson, A.R., Lang, F.M.: An overview of MetaMap: historical perspective and recent advances. J. Am. Med. Inform. Assoc. JAMIA **17**(3), 229–236 (2010). https://doi.org/10.1136/jamia.2009.002733
2. Buonocore, T.M., Crema, C., Redolfi, A., Bellazzi, R., Parimbelli, E.: Localising In-Domain Adaptation of Transformer-Based Biomedical Language Models, December 2022. https://doi.org/10.48550/arXiv.2212.10422, http://arxiv.org/abs/2212.10422, arXiv:2212.10422 [cs]
3. Devlin, J., Chang, M.W., Lee, K., Toutanova, K.: BERT: pre-training of deep bidirectional transformers for language understanding. In: Proceedings of the 2019 Conference of the North American Chapter of the Association for Computational Linguistics: Human Language Technologies, Volume 1 (Long and Short Papers), pp. 4171–4186. Association for Computational Linguistics, Minneapolis, Minnesota, June 2019. https://doi.org/10.18653/v1/N19-1423, https://www.aclweb.org/anthology/N19-1423
4. Mulyar, A., Uzuner, O., McInnes, B.: MT-clinical BERT: scaling clinical information extraction with multitask learning. J. Am. Med. Inform. Associ. **28**(10), 2108–2115 (2021). https://doi.org/10.1093/jamia/ocab126, https://doi.org/10.1093/jamia/ocab126
5. Percha, B., Pisapati, K., Gao, C., Schmidt, H.: Natural language inference for curation of structured clinical registries from unstructured text. J. Am. Med. Inform. Assoc. **29**(1), 97–108 (2022). https://doi.org/10.1093/jamia/ocab243
6. Savova, G.K., et al.: Mayo clinical text analysis and knowledge extraction system (cTAKES): architecture, component evaluation and applications. J. Am. Med. Inform. Assoc. JAMIA **17**(5), 507–513 (2010). https://doi.org/10.1136/jamia.2009.001560, https://www.ncbi.nlm.nih.gov/pmc/articles/PMC2995668/
7. Viani, N., et al.: Information extraction from Italian medical reports: an ontology-driven approach. Int. J. Med. Inform. **111** (2017). https://doi.org/10.1016/j.ijmedinf.2017.12.013
8. Viani, N., et al.: Supervised methods to extract clinical events from cardiology reports in Italian. J. Biomed. Inform. **95**, 103219 (2019). https://doi.org/10.1016/j.jbi.2019.103219
9. Viviani, L., Zolin, A., Mehta, A., Olesen, H.V.: The European cystic fibrosis society patient registry: valuable lessons learned on how to sustain a disease registry. Orphanet J. Rare Diseases **9**(1), 81 (2014). https://doi.org/10.1186/1750-1172-9-81
10. Wang, Y., et al.: Clinical information extraction applications: a literature review. J. Biomed. Inform. **77**, 34–49 (2018). https://doi.org/10.1016/j.jbi.2017.11.011, https://www.sciencedirect.com/science/article/pii/S1532046417302563
11. Wei, Q., et al.: Relation extraction from clinical narratives using pre-trained language models. In: AMIA Annual Symposium Proceedings 2019, pp. 1236–1245, March 2020. https://www.ncbi.nlm.nih.gov/pmc/articles/PMC7153059/
12. Yang, X., Bian, J., Hogan, W.R., Wu, Y.: Clinical concept extraction using transformers. J. Am. Med. Inform. Assoc. **27**(12), 1935–1942 (2020). https://doi.org/10.1093/jamia/ocaa189

Classification of Fall Types in Parkinson's Disease from Self-report Data Using Natural Language Processing

Jeanne M. Powell[1]([✉]) [iD], Yuting Guo[1] [iD], Abeed Sarker[1,2] [iD],
and J. Lucas McKay[1,2] [iD]

[1] Emory University, Atlanta, GA 30322, USA
{jeanne.marie.powell,yuting.guo,abeed.sarker,
j.lucas.mckay}@emory.edu
[2] Georgia Institute of Technology, Atlanta, GA 30332, USA

Abstract. Falls are a leading cause of injury globally, and people with Parkinson's disease are particularly at risk. An important step in reducing the probability of falls is to identify their causes, but manually classifying fall types is laborious and requires expertise. Natural language processing (NLP) approaches hold potential to automate fall type identification from descriptions. The aim of this study was to develop and evaluate NLP–based methods to classify fall types from Parkinson's disease patient self-report data. We trained supervised NLP classifiers using an existing dataset consisting of both structured and unstructured data, including the age, gender, and duration of Parkinson's disease of the faller, as well as the fall location, free-text fall description, and fall class of each fall. We trained supervised classification models to predict fall class based on these attributes, and then performed an ablation study to determine the most important factors influencing the model. The best performing classifier was a hard voting ensemble model that combined the Adaboost, unweighted decision tree, weighted k-nearest neighbor, naïve Bayes, random forest, and support vector machine classifiers. On the testing set, this ensemble classifier achieved an F_1-macro of 0.89. We also experimented with a transformer-based model, but its performance was subpar compared to that of the other models. Our study demonstrated that automatic fall type classification in Parkinson's disease patients is possible via NLP and supervised classification.

Keywords: Falls · Natural Language Processing · Text Classification · Parkinson's disease

1 Introduction

Falls are unintentional events where a person lands on a lower level [1, 2], which can result in significant personal, financial, and health costs [3–6]. For example, falls were the leading cause of injury-related death in the United States between 2007 and 2016 [3]. These high costs could be minimized with a better understanding of the causes of falls and subsequent implementation of preventative measures.

© The Author(s), under exclusive license to Springer Nature Switzerland AG 2023
J. M. Juarez et al. (Eds.): AIME 2023, LNAI 13897, pp. 163–172, 2023.
https://doi.org/10.1007/978-3-031-34344-5_20

Many studies report overall fall frequency without accounting for the circumstances surrounding a fall, which limits our understanding of their etiology [7]. Falls are heterogeneous and can result from multiple types of biomechanical perturbations, including perturbations to an individual's base of support (BoS; e.g., trips) or center of mass (CoM; e.g., overextension during bending) [2]. BoS falls are more common in healthy older adults compared to CoM falls [2]. However, the opposite holds true in subpopulations of people, such as people with Parkinson's disease, where disease-related postural instability results in more CoM falls [1].

People with Parkinson's disease are more likely to fall and be frequent fallers than healthy older adults [1, 6]. Falling in this population can be incapacitating, often resulting in soft tissue injuries, and disabling even early in disease progression [1]. Therefore, it is of particular importance to predict and prevent falls in this population.

A necessary step in this pursuit is to track falls and fall circumstances because risk factors for trips and slips might differ from those for falls due to impaired self-motion or other causes. To determine the cause of a fall, one must collect free-text information about the circumstances of the fall from the faller [6]. Historically, fall classes have been manually coded from these free-text descriptions [6, 8–10], but this practice is subjective, resource intensive, and difficult to scale. Recent advances in the field of natural language processing (NLP) hold exciting promise to automate processes such as fall type classification from free-text fall descriptions.

NLP techniques have been used in many biomedical domains, including mining unstructured electronic health records [11]. For example, Tohira and colleagues trained support vector machine (SVM) and random forest (RF) classifiers to detect falls from ambulance services provider reports [12]. Electronic health records and patient self-reports provide rich data that can capture nuances that structured medical data may miss [13]. NLP techniques may aid in Parkinson's disease diagnosis, given its impact on language production. Pérez-Toro and colleagues demonstrated that NLP techniques could be leveraged to distinguish people with Parkinson's disease from healthy older adults based on differences transcribed in speech patterns [14]. Given that Parkinson's disease is the second most common neurodegenerative disease worldwide, affecting over six million people globally [15], it is extremely important to gain a nuanced understanding of the disease.

Here, we aimed to develop an NLP classification model to distinguish CoM falls from other fall types in people with Parkinson's disease based on patient-provided free-text descriptions. Our particular focus is on CoM falls as they are more common in people with Parkinson's disease [1].

2   Methods

2.1   Fall Self-report Dataset

In a recent study [16], we followed patients with Parkinson's disease for 12-months to correlate fall risk with biomarkers of balance control. Participants tracked falls on monthly "fall calendars" and missing data and fall details were acquired over telephone interviews. Our dataset consisted of 124 fall self-reports collected from 23 individuals. Demographic information about those individuals is provided in Table 1.

Each fall self-report included structured data (i.e., age, gender, and time since Parkinson's disease diagnosis) and unstructured data (free-text description of the fall and its location). Dr. McKay, an expert in the biomechanics of falls, classified fall causes (i.e., CoM, BoS, or 'Other') based on fall descriptions, and Powell classified falls as occurring inside the home or not using location descriptions. Three fall descriptions were modified because the patient described the fall as "same as fall #1". In these cases, the full description used for #1 was copied. The fall descriptions were also manually checked for spelling errors.

The average length of the fall descriptions was 38 words, with the shortest consisting of only 3 words and the longest of 170 words. There were 922 unique words in the dataset before processing and 731 unique words after processing (i.e., stemming and removal of stop words).

2.2 Data Pre-processing

All binary categorical variables (i.e., gender, fall class, and fall location) were one-hot encoded. Our numerical factors age and disease duration, were scaled. Free-text fall descriptions were pre-processed by lowercasing all text, removing English stop words and punctuations, and tokenizing the remaining text. Each word token was stemmed using the Porter stemmer [17]. Fall descriptions were then vectorized as follows. We generated two sets of features from the pre-processed description texts—n-grams and word clusters. A word n-gram is a sequence of contiguous n words in a text segment. This feature enabled us to represent a document using the union of its terms. We used 1-, 2-, and 3-grams as features with the max number of features set to 150. The n-grams were vectorized so that each n-gram was represented by a numeric value in the feature vector indicating its frequency within a given instance. To enable a more generalized representation of the terms, we used the CMU word clusters [18]. The word clusters were generated via a two-step process—dense vector representations of words were first learned from large unlabeled data using the method described in [18] so that similar terms were close together in vector space, and then the words were grouped via hierarchical clustering. The word clusters were represented as unigrams during training.

2.3 Fall Class Classification

We modeled the discrimination between CoM- and Other-class falls as a binary classification problem, using both structured and unstructured features. Because the dataset was relatively small, we applied a predefined 3-fold cross validation for (80% of data) training and (20% of data) evaluation. We experimented with multiple classifiers, specifically: naïve Bayes (NB), K-Nearest Neighbors (KNN), SVM, RF, Adaboost with single split trees as base classifiers, a Decision Tree (DT) classifier, and a hard-voting ensemble classifier with contributions from each of the previously mentioned classifiers. We experimented with both weighted and unweighted KNN and DT classifiers to account for our unbalanced classes. The performances of the classifiers were compared using the F1-macro score on the test data because that metric is more appropriate when classes are unbalanced. We then performed an ablation study to determine the individual impact of each factor on model performance, as well as model performance when only trained on the text.

We also modeled the discrimination between CoM- and Other-class falls using the RoBERTa transformer model [19]. For this experiment, we utilized the unprocessed, free-text fall descriptions to predict fall-class labels and did not include other factors in the model. We applied a 3-fold cross validation for (80% of data) training and (20% of data) evaluation. The model was trained for 2, 5, and 10 epochs. All model parameters were fine-tuned during training. Performance was measured by taking the median of the F_1-macro score.

Confidence intervals were calculated using bootstrapping with samples taken from the test prediction and ground truth datasets with replacement over 1000 iterations.

3   Results

3.1   Demographic Information

Our fall dataset included 23 individuals with an average age of 67.3 years and an average Parkinson's disease duration of 8.9 years (Table 1). There was no significant difference between the overall fall frequency by gender ($p = 0.203$), nor in fall frequency between gender within each fall class (CoM: $p = 0.267$; BoS: $p = 0.662$; Other: $p = 0.614$).

Table 1. Demographic information and fall frequency of patients

	Overall	Women	Men	P-Value
n	23	8	15	
Age, mean (SD)	67.3 (7.1)	65.4 (6.4)	68.3 (7.5)	0.333
Duration, mean (SD)	8.9 (5.1)	8.6 (5.5)	9.1 (5.1)	0.838
Total, mean (SD)	5.4 (4.9)	7.5 (6.1)	4.3 (4.0)	0.203
CoM, mean (SD)	3.8 (4.5)	5.5 (5.5)	2.9 (3.8)	0.267
BoS, mean (SD)	1.1 (1.0)	1.2 (1.5)	1.0 (0.7)	0.662
Other, mean (SD)	0.5 (1.5)	0.8 (2.1)	0.3 (1.0)	0.614

Of the 124 unique falls, 88 falls occurred due to perturbations to the individuals' center of mass (CoM), 25 falls occurred due to perturbations to the individuals' base of support (BoS), and the remaining 11 falls occurred for other reasons, including falling during exercise (n = 9), low blood pressure (n = 1), and rolling out of bed (n = 1). Because of the relatively low number of BoS- and Other-class falls in our dataset, we collapsed the data into CoM- and Other-class falls. There were 88 CoM-class falls and 36 Other-class falls.

3.2   Binary Classification Hyperparameter Selection

For each classifier, we iterated through a wide range of hyper-parameter values and calculated the mean squared error between the actual fall class and predicted fall class at each hyperparameter value. Mean squared error is inversely proportional to model performance. Therefore, the hyperparameter value that resulted in the lowest mean squared error for each model was chosen (Table 2).

3.3 Binary Classifier Performance

Each classifier was trained on the same 80% of the dataset using 3-fold cross validation with the same folds. Then, each classifier was tested on the same 20% of data that had been excluded from the training phase, and we report the resulting F_1-macro score and its 95% confidence interval (CI). The ensemble classifier had the highest performance out of all trained classifiers (F_1-macro = 0.89, 95% CI: [0.67-1]; Table 2).

Adaboost. We determined the optimal number of estimators for the Adaboost model to be 27 by training on all values of n between 1 and 100, inclusive, and selecting the value of n with the lowest mean squared error between the true and predicted fall classes. The model achieved an F_1-macro of 0.80 (95% CI: [0.55–0.96]; Table 2).

Decision Tree. We trained two DT classifiers, one with weights set to the inverse frequency of each class in the training set (DTa) and the other with equal weights for both classes (DTb). We determined the optimal maximum depth of DTa to be 23 and the optimal maximum depth of DTb to be 18. The weighted model DTa achieved an F_1-macro of 0.67 (95% CI: [0.45–0.85]; Table 2). The unweighted model DTb performed better, achieving an F_1-macro of 0.72 (95% CI: [0.50–0.90]; Table 2).

K-Nearest Neighbor. We trained two KNN classifiers, one with the weight function equal to 'distance' (KNNa) and the other with the weight function equal to 'uniform' (KNNb). For the weighted model, KNNa, the optimal value of k was 6 and for the unweighted model, KNNb, the optimal value of k was 10. The weighted model KNNa achieved an F_1-macro of 0.84 (95% CI: [0.66–1; Table 2). The unweighted model KNNb achieved an F_1-macro of 0.78 (95% CI: [0.53–0.95]; Table 2).

Naïve Bayes. We trained a Gaussian Naïve bayes classifier. The model achieved an F_1-macro of 0.84 (95% CI: [0.63–1]; Table 2).

Random Forest. We determined the optimal number of estimators for the RF model to be 25 by training each model on all values of n between 1 and 100, inclusive, and selecting the value of n that resulted in the lowest mean squared error between the true fall classes and predicted fall classes. The RF model achieved an F_1-macro of 0.78 (95% CI: [0.53–0.95]; Table 2).

Support Vector Machine. We trained the SVM model with gamma set to 'scale' and kernel set to 'rbf' on all values of C between 1 and 100 inclusive. We found that the mean squared error between the true and predicted fall class was at its lowest when C was 4. The model achieved an F_1-macro of 0.78 (95% CI: [0.53–0.95]; Table 2).

Ensemble. The ensemble was composed of one of each type of the classifiers described above. The weighted KNN model KNNa and unweighted DT model DTa were included in the ensemble because they outperformed their complementary model. We set the voting for the ensemble equal to hard. The ensemble achieved an F_1-macro of 0.89 (95% CI: [0.67-1]; Table 2).

Table 2. Classifier performance at predicting fall type

Classifier	Hyperparameters	F_1-macro	95% CI
Adaboost	n_estimators = 27	0.80	0.55–0.96
DTa	max_depth = 23, class_weight = {0:70.0, 1:29.0}	0.67	0.45–0.85
DTb	max_depth = 18	0.72	0.50–0.90
KNNa	k = 6, weights = 'distance'	0.84	0.66–1
KNNb	k = 10, weights = 'uniform'	0.78	0.53–0.95
NB	N/A	0.84	0.63–1
RF	n_estimators = 25	0.78	0.53–0.95
SVM	Gamma = 'scale', kernel = 'rbf', C = 4	0.78	0.53–0.95
Ensemble	{NB, KNNa, SVM, RF, Adaboost, DTb} voting = 'hard'	**0.89**	**0.67–1**

3.4 Ablation Study of Features in the Ensemble Model

Table 3. Results of ablation study on the ensemble model

Dropped Feature	F_1-macro	95% CI
Sex	0.80	0.59–0.96
Location	0.84	0.66–1
PD Duration	0.80	0.59–0.95
Age	**0.89**	**0.70–1**
Word Clusters	0.82	0.63–0.96
1,2,3-g	0.78	0.55–0.95
Sex, Location, PD Duration, & Age	0.83	0.58–1

We chose to perform an ablation study on the ensemble model because it achieved the highest F_1-macro score out of all trained classifiers. The F_1-macro score of the model decreased with the removal of each factor, except age (Table 3).

3.5 Binary Classification Using RoBERTa

The RoBERTa transformer model was trained on 80% of the dataset using 3-fold cross validation. Classifier performance was measured using the F_1-macro metric. The RoBERTa model had equal performance across epochs (F_1-macro = 0.42; Table 4).

We also trained the machine learning models on the vector representations generated by RoBERTa using the same hyperparameters in Table 2. The results show that using RoBERTa as a feature generator underperformed compared to using the n-grams and word clusters as features (Table 5).

Table 4. Performance of RoBERTa model

Epochs	F_1-macro	95% CI
2	0.42	0.37–0.46
5	0.42	0.37–0.45
10	0.42	0.37–0.45

Table 5. Classifier performance using RoBERTa-generated vector representations

Model	F_1-macro	95% CI
Adaboost	0.61	0.40–0.81
DTa	0.70	0.49–0.88
DTb	0.58	0.38–0.77
KNNa	0.71	0.44–0.90
KNNb	0.44	0.37–0.48
NB	0.48	0.29–0.66
RF	0.61	0.39–0.81
SVM	0.41	0.34–0.46

4 Discussion

We trained multiple machine learning classifiers to perform a binary classification task to categorize falls as CoM- or Other-type falls based on patient-provided free-text descriptions of the circumstances surrounding each fall. We found that a hard-voting ensemble classifier performed better than individual classifiers and transformer-based models on this task, achieving an F_1-macro of 0.89 (95% CI: [0.67-1]). This serves as a proof of concept that the historically resource intensive task of manual fall type classification can be automated using NLP models. Importantly, our ensemble approach obtained high performance despite the relatively small size of the annotated dataset, which is often the limiting factor for supervised classification tasks.

Current clinical best practices for tracking falls in Parkinson's disease involve retrospective patient reports during regular Neurologist visits, which may lead to recall bias. Although standardized instruments exist [20], they are very uncommon in clinical practice due to the burden on patients and providers. A technology for classification of falls circumstances and causes based on patient reports may enable future "online trials" and reduce misclassification errors in research studies, which might contribute to the high variability across studies applying exercise on fall risk in Parkinson's disease [21].

4.1 Better Performance from ML Classifiers

Unexpectedly, traditional machine learning classifiers and their ensemble outperformed the RoBERTa transformer model in our study. We initially expected the transformer-based architecture to perform better, but it categorized all falls as CoM falls, suggesting that the small size and class imbalance of our data were not ideal for this model. Furthermore, our ablation study revealed that non-text factors were important contributors to model performance, suggesting that using only free-text descriptions may have hindered the transformer model's performance.

4.2 A Lack of Gender Differences in Fall Frequency and Type

In people with Parkinson's disease, there is no gender difference in fall frequency, unlike healthy older adults [1]. Our study found no difference in falls between genders. Future research could explore if the circumstances surrounding falls differ by gender in people with Parkinson's disease, similar to healthy older adults [22].

4.3 Future Work

To improve our study's generalizability, we plan to expand our small, unbalanced dataset with limited vocabulary by scraping social media profiles of people with Parkinson's disease. This will allow us to test and retrain our ensemble model for better performance. We also plan to explore other transformer-based models and strategies for integrating structured data.

5 Conclusion

Our study demonstrated that it is possible to automate the laborious process of fall type identification by using supervised classification methods that integrate structured and unstructured data. An ensemble classification approach produced excellent results, outperforming a state-of-the-art transformer model despite the small size of annotated data. Find the dataset and code at the following repository: https://github.com/jeanne mpowell/PD_Falls_NLP.

References

1. Bloem, B.R., Grimbergen, Y.A.M., Cramer, M., Willemsen, M., Zwinderman, A.H.: Prospective assessment of falls in Parkinson's disease. J. Neurol. **248**, 950–958 (2001). https://doi.org/10.1007/s004150170047
2. Maki, B.E., Holliday, P.J., Topper, A.K.: A prospective study of postural balance and risk of falling in an ambulatory and independent elderly population. J. Gerontol. **49**, M72–M84 (1994). https://doi.org/10.1093/geronj/49.2.M72
3. Burns, E., Kakara, R.: Deaths from Falls Among Persons Aged ≥65 Years — United States, 2007–2016. MMWR Morb. Mortal. Wkly. Rep. **67**, 509–514 (2018). https://doi.org/10.15585/mmwr.mm6718a1

4. Florence, C.S., Bergen, G., Atherly, A., Burns, E., Stevens, J., Drake, C.: Medical costs of fatal and nonfatal falls in older adults: medical costs of falls. J. Am. Geriatr. Soc. **66**, 693–698 (2018). https://doi.org/10.1111/jgs.15304

5. Haddad, Y.K., Bergen, G., Florence, C.S.: Estimating the economic burden related to older adult falls by state. J. Public Health Manag. Pract. **25**, E17–E24 (2019). https://doi.org/10.1097/PHH.0000000000000816

6. Stack, E., Ashburn, A.: Fall events described by people with Parkinson's disease: implications for clinical interviewing and the research agenda. Physiother. Res. Int. **4**, 190–200 (1999). https://doi.org/10.1002/pri.165

7. Ross, A., Yarnall, A.J., Rochester, L., Lord, S.: A novel approach to falls classification in Parkinson's disease: development of the Fall-Related Activity Classification (FRAC). Physiotherapy **103**, 459–464 (2017). https://doi.org/10.1016/j.physio.2016.08.002

8. Ashburn, A., Stack, E., Ballinger, C., Fazakarley, L., Fitton, C.: The circumstances of falls among people with Parkinson's disease and the use of Falls Diaries to facilitate reporting. Disabil. Rehabil. **30**, 1205–1212 (2008). https://doi.org/10.1080/09638280701828930

9. Pelicioni, P.H.S., Menant, J.C., Latt, M.D., Lord, S.R.: Falls in Parkinson's disease subtypes: risk factors, locations and circumstances. Int. J. Environ. Res. Public. Health. **16**, 2216 (2019). https://doi.org/10.3390/ijerph16122216

10. Magnani, P.E., et al.: Use of the BESTest and the Mini-BESTest for fall risk prediction in community-dwelling older adults between 60 and 102 years of age. J. Geriatr. Phys. Ther. **43**, 179–184 (2020). https://doi.org/10.1519/JPT.0000000000000236

11. Houssein, E.H., Mohamed, R.E., Ali, A.A.: Machine learning techniques for biomedical natural language processing: a comprehensive review. IEEE Access. **9**, 140628–140653 (2021). https://doi.org/10.1109/ACCESS.2021.3119621

12. Tohira, H., Finn, J., Ball, S., Brink, D., Buzzacott, P.: Machine learning and natural language processing to identify falls in electronic patient care records from ambulance attendances. Inform. Health Soc. Care. **47**, 403–413 (2022). https://doi.org/10.1080/17538157.2021.2019038

13. Guetterman, T.C., Chang, T., DeJonckheere, M., Basu, T., Scruggs, E., Vydiswaran, V.V.: Augmenting qualitative text analysis with natural language processing: methodological study. J. Med. Internet Res. **20**, e231 (2018). https://doi.org/10.2196/jmir.9702

14. Pérez-Toro, P.A., Vásquez-Correa, J.C., Strauss, M., Orozco-Arroyave, J.R., Nöth, E.: Natural language analysis to detect Parkinson's disease. In: Ekštein, K. (ed.) Text, Speech, and Dialogue, pp. 82–90. Springer, Cham (2019). doi.https://doi.org/10.1007/978-3-030-27947-9_7.

15. Dorsey, E.R., et al.: Global, regional, and national burden of Parkinson's disease, 1990–2016: a systematic analysis for the Global Burden of Disease Study 2016. Lancet Neurol. **17**, 939–953 (2018). https://doi.org/10.1016/S1474-4422(18)30295-3

16. McKay, J.L., Lang, K.C., Bong, S.M., Hackney, M.E., Factor, S.A., Ting, L.H.: Abnormal center of mass feedback responses during balance: a potential biomarker of falls in Parkinson's disease. PLoS ONE **16**, e0252119 (2021). https://doi.org/10.1371/journal.pone.0252119

17. Porter, M.F.: An algorithm for suffix stripping. Program **14**, 130–137 (1980). https://doi.org/10.1108/eb046814

18. Owoputi, O., O'Connor, B., Dyer, C., Gimpel, K., Schneider, N., Smith, N.A.: Improved part-of-speech tagging for online conversational text with word clusters. Presented at the Proceedings of the 2013 Conference of the North American Chapter of the Association for Computational Linguistics: Human Language Technologies June (2013)

19. Liu, Y., et al.: RoBERTa: A Robustly Optimized BERT Pretraining Approach

20. Harris, D.M., et al.: Development of a Parkinson's disease specific falls questionnaire. BMC Geriatr. **21**, 614 (2021). https://doi.org/10.1186/s12877-021-02555-6

21. Allen, N.E., et al.: Interventions for preventing falls in Parkinson's disease. Cochrane Database Syst. Rev. **2022** (2022). https://doi.org/10.1002/14651858.CD011574.pub2

22. Duckham, R.L., Procter-Gray, E., Hannan, M.T., Leveille, S.G., Lipsitz, L.A., Li, W.: Sex differences in circumstances and consequences of outdoor and indoor falls in older adults in the MOBILIZE Boston cohort study. BMC Geriatr. **13,** 133 (2013). https://doi.org/10.1186/1471-2318-13-133

BERT for Complex Systematic Review Screening to Support the Future of Medical Research

Marta Hasny[1]([⊠]) [ID], Alexandru-Petru Vasile[1], Mario Gianni[3] [ID], Alexandra Bannach-Brown[4] [ID], Mona Nasser[5] [ID], Murray Mackay[1] [ID], Diana Donovan[1] [ID], Jernej Šorli[1], Ioana Domocos[1], Milad Dulloo[1], Nimita Patel[1], Olivia Drayson[1], Nicole Meerah Elango[5] [ID], Jéromine Vacquie[1] [ID], Ana Patricia Ayala[6] [ID], and Anna Fogtman[1,2] [ID]

[1] Space Medicine Team, European Astronaut Centre, European Space Agency, Cologne, Germany
martahasny@gmail.com
[2] SciSpacE, European Astronaut Centre, European Space Agency, Cologne, Germany
[3] School of Engineering, Computing and Mathematics, University of Plymouth, Plymouth, UK
[4] Center for Transforming Biomedical Research, Berlin Institute of Health, Charité, Universitätsmedizin Berlin, Berlin, Germany
[5] Peninsula Dental School, University of Plymouth, Plymouth, UK
[6] University of Toronto, Gerstein Science Information Centre, Toronto, ON, Canada

Abstract. This work presents a Natural Language Processing approach to screen complex datasets of medical articles to provide timely and efficient response to pressing issues in medicine. The approach is based on the Bidirectional Encoder Representation from Transformers (BERT) to screen the articles using their titles and abstracts. Systematic review screening is a classification task aiming at selecting articles fulfilling the criteria for the next step of the review. A number of BERT models are fine-tuned for this classification task. Two challenging space medicine systematic review datasets that include human, animal, and in-vitro studies are used for the evaluation of the models. Backtranslation is used as a data augmentation technique to handle the class imbalance and a performance comparison of the models on the original and augmented data is presented. The BERT models provide an accessible solution for screening systematic reviews, which are considered complex and time-consuming. The proposed approach can change the workflow of conducting these types of reviews, especially in response to urgent policy and practice questions in medicine. The source code and datasets are available on GitHub: https://github.com/ESA-RadLab/BERTCSRS.

Keywords: Machine learning · BERT · Natural Language Processing · Automated Systematic Review · Space Medicine

---

M. Hasny and A.-P. Vasile—These authors contributed equally to this work.

1   Introduction and Motivations

Systematic reviews (SR) are an important tool in biomedicine. They are, however, very time-consuming and require a lot of resources. An experienced reviewer can take between 30 s to several minutes to evaluate a citation [26]. Thus the work involved in screening 10,000 citations is considerable (and the screening burden in some reviews is considerably higher than this). Cochrane Handbook for Systematic Reviews of Interventions [12] recommends two independent reviewers assess each article, and the last database search of relevant articles to be completed 6 months before the publication. However, the average time from protocol registration to publication is around 67.3 weeks [6]. Moreover, conducting SR is a highly resource-intensive task; e.g., a review of 10,000 citations is estimated to cost over $300 000.00 [16].

Computational tools can alleviate the burdens of conducting SR, enabling quicker and more economical research efforts [4,8]. Machine Learning (ML) techniques have been gathering more popularity to expedite research efforts [3,13] by reducing the work needed for eligibility screening for SR. An estimate from 2015 [20] evaluates that the work on particular SR could be lessened by between 30–70% by using ML, if a loss of at most 5% of studies is allowed. Despite some work completed on the topic, limited work is presented around complex datasets involving human, animal, and in-vitro studies that policy makers often require. This complexity, along with the variability in how evidence is reported in articles describing these three populations, contributes to a more challenging classification task. Additionally, there is lack of research completed on using state-of-the-art NLP models, such as BERT [10], to tackle the task.

Limited work has been completed on leveraging BERT to automatize the screening process and decrease the amount of work required to perform a SR. Previous work utilizing BERT focused on extensive pretraining of the model [2]. It is a computationally expensive task to pretrain the model from scratch, while a number of BERT models pretrained on corpora of scientific and medical data are publicly available. Additionally, it is a common issue in SR datasets that the training data is highly imbalanced, with a much higher number of excluded articles than included. In the past, this issue has been tackled by manually paraphrasing the titles and abstracts [2].

The contributions of our work are as following: (1) We show that publicly available pretrained BERT models can achieve high recall values on the task of title and abstract screening for complex SR biomedical datasets, without the need for extensive pretraining. (2) We introduce a data augmentation framework for SR datasets, using the technique of backtranslation to generate samples of the underrepresented class. We compare the results between original and augmented datasets, proving backtranslation to be a powerful technique for handing class imbalance and augmenting scientific biomedical text. (3) We publish the datasets to serve as a benchmark for future research on SR screening task.

2   Related Work

The use of ML to facilitate abstract review in SR is not a novel idea, with early works of Aphinyanaphongs et al. [1] and Cohen et al. [9] from as early as 2005. Jaspers et al. [14] published an in-depth report that discusses useful ML techniques to automate SR. Automation through the use of ML techniques was proposed for three different steps: the screening of abstracts, data extraction, and critical appraisal. SVMs, gradient boosting machines, neural networks and random forests classifiers were tested. Moreover, due to unbalanced data, the SMOTE [7] and ROSE [18] sampling adjustments were applied during training. The results showed that the classifiers were able to achieve high precision scores. However, the recall values were low, meaning many relevant articles are incorrectly classified as irrelevant even after the sampling adjustment.

Previous work on the SR screening tasks focuses on the traditional ML methods. Those methods yield good results. However, there is a lack of literature focusing on using newer deep learning methods, such as transformers [25] on the task. One of the current state-of-the-art transformer models that deserves attention is BERT [10]. The original BERT model was pretrained on the Wikipedia and BooksCorpus. This general corpora of pretrained data can be a limitation in achieving high results on domain specific datasets. Thus, domain specific models such as BioBERT [17], PubMedBERT [11], MedBERT [24], and SciBERT [5] have been evaluated and published to aid domain specific tasks. srBERT [2] showed that through pretraining BERT on the domain specific dataset and then fine-tuning, one is able to achieve a recall value of above 90% on the article screening task. However, even with the pretraining, the model was able to achieve such results only after manual data augmentation was performed by paraphrasing the excluded articles to fulfill the inclusion criteria.

Another important aspect which seems to not have received adequate attention in the literature is pertaining the homogeneity of reporting of the data used for training the classifiers in the first place. Bannach-Brown et al. [3] hints at this issue while discussing the lower than expected precision of their algorithms while working with data coming from preclinical animal studies, these types of studies having knowingly a higher degree of reporting variability, when compared to human studies. Therefore, we are interested to evaluate whether the new developments in NLP can be used in screening these more complex datasets that include animal, human and in-vitro studies to a level of recall and precision that is acceptable to the SR community. Automated approaches are intended to have highly recall (95–100%) to not miss any potentially relevant evidence, and precision maximised thereafter. As this automated approaches are intended to mimic the human process, it is estimated that there may be approximately 5% human error in this task, although this has not been formally investigated [3].

3   Methodology

Figure 1 shows all the steps of the SR screening automatization procedure used in this work.

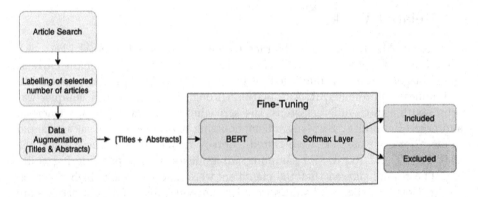

Fig. 1. The steps of our framework. First, articles that fulfil the search criteria are obtained from a database. It is necessary to manually label a set of articles for training of the model. The articles used for training go through data augmentation to account for the data imbalance. Then, they are inputted into the BERT model for fine-tuning of the model parameters. The prediction of the model states whether the articles should be included for the next screening step.

3.1   Dataset Creation

The datasets have been developed using standardized methods for designing and conducting search strategies in electronic databases and screening the data in a structured and systematic way as outlined in the Cochrane Handbook for Systematic reviews [12]. Two independent reviewers assessed each article in the training set and where disagreements arose a third reviewer was employed. The methods used to generate our SR datasets are documented and shared on the Open Science Framework repository for each review respectively [19,30]. All the collected articles are in English.

Unlike most datasets developed as part of SR on health care which focus on specific types of study designs and populations, the datasets here include human, animal, and in-vitro studies. To the best of our knowledge, there are a few other groups that developed similar datasets for regulatory purposes. However, most of them are unavailable or confidential. The datasets used in the paper are published together with the paper. They are available on the GitHub repository of the project.

3.2   Data Augmentation

It is a common issue in SR datasets that the data is highly imbalanced, with only a small percentage of the data being labeled as included. The technique of backtranslation is used here to generate more samples of the titles and abstracts labeled as included in the training set. Backtranslation is often used to support translation task [22,27]. However, its ability to paraphrase text has been shown to be a powerful data augmentation technique [29]. Backtranslation is a method of augmenting text by translating it from the original language to a different language. The translated language is then translated back to the original language. Due to the inaccuracies in the translations, it is common the text translated back to the initial language is a paraphrased version of the original text. The augmentation process is presented in Fig. 2. MarianMT [15] models are used for the translations.

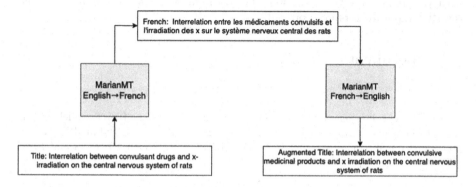

Fig. 2. Data augmentation procedure using backtranslation.

The first step is translating the text from English to French. The resulted translation is fed to the second model to be translated from French to English. The training set consists of titles and abstracts. The abstracts are split into sentences, augmenting two sentences at a time and combining them with the remaining original text to generate an augmented abstract. Titles are augmented by themselves. After generating the augmentations, a duplicate removal is completed to remove the instances where backtranslation did not paraphrase the text.

3.3 BERT for Text Classification

SR screening is a classification task predicting the inclusion of the article in the next stages of the review. We experiment with a number of BERT models pretrained on datasets coming from varying domains and sources. The domain of the corpus on which BERT was pretrained, can results in big differences in the performance on the task. Only publicly available pretrained BERT models are used in this paper. Most of the available models are based on the BERT-base architecture with 12 Transformers blocks, 12 self-attention heads, and the hidden size of 768. A typical BERT tokenization process is followed [10]. The input to the model is a sequence of at most 512 tokens, with padding for texts with lower amount of tokens. For our task, titles and abstracts are concatenated together and serve as input. The first token of the sequence is always [CLS] which contains classification embeddings. These embeddings are numerical vectors that contain the meaning of the words in the presented context.

In the classification tasks, the final hidden state h of the [CLS] token is used as a representation of the entire sequence. A softmax layer with a dropout probability of 0.2 is added on top of BERT to output the classification prediction. The output is a probability of label c.

$$P(c|h) = softmax(Wh), \tag{1}$$

with W representing the set of weights. In the fine-tuning process, both W and the BERT parameters are updated.

4   Experimental Results

Table 1. The following table represents the results of different BERT models on the original datasets. A stands for dataset A and B for dataset B. PubMedBERT was pretrained on abstracts only, while PubMEDBERT (PMC) on abstracts and PMC fulltexts.

Model	Recall (%)		F1-Score (%)		Screening Reduction (%)		AUC	
	A	B	A	B	A	B	A	B
BERT-Tiny	74.16	76.14	33.67	34.01	69.66	69.36	0.74	0.75
BERT-Mini	69.66	59.09	38.87	42.11	76.97	84.1	0.76	0.74
BERT-Small	71.91	73.86	37.65	41.01	74.88	77.07	0.76	0.78
BERT-Medium	66.29	80.68	40.97	38.38	80.1	71.77	0.76	0.79
BERT-Base	69.66	64.77	43.66	47.5	80.48	84.79	0.77	0.77
SciBERT	28.09	60.22	14.08	38.69	67.07	81.38	0.51	0.73
BioBERT	77.53	77.27	47.75	41.21	79.98	75.78	0.82	0.79
PubMedBERT	86.52	71.59	44.13	43.6	73.97	86.59	0.83	0.78
PubMedBERT (PMC)	62.92	63.64	54.37	50.45	88.29	79.88	0.78	0.78
MedBERT	67.42	78.41	48.58	49.64	84.18	80.98	0.78	0.83

4.1   Dataset

Two datasets are used for the evaluation of performance of BERT models on the SR screening task in medicine. Dataset A on sex differences in susceptibility to the ionizing radiation and dataset B on non-neoplastic effects of ionizing radiation on central nervous system. Both datasets are highly imbalanced, with 6% of the data being labeled "included" in dataset A and 10% in dataset B. 3257 samples is used for training of dataset A and 3806 for dataset B. The validation set for dataset A consists of 361 samples and 422 samples for dataset B. The testing set consists of 999 samples for both datasets.

4.2   Implementation Details

The models are implemented and fine-tuned using PyTorch [21]. The pretrained models are acquired from HuggingFace Transformer library [28]. Each model is trained for 5 epochs and the one with best validation results is saved for testing. ADAM optimizer is used with a learning rate of 2e-5. Due to the imbalance of the data, a weighted cross-entropy loss is selected as a loss function to favor the samples labeled as included. The weights are 1:50 for the original datasets, 1:18 for augmented dataset A and 1:8 for augmented dataset B. All the networks are trained on a batch size of 8 using the standard GPU of Google Colaboratory.

4.3   Data Augmentation

Backtranslation is used as a data augmentation technique. Only articles from the training set labeled as "included" are used for augmentation. Initially, dataset A consisted of 6.7% included articles and dataset B of 10.8%. After augmentation, the percentage of included articles grew up to 31.5% in dataset A and 41.7% in dataset B.

4.4   Model Performance

Table 2. The following table represents the results of different BERT models on the augmented datasets. A stands for dataset A and B for dataset B. PubMedBERT was pretrained on abstracts only, while PubMED (PMC) on abstracts and PMC fulltexts.

Model	Recall (%)		F1-Score (%)		Screening Reduction (%)		AUC	
	A	B	A	B	A	B	A	B
BERT-Tiny	77.53	100.0	26.19	16.19	56.16	0.0	0.69	0.5
BERT-Mini	89.89	77.27	23.56	34.09	40.94	68.86	0.67	0.75
BERT-Small	89.89	69.32	25.16	34.56	45.25	73.47	0.69	0.74
BERT-Medium	85.39	96.6	25.59	20.31	49.45	25.03	0.69	0.62
BERT-Base	100.0	72.73	16.36	33.86	0.0	70.97	0.5	0.74
SciBERT	82.02	100.0	31.2	16.19	62.06	0.0	0.74	0.5
BioBERT	84.27	93.18	24.92	37.36	48.65	64.86	0.68	0.82
PubMedBERT	91.01	92.05	27.98	31.46	50.95	57.26	0.73	0.77
PubMedBERT (PMC)	74.16	90.91	43.85	36.53	78.78	64.97	0.79	0.81
MedBERT	77.53	89.77	42.2	35.67	76.18	64.47	0.8	0.8

We evaluate the performance of multiple BERT models on the SR tasks using two datasets. The goal of the task is to reduce the amount of articles passed for human screening while preserving a high recall value, as it is crucial to not miss on relevant data. The models are evaluated using recall, F-1 score, Area under the ROC Curve (AUC), and screening reduction, which is the percentage of articles that remains to be screened in the next step of the SR. A number of models pretrained on medical data only is selected, including BioBERT [17], PubMedBERT [11], and MedBERT [24]. SciBERT [5], pretrained partially on medical data and partially on computer science data, is also used. Additionally, the general BERT models of different sizes are included in the study to see whether there is a need for domain specific pretraining on the screening task. Using smaller models [23] could lower the computational cost even more.

Table 1 represents the results of training the models on the original data. For dataset A the PubMedBERT achieved the best recall score. MediumBERT is a distilled BERT model composed of hidden size of 512, 8 transformers blocks, and 8 self-attention heads. The model obtained best recall score for dataset B, while reducing the amount of articles selected for further screening by 70%.

The same models are evaluated on the augmented data from datasets A and B. The results are presented in table 2. We can see a clear improvement in the recall of the models while trained on the augmented data. BERT models pretrained on medical corpus of data score the best performances on both datasets. MediumBERT achieve the highest recall for dataset B. However, the reduction of workload is very low, scoring only 25%, showing the model predicts most of the articles as included. On the other hand, BioBERT obtains a high recall value, while reducing the workload by nearly 65%, scoring an overall better performance. For dataset A, PubMedBERT once again achieve highest values, improving its results from the original data. Models pretrained on medical corpora of data, show to be superior compared to the general corpora.

5   Conclusions

This paper presents a comprehensive review of BERT models pretrained on varying corpora of data, fine-tuned for classification of SR articles. We evaluate the models on two challenging SR datasets composed of human, in-vitro, and animal studies. Inclusion of multiple types of studies poses an additional challenge for the classification model due to reporting variability. Fine-tuning publicly available models, without the need for computationally expensive pretraining, scores recall values of at least 90% while reducing the human workload by at least 50%. All the models were trained on a Google Colaboratory's standard GPU version in under an hour, showing the accessibility of our method. Additionally, back-translation shows to be a powerful data augmentation technique for scientific text, improving the final results for both datasets. The developments of NLP domain can accelerate the process of SR, while preserving the important articles. Inclusion of such systems in the process can ensure the results of SR are delivered in a timely manner, reduce the human workload, and lower the cost of performing such studies.

Acknowledgements. This work was supported by the internship programme at the European Astronaut Centre (EAC), European Space Agency (ESA). The authors would like to thank dr. Guillauime Weerts and dr. Sergi Vaquer Araujo as the Space Medicine Team Leads and dr. Ulrich Straube for their support throughout this project. Mona Nasser's research is partially supported by the National Institute for Health Research Applied Research Collaboration South West Peninsula. The views expressed in this publication are those of the author(s) and not necessarily those of the National Institute for Health Research or the Department of Health and Social Care or the European Space Agency.

References

1. Aphinyanaphongs, Y., et al.: Text categorization models for high-quality article retrieval in internal medicine. J. Am. Med. Inform. Assoc. **12**(2), 207–216 (2005). https://doi.org/10.1197/jamia.M1641
2. Aum, S., Choe, S.: srBERT: automatic article classification model for systematic review using BERT. Syst. Rev. **10**(1), 1–8 (2021)
3. Bannach-Brown, A., et al.: Machine learning algorithms for systematic review: reducing workload in a preclinical review of animal studies and reducing human screening error. Syst. Rev. **8**(1), 23 (2019). https://doi.org/10.1186/s13643-019-0942-7
4. Beller, E., et al.: Making progress with the automation of systematic reviews: principles of the international collaboration for the automation of systematic reviews (ICASR). Syst. Rev. **7**(1), 77 (2018). https://doi.org/10.1186/s13643-018-0740-7
5. Beltagy, I., Lo, K., Cohan, A.: SciBERT: a pretrained language model for scientific text. In: Proceedings of the 2019 Conference on Empirical Methods in Natural Language Processing and the 9th International Joint Conference on Natural Language Processing (EMNLP-IJCNLP). Association for Computational Linguistics, Stroudsburg, PA, USA (2019)
6. Borah, R., Brown, A.W., Capers, P.L., Kaiser, K.A.: Analysis of the time and workers needed to conduct systematic reviews of medical interventions using data from the prospero registry. BMJ Open **7**(2), e012545 (2017)
7. Chawla, N.V., Bowyer, K.W., Hall, L.O., Kegelmeyer, W.P.: SMOTE: synthetic minority over-sampling technique. J. Artif. Intell. Res. **16**, 321–357 (2002)
8. Clark, J., et al.: A full systematic review was completed in 2 weeks using automation tools: a case study. J. Clin. Epidemiol. **121**, 81–90 (2020). https://doi.org/10.1016/j.jclinepi.2020.01.008, publisher: Elsevier
9. Cohen, A., Hersh, W., Peterson, K., Yen, P.Y.: Reducing workload in systematic review preparation using automated citation classification. J. Am. Med. Inform. Assoc. **13**(2), 206–219 (2006). https://doi.org/10.1197/jamia.M1929
10. Devlin, J., et al.: BERT: pre-training of deep bidirectional transformers for language understanding (2018). https://doi.org/10.48550/ARXIV.1810.04805
11. Gu, Y., et al.: Domain-specific language model pretraining for biomedical natural language processing (2020)
12. Higgins, J., et al.: Cochrane handbook for systematic reviews of interventions version 6.3 (updated February 2022). Cochrane, www.training.cochrane.org/handbook (2022)
13. Howard, B., et al.: SWIFT-review: a text-mining workbench for systematic review. Syst. Rev. 5 (2016). https://doi.org/10.1186/s13643-016-0263-z

14. Jaspers, S., De Troyer, E., Aerts, M.: Machine learning techniques for the automation of literature reviews and systematic reviews in EFSA. EFSA Support. Publ. **15**(6), 1427E (2018). https://doi.org/10.2903/sp.efsa.2018.EN-1427

15. Junczys-Dowmunt, M., et al.: Marian: fast neural machine translation in C++. In: Proceedings of ACL 2018, System Demonstrations, pp. 116–121. Association for Computational Linguistics, Melbourne, Australia (2018). http://www.aclweb.org/anthology/P18-4020

16. Lau, J.: Editorial: systematic review automation thematic series. Syst. Rev. **8**(1), 70 (2019). https://doi.org/10.1186/s13643-019-0974-z

17. Lee, J., et al.: BioBERT: a pre-trained biomedical language representation model for biomedical text mining. Bioinformatics **36**(4), 1234–1240 (2020)

18. Lunardon, N., Menardi, G., Torelli, N.: Rose: a package for binary imbalanced learning. R J. 6, 79–89 (06 2014). https://doi.org/10.32614/RJ-2014-008

19. Nasser, M., et al.: Are there sex differences in susceptibility to ionised radiation. Open Science Framework (2021). https://doi.org/10.17605/OSF.IO/23TKV, https://osf.io/23tkv/, publisher: OSF

20. O'Mara-Eves, A., Thomas, J., McNaught, J., Miwa, M., Ananiadou, S.: Using text mining for study identification in systematic reviews: a systematic review of current approaches. Syst. Rev. **4**(1), 5 (2015). https://doi.org/10.1186/2046-4053-4-5

21. Paszke, A., et al.: Automatic differentiation in PyTorch. In: NIPS-W (2017)

22. Sennrich, R., Haddow, B., Birch, A.: Improving neural machine translation models with monolingual data. arXiv preprint arXiv:1511.06709 (2015)

23. Turc, I., et al.: Well-read students learn better: On the importance of pre-training compact models. arXiv preprint arXiv:1908.08962v2 (2019)

24. Vasantharajan, C., Tun, K.Z., Thi-Nga, H., Jain, S., Rong, T., Siong, C.E.: Medbert: a pre-trained language model for biomedical named entity recognition. In: 2022 Asia-Pacific Signal and Information Processing Association Annual Summit and Conference (APSIPA ASC), pp. 1482–1488 (2022). https://doi.org/10.23919/APSIPAASC55919.2022.9980157

25. Vaswani, A., et al.: Attention is all you need. CoRR abs/1706.03762 (2017). https://doi.org/10.48550/ARXIV.1706.03762

26. Wallace, B.C., Trikalinos, T.A., Lau, J., Brodley, C., Schmid, C.H.: Semi-automated screening of biomedical citations for systematic reviews. BMC Bioinform. **11**(1), 1–11 (2010)

27. Wieting, J., Mallinson, J., Gimpel, K.: Learning paraphrastic sentence embeddings from back-translated bitext. arXiv preprint arXiv:1706.01847 (2017)

28. Wolf, T., et al.: Transformers: state-of-the-art natural language processing. In: Proceedings of the 2020 Conference on Empirical Methods in Natural Language Processing: System Demonstrations, pp. 38–45. Association for Computational Linguistics, Online (Oct 2020). https://www.aclweb.org/anthology/2020.emnlp-demos.6

29. Yu, A.W., Dohan, D., Luong, M.T., Zhao, R., Chen, K., Norouzi, M., Le, Q.V.: QANet: combining local convolution with global self-attention for reading comprehension (2018)

30. Šorli, J., et al.: Non-neoplastic effects of Ionising radiation on central nervous system - a systematic review. Open Science Framework (2021). https://doi.org/10.17605/OSF.IO/Q8ZV3, https://osf.io/q8zv3/, publisher: OSF

GGTWEAK: Gene Tagging with Weak Supervision for German Clinical Text

Sandro Steinwand[iD], Florian Borchert[✉][iD], Silvia Winkler[iD],
and Matthieu-P. Schapranow[iD]

HPI Digital Health Center, Hasso Plattner Institute, University of Potsdam,
Prof. -Dr. -Helmert -Str. 2 -3, 14482 Potsdam, Germany
sandro.steinwand@student.hpi.uni-potsdam.de, florian.borchert@hpi.de

Abstract. Accurate extraction of biomolecular named entities like genes and proteins from medical documents is an important task for many clinical applications. So far, most gene taggers were developed in the domain of English-language, scientific articles. However, documents from other genres, like clinical practice guidelines, are usually created in the respective language used by clinical practitioners. To our knowledge, no annotated corpora and machine learning models for gene named entity recognition are currently available for the German language.

In this work, we present GGTWEAK, a publicly available gene tagger for German medical documents based on a large corpus of clinical practice guidelines. Since obtaining sufficient gold-standard annotations of gene mentions for training supervised machine learning models is expensive, our approach relies solely on programmatic, weak supervision for model training. We combine various label sources based on the surface form of gene mentions and gazetteers of known gene names, with only partial individual coverage of the training data. Using a small amount of hand-labelled data for model selection and evaluation, our weakly supervised approach achieves an F_1 score of 76.6 on a held-out test set, an increase of 12.4 percent points over a strongly supervised baseline.

While there is still a performance gap to state-of-the-art gene taggers for the English language, weak supervision is a promising direction for obtaining solid baseline models without the need to conduct time-consuming annotation projects. GGTWEAK can be readily applied in-domain to derive semantic metadata and enable the development of computer-interpretable clinical guidelines, while the out-of-domain robustness still needs to be investigated.

Keywords: Clinical NLP · Gene Named Entity Recognition · German Language · Computer Interpretable Guidelines

1 Introduction

Molecular Tumor Boards (MTBs) become increasingly established in cancer care and necessitate time-intensive research for the latest scientific evidence [13].

S. Steinwand and F. Borchert—These authors equally share first authorship.

J. M. Juarez et al. (Eds.): AIME 2023, LNAI 13897, pp. 183–192, 2023.
https://doi.org/10.1007/978-3-031-34344-5_22

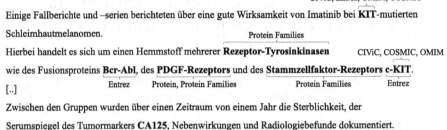

Fig. 1. Examples of gene / protein mentions (**bold**) in German oncology guidelines and labelling functions (blue) matching partially overlapping subsets of these mentions.

Therefore, specialized language technology is needed to extract molecular information from the medical literature and make insights from oncogenetics accessible in a scalable manner. For most downstream processing, Named Entity Recognition (NER) is an essential building block [19]. In this work, we consider the detection of gene and protein mentions, as shown in Fig. 1. We follow common practice in biomedical text mining and treat these entities interchangeably [6].

State-of-the-art approaches for gene NER typically rely on supervised machine learning models, trained on text corpora manually annotated by subject-matter experts. As such strong supervision is expensive to obtain, different sources of external knowledge, heuristics, and other kinds of noisy labels can be exploited in addition. Recently, such weakly supervised methods were successfully applied for clinical NER and can approach the performance of models trained with a comparable amount of strong supervision [9].

In this work, we propose German Gene Tagging with Weak Supervision (GGTWEAK), the first publicly available NER model for genes and proteins in German medical text. It is based on large amounts of unlabelled text in the German Guideline Program in Oncology NLP Corpus (GGPONC) [3] and requires only a minimal amount of hand-labelled data for model selection and evaluation. We use the SKWEAK framework to implement a range of labelling functions (LFs), and combine their predictions to train a Transformer-based NER model [17]. The contributions of this work are: (1) a dataset with novel gold-standard annotations of gene mentions for a subset of GGPONC, (2) an implementation and detailed analysis of various LFs for finding gene mentions and (3) a freely distributable neural model for gene tagging trained on aggregated weak labels. We make the source code and trained model publicly available [12].

The remainder of this work is structured as follows: In Sect. 3, we share our weak supervision methodology and incorporated data. In Sect. 4, we evaluate our LFs and model performance with respect to a small amount of gold-standard annotations. We discuss our findings in Sect. 5 and conclude our work with an outlook in Sect. 6.

Table 1. A selection of biomedical text corpora annotated on the level of molecular entities as well as the performance of recently published gene taggers evaluated on these corpora. Note: The annotation schemes vary considerably and may include more fine-grained distinctions of subclasses than we use.

Corpus	Lang.	Sent.	NER Model	F₁ Score	Year
CRAFT [1]	EN	21K	HUNFLAIR [25]	0.722	2020
BC2GM [23]	EN	20K	DTRANNER [14]	0.845	2020
PROGENE [8]	EN	36K	FLAIR [8]	0.850	2020
JNLPBA [7]	EN	19K	BIOELECTRA [20]	0.802	2021
PHARMACONER [11]	ES	14K	BIOBERT v1.1 [24]	0.899	2021

2   Related Work

In Tab. 1, we give an overview of corpora annotated with biomolecular entities and the respective performance of NER taggers. There are a number of English-language corpora based on scientific articles. In addition to such gold-standard corpora, silver-standard annotations can improve the performance of fine-tuned NER taggers when used for transfer learning [10]. Given the large amount of existing annotated gold-standard corpora for the English language, such silver-standard annotations can be obtained by applying existing NER taggers to large, unlabelled corpora. In contrast, non-English corpora with a clinical focus and annotations of biomolecular entities are scarce, with few exceptions, such as Spanish-language clinical case reports in PHARMACONER [11].

For the German language, there is a general shortage of annotated medical text corpora and none of the few existing ones provides annotations of genes or proteins [27]. We suppose that the genres of clinical texts used so far (often discharge summaries) did not contain any particularly rich molecular information. Moreover, each additional annotation layer complicates annotation and requires specialized domain expertise. Earlier experiments with dictionary-based silver-standard annotations for genes on GGPONC 1.0 resulted in an extremely large number of false positive results [2]. In this work, we aim to alleviate this shortcoming by combining different sources of weak labels instead and train a statistical NER tagger on top of them.

Recently, a number of solutions for integrating multiple label sources as programmatic, weak supervision in structured prediction tasks like NER have been proposed. Fries et al. use the Snorkel framework to integrate labels obtained from a large set of medical terminologies [9,21]. Extensions to the generative label model introduced by Snorkel employ structured probabilistic models, like HMMs, which allow modelling the dependencies of adjacent token labels [17,22]. For this work, we use the SKWEAK framework, as it employs an HMM to model dependencies across labelled tokens. Due to its tight integration with SPACY, the resulting pipeline can be easily shared and integrated into downstream applications [18].

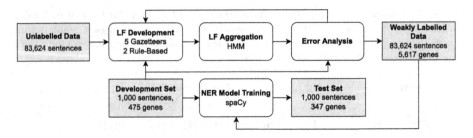

Fig. 2. Overview of GGTWEAK. Rectangular boxes represent datasets, while round-cornered boxes indicate process steps. Starting from a large unlabelled dataset, we apply seven LFs that cover subsets of gene mentions. Their outputs are aggregated using a Hidden Markov Model (HMM), resulting in a weakly labelled dataset, which is used to train a Transformer-based NER model. A small set of manual annotations are used as development and test data for error analysis and model selection. For comparison, we also train a strongly supervised NER model on the gold standard development set and evaluate it on the test set.

3   Methods

This section describes the used dataset and the weak supervision approach based on LFs and their aggregation, outlined in Fig. 2.

3.1   Dataset and Annotation

As a dataset, we use the freely distributable GGPONC corpus. Originally, the complete corpus contains 1,877K tokens in 10.19K documents. For compatibility with SKWEAK, sentence segmentation and tokenization was carried out again using the SPACY model de_core_news_md.

We randomly sampled 2,000 sentences from the subset of documents that contain at least one gene mention according to the silver-standard annotations in GGPONC 1.0 [2]. These sentences were then manually annotated with gene mentions using the INCEPTION tool [15]. Annotation was performed by a single medical student with extensive experience in linguistic annotation. The amount of hand-labelled data was chosen a priori such that a single annotator can annotate it in around one work week. In total, 822 mentions of genes and proteins were annotated. The manually annotated documents are used as development and test sets (1,000 sentences each). The remaining 83,624 sentences are labelled automatically with weak supervision and used as training data.

3.2   Labelling Function Development

We apply the following LFs based on external knowledge bases, naming conventions for gene names, and other heuristics to programmatically annotate the unlabelled part of the corpus with automatically induced labels. For implementation details, please refer to the interactive notebooks in our source code repository [12].

Gazetteer-Based LFs. The following LFs are based on *gazetteers*, i.e., they match tokens to entries in a list of known gene and protein names.

- **CIViC:** Gazetteer (case-sensitive) based on canonical gene names in the Clinical Interpretation of Variants in Cancer database, a community knowledge base of cancer genes and variants (*Examples: SMO, VEGF, TP53*).
- **Entrez:** Gazetteer (case-sensitive) based on the aliases of all genes in CIViC that occur in Entrez Gene, the gene-specific database of the National Center for Biotechnology Information [5]. Since many gene aliases can also occur as common German terms in other contexts (e.g., the pronoun "er"), we further filter by the part-of-speech tag of matched tokens (NOUN, PROPN, or X) (*Examples: p16, B-Raf, HER-2*).
- **OMIM:** Gazetteer (case-insensitive) based on the Online Mendelian Inheritance in Man database, a comprehensive catalogue of human genes and genetic disorders (*Examples: PALB2, TNF, BRCA1*).
- **COSMIC:** Gazetteer (case-sensitive) based on the Catalogue of Somatic Mutations in Cancer database, a knowledge base of somatic mutations and additional information associated with cancer in humans (*Examples: BTK, IGHV, BRAF*).
- **Proteins:** Custom gazetteer (case-sensitive) sourced from the German Wikipedia overview page of proteins, manually refined by exploration of the unlabelled training part of the dataset [26] (*Examples: PD-L1, Cyclooxygenase, Uridin-5'-Diphospho-Glucuronosyltransferase*).

Rule-Based LFs. Another type of LF is based on heuristics that take the surface forms of tokens into account, e.g., a particular composition of uppercase letters and numbers, as well as specific prefixes and suffixes.

- **HGNC:** Heuristic derived from the HUGO Gene Nomenclature Committee naming conventions for genes, using regular expression. As matching short gene names based on this convention would lead to many false positives, we instead rely on a case-insensitive lookup in CIViC for these genes (*Examples: CA125, CYP19, mTORC1*).
- **Protein Families:** Heuristic based on common suffixes describing groups of proteins, e.g., "-rezeptor", "-kinase" or the "-RAS" family (*Examples: Rezeptor-Tyrosinkinasen, MAP-Kinase, k-ras*).

3.3   Labelling Function Aggregation

The LFs were designed such that they cover specific subsets of gene mentions in our corpus (as shown in Fig. 1). Therefore, the partial and potentially conflicting outputs of these LFs are aggregated using the HMM label model from SKWEAK, which emits a single label per token, accounting for correlations and conflicts among the LFs. We fit the HMM on the LF outputs on the training set. The HMM predictions on the 83.6K sentences of the training set result in more than 5,617 automatic annotations of gene mentions.

3.4   Named Entity Recognition Models

We can use the trained HMM to predict labels for unseen instances, which we do for comparison on the development and test set. However, in order to obtain a model that can potentially generalize beyond our LFs, we train another Transformer-based NER model with the SPACY framework on top of the HMM output. To this end, we use the aggregated, weakly labelled data for model training and the gold-standard development set for model selection. The NER model's encoder is initialized from the BERT checkpoint `bert-base-german-cased`, which was pre-trained on German general domain and legal text. We use an initial learning rate of 10^{-5}, with 250 warmup steps and linear learning rate decay. For all other hyperparameters, we use the default values provided by SPACY. The model was trained for 20,000 optimization steps, which takes about 2 h on a single NVIDIA A40 GPU. As the final model, we choose the checkpoint that achieves the maximal F_1 score on the development set.

For comparison with the traditional setting of building NER taggers, we train another model with the same architecture and hyperparameters using the development set as training data, i.e., with just the small amount of available strongly supervised data. The final evaluation of both models is performed on our initially defined test set.

4   Results

The results of the incorporated LFs, the HMM and NER models are presented in Table 2. Since we do not have access to ground truth labels on the training set, we estimate the contributions made by each LFs through coverage and overlap. For the development and test set, we can compare all LFs and aggregated models to gold-standard labels.

4.1   Labelling Function Analysis

All LFs achieve high levels of precision and a coverage of up to 40.6% of the targetted labels. The rule-based LFs show small overlap (38.2% and 36.9%), i.e., more than 60% of the mentions they label are unique to these LFs. While the coverage of the suffix-based LF for protein families is low, it has a non-negligible recall on the development set (7.6%), that, combined with the uniqueness of its labels, has a positive impact on the final model.

Considering synonyms from Entrez Gene drastically improves coverage on the training set compared to CIVIC, at the expense of a small decrease in precision. Likewise, OMIM as the biggest database has high coverage and only 50.0% overlap with other LFs. In contrast, CIVIC and COSMIC both share high overlap but rather low coverage. After aggregation, the combined labels from the HMM result in a slightly lower precision compared to the individual LFs, but provide a better recall and F_1 score.

Table 2. Performance metrics of each LF and derived statistical models (coverage = number of tokens labelled by one LF divided by the number of tokens labelled by all LFs, overlap = number of tokens labelled by one LF that are also labelled by any other LF divided by the total number of tokens labelled by this LF). The strongly supervised model was trained on the development set and is therefore only evaluated on the test set.

	Training		Development			Test		
	Coverage	Overlap	Pr.	Rec.	F_1	Pr.	Rec.	F_1
Gazetteers								
CIViC	.210	.980	.944	.465	.624	.841	.473	.606
Entrez	.406	.686	.902	.503	.646	.890	.608	.722
OMIM	.344	.500	.926	.524	.670	.818	.493	.616
COSMIC	.223	.930	.928	.436	.594	.854	.473	.608
Proteins	.137	.699	.934	.120	.212	**.975**	.112	.200
Rule-based								
HGNC	.365	.382	.833	.305	.446	.836	.280	.420
Protein Families	.018	.369	**1.000**	.076	.142	.250	.012	.022
HMM	-	-	.841	.680	.752	.789	.689	.736
GGTWEAK	-	-	.855	**.720**	**.782**	.819	.718	**.766**
Strong Supervision	-	-	-	-	-	.558	**.758**	.642

4.2   Evaluation Against Gold-Standard Annotations

The final GGTWEAK NER model achieves an F1 score of 78.2% on the development set and 76.6% on the held-out test set. Moreover, GGTWEAK performs 12.4 percent points better than the model trained with strong supervision in terms of F_1 score. While the strongly supervised model has slightly higher recall (+4 pp.), GGTWEAK shows dramatically higher precision (+26.1 pp.). GGTWEAK also outperforms the HMM consistently by a margin of 3 pp., highlighting the added value of transfer learning through pre-trained Transformer weights.

5   Discussion and Limitations

The foundation of our work is GGPONC, a corpus of German oncology guidelines. While extensive in volume, it is imbalanced regarding the presence of molecular entities, i.e., most sentences do not contain mentions of genes or proteins. We note that the HMM implementation provided by SKWEAK is particularly sensitive to false positives. For these reasons, it is challenging to develop high-precision LFs while maintaining high coverage. Although we have not performed exhaustive ablation experiments, we notice that additional LFs increase the recall of the final model, usually at the expense of decreased precision.

Interestingly, the performance of the final model drops only slightly when evaluating it on the test set in comparison to the development set, although the latter was used during LF development. This indicates a certain generalizability of the model beyond the scope of the LFs. However, we note that the considered text genre provides only a partial representation of the different notations for gene names that may occur in clinical documents. Both aspects impact the generalization performance of our model and need to be further investigated.

There still remains a performance gap to state-of-the-art gene taggers for English biomedical literature, which often achieve F_1 scores significantly larger than 80% (see Table 1). However, we have to bear in mind that the research community has had access to annotated Englishlanguage corpora for a much longer time. Furthermore, the underlying problem might be intrinsically harder for the German language due to its grammatical intricacies, such as the prevalence of compound nouns. We rely on several upstream components in SPACY for basic linguistic tasks, such as tokenization and POS tagging. Although these general-domain solutions appear to work reasonably well on GGPONC, errors introduced by them might influence the performance of our LFs. Lastly, we have not performed any optimization of hyperparameters of both the HMM label model and the Transformer-based NER training, which would likely have a positive impact on model performance.

6   Conclusion and Future Work

In this work, we presented a novel approach for gene tagging in German medical text. With an F_1 score of 76.6%, we could demonstrate the viability of weak supervision for this task with substantially decreased demand for labels from human experts. Importantly, GGTWEAK outperforms a model that was trained on the same amount of gold-standard labels that we used for model selection only.

As future work, we plan to add more diversely targetting LFs and explore other Transformer checkpoints, e.g., domain-specialized models for the German [4,16] or other languages, as shown by Sun et al. [24]. An important downstream task is the normalization of gene mentions to identifiers in knowledge bases, such as Entrez Gene [5]. We expect that this will be challenging, as German terms relating to genes and more generally to groups of genes might not have easily identifiable aliases in such knowledge bases.

We believe that more annotated language resources in conjunction with weak supervision can support the development of high-quality gene taggers for clinical documents. Our findings should be readily applicable to other languages, as clinical guidelines are a widely available text genre and most of our LFs do not rely on language-specific resources.

Acknowledgements. Parts of this work were generously supported by a grant of the German Federal Ministry of Research and Education (01ZZ1802H).

References

1. Bada, M., et al.: Concept annotation in the CRAFT corpus. BMC Bioinform. **13**(1), 1–20 (2012)
2. Borchert, F., et al..: GGPONC: A corpus of German medical text with rich metadata based on clinical practice guidelines. In: Proceedings of the 11th International Workshop on Health Text Mining and Information Analysis. pp. 38–48 (2020)
3. Borchert, F., et al.: GGPONC 2.0 - the German clinical guideline corpus for oncology: Curation workflow, annotation policy, baseline ner taggers. In: Proceedings of the Thirteenth Language Resources and Evaluation Conference, pp. 3650–3660 (2022)
4. Bressem, K.K., et al.: MEDBERT.de: A comprehensive German BERT model for the medical domain. arXiv (2023). https://doi.org/10.48550/ARXIV.2303.08179
5. Brown, G.R., et al.: Gene: a gene-centered information resource at NCBI. Nucleic Acids Res. **43**(D1), D36–D42 (2015)
6. Cohen, A.M., Hersh, W.R.: A survey of current work in biomedical text mining. Briefings Bioinform. **6**(1), 57–71 (03 2005)
7. Collier, N., Ohta, T., Tsuruoka, Y., Tateisi, Y., Kim, J.D.: Introduction to the bio-entity recognition task at JNLPBA. In: Proceedings of the International Joint Workshop on Natural Language Processing in Biomedicine and its Applications (NLPBA/BioNLP). pp. 73–78. Geneva, Switzerland (2004)
8. Faessler, E., Modersohn, L., Lohr, C., Hahn, U.: ProGene - a large-scale, high-quality protein-gene annotated benchmark corpus. In: Proceedings of the 12th Language Resources and Evaluation Conference, pp. 4585–4596 (2020)
9. Fries, J.A., et al.: Ontology-driven weak supervision for clinical entity classification in electronic health records. Nat. Commun. **12**(1), 1–11 (2021)
10. Giorgi, J.M., Bader, G.D.: Transfer learning for biomedical named entity recognition with neural networks. Bioinformatics **34**(23), 4087–4094 (2018)
11. Gonzalez-Agirre, A., Marimon, M., Intxaurrondo, A., Rabal, O., Villegas, M., Krallinger, M.: PharmaCoNER: Pharmacological substances, compounds and proteins named entity recognition track. In: Proceedings of the 5th Workshop on BioNLP Open Shared Tasks, pp. 1–10. Association for Computational Linguistics, Hong Kong, China (2019)
12. Hasso Plattner Institute's Digital Health Center on GitHub: GGTweak source code repository. https://github.com/hpi-dhc/ggponc_molecular (2023)
13. Henkenjohann, R., et al.: An engineering approach towards multi-site virtual molecular tumor board software. In: International Conference on ICT for Health, Accessibility and Wellbeing, pp. 156–170. Springer (2021)
14. Hong, S., Lee, J.G.: DTranNER: biomedical named entity recognition with deep learning-based label-label transition model. BMC Bioinform. **21**(1), 1–11 (2020)
15. Klie, J.C., Bugert, M., Boullosa, B., Eckart de Castilho, R., Gurevych, I.: The INCEpTION platform: machine-assisted and knowledge-oriented interactive annotation. In: COLING 2018 – Proceedings of the 27th International Conference on Computational Linguistics: System Demonstrations, pp. 5–9 (2018)
16. Lentzen, M., et al.: Critical assessment of transformer-based ai models for German clinical notes. JAMIA open **5**(4), ooac087 (2022)
17. Lison, P., Barnes, J., Hubin, A.: skweak: Weak supervision made easy for NLP. In: Proceedings of the 59th Annual Meeting of the Association for Computational Linguistics, pp. 337–346. Association for Computational Linguistics, Online (2021)

18. Montani, I.,et al.Flusskind: explosion/ spaCy: v3.4.1: Fix compatibility with CuPy v9.x (Jul 2022). https://doi.org/10.5281/zenodo.6907665

19. Perera, N., Dehmer, M., Emmert-Streib, F.: Named entity recognition and relation detection for biomedical information extraction. Frontiers in cell and developmental biology, p. 673 (2020)

20. Raj Kanakarajan, K., Kundumani, B., Sankarasubbu, M.: BioELECTRA: pre-trained biomedical text encoder using discriminators. In: Proceedings of the 20th Workshop on Biomedical Language Processing, pp. 143–154 (2021)

21. Ratner, A., Bach, S.H., Ehrenberg, H., Fries, J., Wu, S., Ré, C.: Snorkel: Rapid training data creation with weak supervision. In: Proceedings of the VLDB Endowment. International Conference on Very Large Data Bases. vol. 11, p. 269 (2017)

22. Safranchik, E., Luo, S., Bach, S.: Weakly supervised sequence tagging from noisy rules. In: Proceedings of the AAAI Conference on Artificial Intelligence. vol. 34, pp. 5570–5578 (2020)

23. Smith, L., et al.: Overview of BioCreative II gene mention recognition. Genome Biol. **9**(2), 1–19 (2008)

24. Sun, C., Yang, Z., Wang, L., Zhang, Y., Lin, H., Wang, J.: Deep learning with language models improves named entity recognition for PharmaCoNER. BMC Bioinform. **22**(1), 1–16 (2021)

25. Weber, L., Sänger, M., Münchmeyer, J., Habibi, M., Leser, U., Akbik, A.: Hun-Flair: an easy-to-use tool for state-of-the-art biomedical named entity recognition. Bioinformatics **37**(17), 2792–2794 (2021)

26. Wikipedia: Kategorie:Protein. https://de.wikipedia.org/wiki/Kategorie: Protein (2023)

27. Zesch, T., Bewersdorff, J.: German medical natural language processing-a data-centric survey. In: Applications in Medicine and Manufacturing, pp. 137–142 (2022)

Soft-Prompt Tuning to Predict Lung Cancer Using Primary Care Free-Text Dutch Medical Notes

Auke Elfrink[1,2]([envelope]) [ID], Iacopo Vagliano[2] [ID], Ameen Abu-Hanna[2] [ID], and Iacer Calixto[2] [ID]

[1] Informatics Institute, University of Amsterdam, Amsterdam, Netherlands
auke.elfrink@gmail.com
[2] Amsterdam UMC, Department of Medical Informatics, Amsterdam Public Health Research Institute, University of Amsterdam, Meibergdreef 9, Amsterdam, Netherlands
{i.vagliano,a.abu-hanna}@amsterdamumc.nl

Abstract. We examine the use of large Transformer-based pretrained language models (PLMs) for the problem of early prediction of lung cancer using free-text patient medical notes of Dutch primary care physicians. Specifically, we investigate: 1) how *soft prompt-tuning* compares to standard model fine-tuning; 2) whether simpler static word embedding models (WEMs) can be more robust compared to PLMs in highly imbalanced settings; and 3) how models fare when trained on notes from a small number of patients. All our code is available open source in https://bitbucket.org/aumc-kik/prompt_tuning_cancer_prediction/.

Keywords: Prediction models · Natural Language Processing · Cancer

1 Introduction

In the Netherlands, a patient's visits to their general practitioner (GP) is registered in narrative form as *free-text notes*, and collectively these provide a unique lens into years of a patient's health and medical history. In previous work, Luik et al. [4] have used free-text patient notes in the early prediction of lung cancer with *context-independent* word embedding methods (WEMs) and a simple logistic regression objective, with promising preliminary results.

Ideally, one would like to use state-of-the-art contextualised pretrained language models (PLMs) instead of simpler context-independent WEMs. In this work, we use PLM's to address the following research questions. **RQ1.** *How does soft-prompt tuning compare to standard model fine-tuning in terms of discrimination and calibration?* We use *soft prompt-tuning* [3], an NLP technique used to adapt PLMs using small amounts of training data, and compare it to standard model fine-tuning. **RQ2.** *How does the number of patients used for model training affect model performance in terms of discrimination and calibration?* We empirically evaluate models trained on very small numbers of patients and models trained

J. M. Juarez et al. (Eds.): AIME 2023, LNAI 13897, pp. 193–198, 2023.
https://doi.org/10.1007/978-3-031-34344-5_23

on datasets with different degrees of class imbalance. **RQ3.** *How do PLMs compare to simpler static WEMs in terms of discrimination and calibration?* Overall, contextualised PLMs outperform static WEMs in terms of discrimination, but WEMs show better calibration in both balanced and imbalanced classification settings. Results we obtain for few-shot experiments show mixed results, suggesting that WEMs are an alternative to PLMs in this setting.

2   Methodology

Data. We have unique access to data from patients of General Practitioners (GPs) associated with the two hospitals of the Amsterdam University Medical Centers, the Free University Medical Center (VUMC) and the Academic Medical Center of the University of Amsterdam (AMC). We establish lung cancer diagnoses for the patients in our cohort by linking to a central database maintained by *Integraal Kankercentrum Nederland* (IKNL). Henceforth, we refer to patients with a lung cancer diagnosis as 'positive', and all other patients as 'negative'. We include patients with entry date of 01/01/2002 and an exit date of 31/12/2020 who: *i)* are at least 40 years-old, *ii)* have at least one valid free-text note.[1] For the note of a positive patient to be considered valid, it must be dated at least 150 days ($\sim$ 5 months) and at most 730 days ($\sim$ 2 years) before the date of the diagnosis. For negative patients, notes must be dated at most 730 days ($\sim$ 2 years) before the date of the patient's last available note. We denote by $\mathcal{D}$ the set of patients after we apply all the above mentioned procedures.

We first create the subset $\mathcal{D}^{bal} \subset \mathcal{D}$ with the same number of positive and negative patients. In $\mathcal{D}^{bal}$ we include all $1,733$ positive patients in $\mathcal{D}$ and randomly subsample negative patients from $\mathcal{D}$ to have a balanced dataset. To study the effect of class imbalance in our models, we propose different *test sets* $\mathcal{D}^{pos:neg}$ with different positive-to-negative ratios, and build the test sets $\mathcal{D}^{1:10}$, $\mathcal{D}^{1:100}$, and $\mathcal{D}^{1:250}$. Finally, we also build few-shot datasets $\mathcal{F}^k \subset \mathcal{D}^{bal}$ where k is the number of patients in *each* class (positive and negative). We use $k \in \{2, 4, 8, 16, 32, 64, 128\}$. In Table 1 we show key characteristics for these datasets. Note that train, validation, and test splits are always *stratified*, i.e., they inherit the same ratio of patients with/without cancer as the original dataset.

Models. We use two Dutch-language PLMs: RobBERT [2] and MedRoBERTa.nl [6], which are trained on general and medical texts, respectively. We adapt the models with *soft-prompt tuning*: we append *continuous, trainable embeddings* to the input sequence which are optimised via backpropagation, while the original model parameters remain frozen and are not updated. Please see [3] for details.

We construct each training instance $\{X_n^P, Y^P\}_{n=1}^N$ by propagating the binary label Y^P of patient P to each of the N notes available for that patient. After the model is trained, we aggregate per-note into per-patient predictions by:

$$P(Y^P = 1|\{X_n^P\}_{n=1}^N) = P_{min}, \tag{1}$$

where P_{min} is the lowest per-note probability predicted across all notes $\{X_n^P\}_{n=1}^N$.

---

[1] All notes were anonymized with a modified version of the DEDUCE algorithm [5].

Table 1. Dataset statistics for our balanced dataset split $\mathcal{D}^{bal}$ and our imbalanced test sets $\mathcal{D}^{pos:neg}$. Overall, there are a total of *1,733* positive patients in our cohort (692+259+85+697).

Dataset name	# patient notes mean (min, max)	total patients (#positive, #negative)			
		train	valid	test_1	test_2
$\mathcal{D}^{bal}$	32.2 (1, 284)	1384 (692, 692)	518 (259, 259)	170 (85, 85)	1394 (697, 697)
$\mathcal{D}^{1:10}$	28.1 (1, 293)	—	—	—	7667 (697, 6970)
$\mathcal{D}^{1:100}$	29.2 (1, 572	—	—	—	70396 (697, 69699)
$\mathcal{D}^{1:250}$	29.5 (1, 1403)	—	—	—	174946 (697, 174249)
$\mathcal{F}^2$	30.7 (1, 284)	2 (1, 1)	2 (1, 1)	—	—
$\mathcal{F}^4$	30.7 (1, 284)	4 (2, 2)	4 (2, 2)	—	—
$\mathcal{F}^8$	30.9 (1, 284)	8 (4, 4)	8 (4, 4)	—	—
$\mathcal{F}^{16}$	30.9 (1, 284)	16 (8, 8)	16 (8, 8)	—	—
$\mathcal{F}^{32}$	30.9 (1, 284)	32 (16, 16)	32 (16, 16)	—	—
$\mathcal{F}^{64}$	31.1 (1, 284)	64 (32, 32)	64 (32, 32)	—	—
$\mathcal{F}^{128}$	31.5 (1, 284)	128 (64, 64)	128 (64, 64)	—	—

Table 2. Results on $\mathcal{D}^{bal}$ test_1 for models fine-tuned (FT) to predict lung cancer vs. using soft-prompt tuning (ST). We highlight best results in **bold** and second-best by underscoring. **N** denotes per-note and **P** per-patient results.

Model	AUROC ($\uparrow$)		AUPRC ($\uparrow$)		Brier score ($\downarrow$)	
	N	P	N	P	N	P
FastText (FT)	—	90.8	—	90.4	—	**44.7**
RobBERT (FT)	81.8	90.2	65.7	84.8	18.2	54.3
MedRoBERTa.nl (FT)	82.4	90.2	68.6	82.7	17.6	56.8
RobBERT (ST)	**84.5**	**94.3**	72.3	**91.4**	16.7	55.9
MedRoBERTa.nl (ST)	84.3	92.7	**75.5**	86.5	**15.9**	58.0

As our static word embedding model (WEM) we use FastText [1], a WEM where word representations encode *subword information*. We follow two steps: 1) We pretrain FastText from scratch using all free-text notes available for all patients from all splits in $\mathcal{D}^{bal}$; 2) We use the training notes in $\mathcal{D}^{bal}$ to fine-tune FastText to predict risk of lung cancer for that dataset.

3   Results and Discussion

In Table 2, we report results on our experiments where we use the $\mathcal{D}^{bal}$ dataset. We compare models when directly *fine-tuned* or when *soft-prompt tuned* on the task of predicting whether a patient has lung cancer. We use the *train, valid*, and *test_1* splits for model training, model selection, and testing, respectively.

We first highlight that per-patient predictions consistently improve on per-note predictions, which suggests that the offline aggregation method we use

Table 3. Results on imbalanced/balanced test datasets with the best performing fine-tuning (FT) and soft prompt-tuning (ST) model trained on $\mathcal{D}^{bal}$ *train* split. **N** denotes per-note and **P** per-patient results. **RB**: RobBERT. **FaT**: Fast-text.

Test set $\mathcal{D}$?	AUROC (↑)				AUPRC (↑)				Brier score (↓)			
	bal	1:10	1:100	1:250	*bal*	1:10	1:100	1:250	*bal*	1:10	1:100	1:250
Random	50.0	50.0	50.0	50.0	50.0	10.0	1.0	0.4	—	—	—	—
Per-note predictions												
RB (FT)	76.8	78.3	61.3	54.4	63.9	15.9	1.2	**0.4**	19.9	18.7	28.3	32.4
RB (ST)	**80.1**	**81.4**	**64.5**	**57.8**	**70.6**	**22.0**	**1.5**	0.4	**18.3**	**16.8**	**27.4**	**32.3**
Per-patient predictions												
FaT (FT)	78.7	81.0	60.1	56.1	77.2	32.3	1.4	0.5	**44.8**	**8.2**	**1.2**	**0.7**
RB (FT)	88.2	90.1	70.3	65.4	81.9	40.0	1.8	0.6	54.0	67.7	24.3	18.0
RB (ST)	**89.7**	**91.1**	**71.2**	**67.1**	**86.2**	**47.0**	**1.9**	**0.7**	56.3	71.0	25.1	17.8

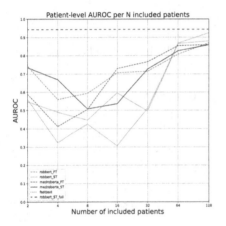 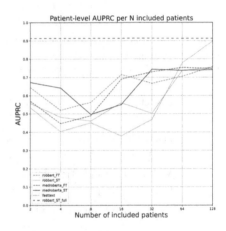

(a) Number of training patients vs. AUROC scores on $\mathcal{D}^{bal}$ *test_1*.

(b) Number of training patients vs. AUPRC scores on $\mathcal{D}^{bal}$ *test_1*.

Fig. 1. Results for models trained on $\mathcal{F}^k$ (*train*), $k \in \{2, 4, 8, 16, 32, 64, 128\}$. **ST**: soft-prompt tuning. **FT**: fine-tuning.

(Eq. 1) suppresses possibly noisy per-note predictions well. We note that Fast-Text performs comparably to or better than our *fine-tuned* PLMs according to all metrics; however, when compared to our *soft prompt-tuned* PLMs, Fast-Text is slightly worse (AUROC), comparable (AUPRC) or much better in terms of calibration (Brier score). Moreover, soft-prompt tuning tends to consistently outperform standard model fine-tuning, according to AUROC (by ∼ 2–4%) and AUPRC (by ∼ 4–6%), but with no clear trend according to the Brier score. Finally, per-note Brier scores are considerably better than per-patient Brier scores.

In Table 3, we show results for our FastText baseline and the best performing RobBERT model trained on $\mathcal{D}^{bal}$ (*train*) and selected according to AUROC scores on $\mathcal{D}^{bal}$ (*valid*). We provide results on the *test_2* splits of $\mathcal{D}^{bal}$, $\mathcal{D}^{1:10}$, $\mathcal{D}^{1:100}$, and $\mathcal{D}^{1:250}$. As expected, both AUROC and AUPRC tend to decrease as we increase the number of negative patients, with more pronounced effects on AUPRC. However, *Brier scores improve as test sets become more imbalanced.* This makes sense when taking our per-patient aggregation method into account: since we use the lowest note probability, the prediction for true negatives will likely be much closer to 0 than the prediction for true positives will be to 1.

Few-shot learning results in Fig. 1 show that when models are trained on 32 or less patients, PLMs tend to clearly outperform FastText, but when trained on 128 patients we already note that FastText outperforms all PLMs (according to both AUROC and AUPRC), while the fine-tuned and soft-prompt tuned PLM's have comparable performance. This suggests that soft-prompt tuning is still not clearly more resilient to 'noisy' inputs than fine-tuned models.

4   Conclusions

We find that using contextualised PLM's (RobBERT and MedRoBERTa.nl) outperforms strong static WEM's (FastText) according to AUROC and AUPRC, although FastText shows much better calibration than both PLMs. In few-shot experiments, the difference between PLMs and WEMs is less clear. Soft prompt tuning consistently outperforms standard model fine-tuning with PLMs, which we find promising for further research. When testing our models on datasets with increased class imbalance, performance deteriorates as expected; nonetheless, the best performing PLM still achieves a reasonable 67.1 AUROC when tested on a 1:250 ratio, though the corresponding 0.7 AUPRC shows considerable improvements upon the random baseline (0.4 AUPRC).

References

1. Bojanowski, P., Grave, E., Joulin, A., Mikolov, T.: Enriching word vectors with subword information. arXiv preprint arXiv:1607.04606 (2016)
2. Delobelle, P., Winters, T., Berendt, B.: RobBERT: a dutch RoBERTa-based language model. In: Findings of the Association for Computational Linguistics: EMNLP 2020, pp. 3255–3265. Association for Computational Linguistics, Online (Nov 2020). https://doi.org/10.18653/v1/2020.findings-emnlp.292
3. Liu, X., et al.: P-tuning: prompt tuning can be comparable to fine-tuning across scales and tasks. In: 60th ACL (Volume 2: Short Papers). pp. 61–68. Association for Computational Linguistics, Dublin, Ireland (May 2022). https://doi.org/10.18653/v1/2022.acl-short.8
4. Luik, T.T., Rios, M., Abu-Hanna, A., van Weert, H.C.P.M., Schut, M.C.: The effectiveness of phrase skip-gram in primary care NLP for the prediction of lung cancer. In: Tucker, A., Henriques Abreu, P., Cardoso, J., Pereira Rodrigues, P., Riaño, D. (eds.) AIME 2021. LNCS (LNAI), vol. 12721, pp. 433–437. Springer, Cham (2021). https://doi.org/10.1007/978-3-030-77211-6_51

5. Menger, V., Scheepers, F., van Wijk, L.M., Spruit, M.: DEDUCE: a pattern matching method for automatic de-identification of dutch medical text. Telematics Inf. **35**(4), 727–736 (2018). https://doi.org/10.1016/j.tele.2017.08.002.https://www.sciencedirect.com/science/article/pii/S0736585316307365
6. Verkijk, S., Vossen, P.: Medroberta.nl: a language model for dutch electronic health records. Comput. Ling. Netherlands J. **11**, 141–159 (2021)

Machine Learning Models for Automatic Gene Ontology Annotation of Biological Texts

Jayati H. Jui[✉][ID] and Milos Hauskrecht[ID]

University of Pittsburgh, Pittsburgh, PA 15260, USA
{jaj146,milos}@pitt.edu

Abstract. Gene ontology (GO) is a major source of biological knowledge that describes the functions of genes and gene products using a comprehensive set of controlled vocabularies or terms organized in a hierarchical structure. Automatic annotation of biological texts using gene ontology (GO) terms gained the attention of the scientific community as it helps to quickly identify relevant documents or parts of text related to specific biological functions or processes. In this paper, we propose and investigate a new GO-term annotation strategy that uses a non-parametric k-nearest neighbor model and relies on various vector-based representations of documents and GO terms linked to these documents. Our vector representations are based on machine learning and natural language processing (NLP) models, including singular value decomposition, Word2Vec and topic-based scoring. We evaluate the performance of our model on a large benchmark corpus using a variety of standard and hierarchical evaluation metrics.

Keywords: Gene Ontology (GO) · GO-term text annotation

1 Introduction

Gene Ontology (GO) is the largest and most diverse open-source repository of structured and standardized vocabulary that describes complex biological functions of genes and gene products across different organisms. The GO knowledge base is developed and maintained by the Gene Ontology (GO) Consortium. It defines vocabulary and its structure using functional attributes known as GO terms, and links these to different genes and gene products. GO centers on three main aspects to describe biological knowledge: biological process, cellular component, and molecular function. GO can be used for a variety of purposes. One important problem is the annotation of documents or text with GO terms which can help

Supported by the Defense Advanced Research Projects Agency (DARPA) through Cooperative Agreement D20AC00002 awarded by the U.S. Department of the Interior, Interior Business Center. The content of the article does not necessarily reflect the position or the policy of the Government, and no official endorsement should be inferred.

J. M. Juarez et al. (Eds.): AIME 2023, LNAI 13897, pp. 199–204, 2023.
https://doi.org/10.1007/978-3-031-34344-5_24

researchers to identify articles based on important biological relations mentioned in the articles.

Early efforts at GO annotation relied on manual annotations. Unfortunately, such annotations were tedious and required well-established guidelines to avoid inconsistencies and errors [3,5]. The focus of recent GO-annotation efforts has been gradually shifting towards automatic methods based on Natural Language Processing (NLP) and machine learning (ML) solutions. BioCreAtIvE text mining competitions were among the first attempts to design solutions to facilitate automatic GO annotations of text [2,8]. Multiple strategies have been devised which range from simple keyword matching to drive the annotation in pattern-based methods to complex machine learning models relying on gene concepts and language-based features [4,6,7].

In this study, we use cutting-edge natural language processing (NLP) and machine learning (ML) methods to develop and investigate GO-term annotation solutions for biological text. Briefly, the annotation problem can be seen as a supervised multi-label classification problem with GO terms defining the class labels. We employ non-parametric techniques for label prediction, including featurization and representation of documents with several natural language processing (NLP) vector-based models, including Singular Value Decomposition (SVD), Word2Vec, and topic-based features. To evaluate our solution we used a new benchmark dataset with scientific abstracts and their GO annotations.

2   Methods

Corpus: We have created a benchmark corpus with the latest GO annotations for our model training and assessment. Using the Gene Ontology Annotation (GOA) database from Uniprot[1] (Uniprot-GOA), we retrieved all human GO annotations with references to PubMed articles. After updating the old Uniprot-GOA annotations with the latest GO functional attributes, our final corpus had ∼42k article abstracts with 14707 unique GO annotations. We randomly split the corpus into two disjoint train and test sets with a 90:10 ratio resulting in ∼38k train and ∼4k test documents. The dataset used in this study is available on GitHub[2]. The distribution of the three GO categories in the dataset is available in a supplementary document[3].

Text Processing and Vectorization: We performed text processing of all documents in the corpus using **scyspaCy**[4], a python package built for biomedical, clinical, and scientific text analysis. We utilized scispaCy's "en_core_sci_md" model to conduct Named Entity Recognition (NER). We removed all words from the documents that were not recognized as NER entities in order to shorten the documents and computed the Term frequency-inverse document frequency

---

[1] https://www.ebi.ac.uk/GOA/index.

[2] https://github.com/juijayati/GOA-AIME2023.git.

[3] https://tinyurl.com/ypzttrur.

[4] https://allenai.github.io/scispacy/.

(TF-iDF) of each document. Using document TF-iDFs, we computed two vector representations of each document: SVD and Word2Vec. SVD is a popular dimensionality reduction technique for data with a large number of features. We computed 100-dimensional SVDs of the sparse document TF-iDFs and pre-trained Word2Vec word embeddings were extracted directly from scispaCy's "en_core_sci_md" model. Word2Vec vectors were weighted using the Tf-iDF weights of the words to generate document embeddings.

Prediction Model: We employed a non-parametric method that models the relationships between documents and GO terms for text annotation. Specifically, vector representations of documents in the training data and their associated GO-labels are used to make predictions on the test documents using the k Nearest Neighbor (k-NN) approach applied to their vector representations. Our method can be summarized as follows:

For a test article Q

- Compute vector representations of Q and assign topics to Q
- Extract k most similar documents from the training set using k-NN strategy.
- Build a set of GO terms $\mathcal{G}$ by combining all GO annotations from the top-k articles extracted from the training set.
- Calculate document-based similarity score $\phi_D(Q, t)$ and topic-based similarity score $\phi_T(Q, t)$ for each GO-term $t \in \mathcal{G}$
- For each term t in $\mathcal{G}$, calculate annotation likelihood of term t given Q as:

$$(t|Q) = \phi_D(Q, t) * \phi_T(Q, t) \tag{1}$$

- Annotate Q with top n terms in $\mathcal{G}$ based on the highest likelihoods.

Topic Assignment: For assigning topics to a query article, we used scispaCy's "Gene Ontology" linker that links NER entities to a set of UMLS[5] concepts related to GO functional attributes. The topics of the test articles were determined by direct mapping of the UMLS concept names to GO terms.

Document-Based Score: Document-based similarity score for a GO term $t \in \mathcal{G}$ given a query article Q is calculated using the documents in the list K_Q of top-k most similar documents that annotates t.

$$\phi_D(Q, t) = \left(1 + \sum_{d \in K_Q : t \text{ annotates } d} sim(Q, d)\right)^2 \tag{2}$$

Topic-Based Score: Topic-based similarity score for a GO term $t \in \mathcal{G}$ given a query article Q is calculated based on the maximum semantic similarity between the term t and any topics of Q that is in the same path of the GO hierarchy as the term t. The semantic similarity between two go terms is defined as:

---

[5] https://www.nlm.nih.gov/research/umls/index.html.

Table 1. Model performances on the benchmark test corpus

Vectorization	Vector Dimension	GO Scoring	R_{10}	TREC	BioCreAtIve		
				MMR_{10}	hR_{10}	hP_{10}	hF_{10}
SVD	100	Doc	0.41	0.44	0.19	0.65	0.25
Word2Vec	200	Doc	0.43	0.45	0.19	0.67	0.26
SVD	100	Doc + Topic	0.43	0.45	**0.21**	0.66	0.27
Word2Vec	200	Doc + Topic	**0.45**	**0.46**	**0.21**	**0.69**	**0.28**

$$sim(t_1, t_2) = \begin{cases} 1 & \text{if } t_1 = t_2 \\ \frac{1}{dist(t_1, t_2)} & \text{if } t_2 \in ancestors(t_1) \cup children(t_1) \\ 0 & \text{otherwise} \end{cases} \tag{3}$$

where $dist(t_1, t_2)$ denotes the semantic distance between the terms t_1 and t_2 and is defined as the shortest distance between t_1 and t_2 in GO hierarchy. The topic-based score is then calculated to reflect the maximum semantic similarity between a candidate GO term t and the topic set T_Q of a query Q.

$$\phi_T(Q, t) = \left(1 + \max_{t_Q \in T_Q} sim(t, t_Q)\right)^2 \tag{4}$$

3   Results and Discussion

We evaluated our models on the test corpus consisting of 4034 articles, and 4191 unique GO annotations among which 343 annotations were not present in the training corpus. For evaluation, we used evaluation metrics developed for hierarchical biological ontologies. In particular, we considered Mean Reciprocal Rank (MRR_n) used in TREC question answering track and hierarchical measures of precision (hP_n), recall (hR_n) and F-scores (hF_n) introduced at BioCreAtIve IV competition [1,9]. We also considered Recall at rank n (R_n) that measures the exact recall achieved by the model's top n predictions. The detailed explanations of the evaluation metrics are available in a supplementary document (See Footnote 2).

The classification performances of the proposed machine learning models are summarized in Table 1. All statistics were based on the top 10 GO terms predicted by the models. As can be seen from the results, the Word2Vec model combined with document and topic-based GO terms scoring achieved the best performance across all five evaluation metrics. It is interesting to see that applying a dimensionality reduction technique like SVD on the TF-iDFs was able to achieve comparable performance to Word2Vec models. In contrast to TF-iDF or SVD, Word2Vec captures the context of words and the semantic relationship between words. We note that scispaCy's word vectors were trained on biomedical and clinical corpora and offer vector representations of key biological words and concepts. Since the corpus introduced in this study is built using biological texts, the Word2Vec models provide a better representation of the articles than term frequencies. The Word2Vec models also strictly outperform SVD models except for hierarchical precision. Furthermore, it can be seen that incorporating topic-based similarity scores in addition to document-based similarities enabled improved scoring of the gold standard GO terms. It shows that a rough set of terms related to the actual protein functions can be identified via direct word mentions or textual cues from the NER entity tokens.

Both SVD and Word2Vec models achieved high hierarchical recall (hR_{10}) on the training data. This indicates that the ancestor sets of the predicted terms and the true annotations have a high overlap. Higher hR_{10} were achieved by topic-based models because topic-based similarity scores prioritize semantic similarity between two terms with ancestor-descendent relationships. However, the hierarchical precision of the models remained very low. Hierarchical precision favors predictions of more general GO terms with fewer ancestors. This is contradictory to providing the most specific terms for annotation it is a poor metric for such ontologies. According to MRR_{10}, the first prediction of a true annotation occurs within the top three predicted terms. Finally, 45% of the true annotations were typically included in the top-10 predictions, as indicated by the R_{10} statistics. Additional results regarding the performance of the top Word2Vec model across three GO categories are available in a supplementary document (See Footnote 2).

4   Conclusions

We have proposed and investigated an automated approach for the annotation of biomedical articles with GO terms that represent molecular functions, underlying biological processes, and cellular components mentioned in the text. The annotation of a test article uses k-nearest neighbor matching of training articles using their vector representation. In the future, we plan to investigate additional modern text vectorization methods offered, for example, by BERT or ELMO architectures for biological domains, as well as featurization based on gene or gene products mentioned in the articles.

References

1. Arighi, C., et al.: Proceedings of the fourth biocreative challenge evaluation workshop (2013)
2. Blaschke, C., Leon, E.A., Krallinger, M., Valencia, A.: Evaluation of biocreative assessment of task 2. BMC Bioinform. **6**, 1–13 (2005)
3. Camon, E.B., et al.: An evaluation of go annotation retrieval for biocreative and goa. BMC Bioinf. **6**, 1–11 (2005)
4. Chen, Y.D., Yang, C.J., Li, W.G., Huang, C.Y., Chiang, J.H., et al.: Gene ontology evidence sentence extraction and concept extraction: two rule-based approaches (2013)

5. Faria, D., Schlicker, A., Pesquita, C., Bastos, H., Ferreira, A.E., Albrecht, M., Falcão, A.O.: Mining go annotations for improving annotation consistency. PLoS ONE **7**(7), e40519 (2012)
6. Gobeill, J., Pasche, E., Vishnyakova, D., Ruch, P.: Closing the loop: from paper to protein annotation using supervised gene ontology classification. Database 2014 (2014)
7. Lena, P.D., Domeniconi, G., Margara, L., Moro, G.: Gota: Go term annotation of biomedical literature. BMC Bioinform. **16**, 1–13 (2015)
8. Lu, Z., Hirschman, L.: Biocuration workflows and text mining: overview of the biocreative 2012 workshop track ii. Database 2012 (2012)
9. Voorhees, E.M., Buckland, L.: Overview of the trec 2003 question answering track. In: TREC, vol. 2003, pp. 54–68 (2003)

Image Analysis and Signal Analysis

A Robust BKSVD Method for Blind Color Deconvolution and Blood Detection on H&E Histological Images

Fernando Pérez-Bueno[1]([✉]) [ID], Kjersti Engan[2] [ID], and Rafael Molina[1] [ID]

[1] Dpto. de Ciencias de la Computación e I. A., Universidad de Granada,
Granada, Spain
fpb@ugr.es, rms@decsai.ugr.es
[2] Department of Electrical Engineering and Computer Science,
University of Stavanger, Stavanger, Norway
kjersti.engan@uis.no

Abstract. Hematoxylin and Eosin (H&E) color variation between histological images from different laboratories degrades the performance of Computer-Aided Diagnosis systems. Histology-specific models to solve color variation are designed taking into account the staining procedure, where most color variations are introduced. In particular, Blind Color Deconvolution (BCD) methods aim to identify the real underlying colors in the image and to separate the tissue structure from the color information. A commonly used assumption is that images are stained with and only with the pure staining colors (e.g., blue and pink for H&E). However, this assumption does not hold true in the presence of common artifacts such as blood, where the blood cells need a third color component to be represented. Blood usually hampers the ability of color standardization algorithms to correctly identify the stains in the image, producing unexpected outputs. In this work, we propose a robust Bayesian K-Singular Value Decomposition (BKSVD) model to simultaneously detect blood and separate color from structure in histological images. Our method was tested on synthetic and real images containing different amounts of blood pixels.

Keywords: Stain Separation · Blood Detection · Histological images

1 Introduction

The development of Computer-Aided Diagnosis (CAD) systems for the analysis of Whole Slide Images (WSIs) is not exempt from challenges [9]. Just using data from different hospitals can hamper their performance [11], mostly due to color variation and artifacts in the image [6]. Therefore, preprocessing is often a key step [6] for reliable CAD systems.

Blind Color Deconvolution (BCD) methods [12] estimate the image-specific stain colors and structure (concentration). The separation itself can reduce

This work has been supported by project B-TIC-324-UGR20 FEDER/Junta de Andalucía and Universidad de Granada.

J. M. Juarez et al. (Eds.): AIME 2023, LNAI 13897, pp. 207–217, 2023.
https://doi.org/10.1007/978-3-031-34344-5_25

the impact of color variation [14] and it is often included as a step for other approaches such as color normalization [19] or color augmentation [18]. See [6] for a complete survey. However, the assumption that the image contains only the expected stains (e.g. Blue for nuclei and pink for cytoplasm and connective tissue in H&E images), is not true when artifacts are present. Artifacts such as blood, degrade the quality and diagnosis value of a WSI and introduce additional sources of color variation [6]. Blood artifacts react differently to H&E staining and get stained in a completely different color [3] (usually red), which is often used to detect them [3] but also hampers the performance of BCD methods [2].

Despite the relationship between artifacts and color, artifact detection and color variation have hardly been explored together. The presence of blood and other artifacts is often ignored by BCD methods. Similarly, it is hard to find works that use BCD to detect artifacts. Our work brings together the fields of artifact detection and color variation by focusing on blood, and how its presence affects BCD methods on H&E images. We propose the use of BCD for blood detection using its difference in color with the H&E stains. For this goal, we use the recently proposed Bayesian K-Singular Value Decomposition (BKSVD) [12] for BCD, which is able to identify the colors of the stains in the image but is affected by blood. We extend it to acknowledge the presence of blood, making it possible to perform blood detection and then obtain an estimation of the H&E stains.

The paper is organized as follows. Section 1.1 describes related works on BCD and blood detection. Section 2 provides an overview of the BKSVD method in [12], discusses the limitations, and provides the necessary improvements for its application to robust blood detection and BCD. In Sect. 3 experimentally evaluate the proposed method. Finally, Sect. 4 concludes the paper.

1.1 Related Work

Blind Color Deconvolution. Most BCD methods use the Beer-Lambert law [15], which establishes a linear combination between the stains in the *optical density* (OD) space. Let I be a RGB image $\mathbf{I} \in \mathbb{R}^{3 \times Q}$, where each value $i_{cq} \in \mathbf{I}$ correspond to pixel q and channel $c \in$ RGB. Then, the OD is defined as $y_{cq} = -\log(i_{cq}/i_{cq}^0)$, where $i_{cq}^0 = 255$ denotes the incident light. Then, a WSI $\mathbf{Y}$ stained with S stains follows the equation

$$\mathbf{Y} = \mathbf{MC} + \mathbf{N}, \tag{1}$$

where $\mathbf{M} = [\mathbf{m}_1, \cdots, \mathbf{m}_S] \in \mathbb{R}^{3 \times S}$ is the normalized stains' specific color-vector matrix; $\mathbf{C} \in \mathbb{R}^{S \times Q}$ is the stain concentration matrix, its q-th column, $\mathbf{c}_q = [c_{1,q}, \ldots, c_{S,q}]^T$, represents the contribution of each stain to the q-th pixel value in $\mathbf{Y}$; and, finally, $\mathbf{N} \in \mathbb{R}^{3 \times Q}$ is a Gaussian noise matrix with independent components of variance β^{-1}.

The goal of BCD is to estimate $\mathbf{C}$ and $\mathbf{M}$ from $\mathbf{Y}$. Here we summarize the most relevant approaches in the literature. See [6] for a complete survey. Ruifrok et al. [15] use a given color-vector matrix to separate the stains, which is widely used as a standard. However, the actual color is usually considered to be unknown due to color variation. In the work by Macenko *et al.* [8] the H&E channels are estimated using Singular Value Decomposition (SVD). Vahadane *et al.* [19] estimate the color-vector matrix using Non-Negative Matrix Factorization (NMF) and the assumption that most pixels in the image are stained by a single stain. Alsubaie *et al.* [1] apply Independent Component Analysis (ICA) in the wavelet domain, under the assumption that stains might not be independent. Few works consider the effect of noise and artifacts. The use of the method by Macenko *et al.* [8] is widely extended, although it is known to be affected by noise and artifacts [2]. To speed up the method, Vahadane *et al.* [19] perform a patch sampling of the WSI. They take advantage of their patch-wise stain estimation to calculate the median color-vector and provide a more robust estimation against artifacts. Alsubaie *et al.* [1] included a linear filtering to reduce the noise contamination when estimating stain matrix, but did not consider large artifacts. The presence of artifacts and their effect on color estimation is acknowledged in [2]. The estimation of the color-vector and the robust maximum (99th percentile) of the H&E concentrations were used to identify low-quality images, substituting the color-vector matrix with average estimates from other images when poor quality is detected.

A Bayesian approach is followed in [5] by defining a smoothness prior on the concentrations and a similarity prior on the color-vectors. To improve the quality of the concentration obtained, this work was extended in [13] by using a Total Variation (TV) prior and in [14] with general super-Gaussian priors. These Bayesian methods share a need for a reference color-vector matrix for the similarity prior. The use of a prior on the color [13,14] can reduce the effect of noise and artifacts, but limits the adaptability to different color distributions. To solve this issue, the work in [12] proposes the use of Bayesian K-SVD (BKSVD) to find the color-vector matrix as a dictionary learning problem.

Blood Detection. Detection of blood is frequently formulated as a color-related problem. In [3] the authors classify blood segmentation techniques into (i) RGB segmentation, (ii) segmentation using other color space (such as HSV, Lab and LUV) and additional techniques, (iii) segmentation using one or two channels of a non-RGB color space. In [7] the detection of blood is approached with a combination of staining protocol and image processing, using mathematical morphology and thresholding the RGB channels of the image. Sertel *et al.* [16] use the color to distinguish five major components in the H&E images (i.e. nuclei, cytoplasm, background, blood, and extracellular material). First they threshold the RGB channels to remove blood and background, and then use k-means in the L*a*b color space. A Maximum Likelihood Estimation is implemented in [10] to classify the pixels into four classes (blood, cytoplasm, nuclei, and background) in the RGB color space. In [17] the magenta channel of the CMYK space is used to detect blood areas that are later classified into hemorrhages or vessels using mathematical morphology and a decision tree. Hue and saturation are used in [4] to detect intracerebral hemorrhage.

More complex approaches combine the choice of color space with clustering, mathematical morphology, classification, or DL [20]. See [6] for details.

2   Material and Methods

2.1   Bayesian k-SVD for Blind Color Deconvolution

Although the color in the slide is a combination of both H&E stains, each biological structure will present structure-specific color properties [12,19]. This allows pathologists to distinguish structures based on their color. In [12] a Bayesian framework is used to approach the problem as a dictionary learning problem, with sparse concentrations [19], which encourages the framework to find the color-vector matrix $\mathbf{M}$ that better represents the differential staining of the structures in the image.

Following (1), the OD observed image $\mathbf{Y}$ is modeled with a Gaussian distribution $p(\mathbf{Y}|\mathbf{C}, \mathbf{M}, \beta)$, where β controls the noise precision. That is, $p(\mathbf{Y}|\mathbf{C}, \mathbf{M}, \beta) = \prod_{q=1}^{Q} \mathcal{N}(\mathbf{y}_q|\mathbf{M}\mathbf{c}_q, \beta^{-1}\mathbf{I}_{3\times 3})$. The sparsity of the solution is promoted by a zero-mean Laplace prior on the concentrations, that is $p(\mathbf{c}_q) \propto \exp\left(-\sqrt{\lambda_q}\|\mathbf{c}_q\|_1\right)$, with $\lambda_q > 0$ the scale parameter. A flat prior $p(\mathbf{M})$ on the color-vector matrix is used, and unit norm for each column $\mathbf{m}_s$ is assumed on their posterior estimation.

The true posterior $p(\boldsymbol{\Theta}|\mathbf{Y}) = p(\mathbf{Y}, \boldsymbol{\Theta})/p(\mathbf{Y})$, where $\boldsymbol{\Theta} = \{\beta, \mathbf{M}, \mathbf{C}\}$ is the set of unknowns, is approximated with variational inference [12]. WSIs are usually very large, rendering their processing computationally expensive. The number of pixels necessary to estimate the color-vector matrix $\hat{\mathbf{M}}$ is reduced by using a uniform random sampling of the WSI. The estimations $\hat{\mathbf{M}}$ and $\hat{\mathbf{C}}$ are then initialized to $\underline{\mathbf{M}}$ and $\hat{\mathbf{C}} = \underline{\mathbf{M}}^+\mathbf{Y}$, where $\underline{\mathbf{M}}$ is the Ruifrok's standard matrix [15] and $\underline{\mathbf{M}}^+$ the Moore-Penrose pseudo-inverse of $\underline{\mathbf{M}}$. After iteratively optimizing the model parameters, estimations for $\hat{\mathbf{M}}$ and $\hat{\mathbf{C}}$ are provided. Notice that the method in [12] did not consider the presence of blood. When the image contains blood artifacts $\mathbf{M}$ will be wrongly estimated because it will be forced to have two color vectors (H&E). Here we extend the method and propose its use for robust blood detection and BCD estimation.

2.2   Robust Blind Color Deconvolution and Blood Detection

Blood and other artifacts hamper the estimation of the color-vector matrix and the separation of the stains. Obtaining a robust method to detect blood and perform BCD estimation is closely related to correctly identifying these elements.

BKSVD [12] uses two channels to separate H&E images, but it can be extended to include more channels. A third channel is often used by BCD [8,13,15], which content is considered to be residual when there are only two stains in the image and is often referred to as "background" channel. As blood gets stained in a different color, often used for its detection [3], it can be seen as an additional effective stain, that can be detected using the third channel. However, including a third channel when not necessary, hampers the quality of the H&E estimated color-vectors $\hat{\mathbf{m}}_h$, $\hat{\mathbf{m}}_e$, and the corresponding concentrations [13]. Assuming blood or other elements might appear in the image, we split the BCD process into two stages. The first stage uses three channels to represent all elements in the image. The second stage uses two and focuses on the H&E channels only.

The first stage starts from the H&E vectors from $\underline{\mathbf{M}}$ [15] and a third color-vector $\underline{\mathbf{m}}_b$ which is orthogonal to $\underline{\mathbf{m}}_h$ and $\underline{\mathbf{m}}_e$ [13]. After estimating $\hat{\mathbf{M}}$ and $\hat{\mathbf{C}}$, it is critical to keep the right order of the stain channels. Thus, we calculate the correlation between the columns of $\hat{\mathbf{M}}$ and the initial $\underline{\mathbf{m}}_h$, $\underline{\mathbf{m}}_e$, and choose $\hat{\mathbf{m}}_h$, $\hat{\mathbf{m}}_e$ as those with maximum correlation to their respective references. The remaining channel is selected as the third (blood) channel, and the order of $\hat{\mathbf{C}}$ is modified accordingly.

We utilize the values in the third concentration channel $\hat{\mathbf{C}}_{:,3}$ to detect anomalies. Since, as explained in [12], the dictionary learning approach of BKSVD aims at finding a representation of the image where each pixel is assigned to only one channel whenever possible, pixels assigned or having a higher value in the third channel are not correctly represented by the H&E channels and can be considered an anomaly. Pixels with a blood channel component above a threshold are marked and discarded. The value for the threshold is experimentally determined in Sect. 3.

Finally, utilizing only the blood-free pixels and starting from $\underline{\mathbf{M}}$, the second stage re-estimates the color-vector matrix and concentrations for the remaining pixels. The procedure is summarized in Algorithm 1.[1]

Algorithm 1. Robust BKSVD for blood detection

Require: Observed image $\mathbf{I}$, initial normalized $\underline{\mathbf{M}}$, batch size B, threshold thr.
Ensure: Estimated stain color-vector matrix, $\hat{\mathbf{M}}$, concentrations, $\hat{\mathbf{C}}$, blood mask.
  1: Obtain the OD image $\mathbf{Y}$ from $\mathbf{I}$
  2: First stage: set $S = 3$. Estimate $\hat{\mathbf{M}}$ and $\hat{\mathbf{C}}$ using BKSVD [12]
  3: Sort $\hat{\mathbf{M}}$ and $\hat{\mathbf{C}}$ using the correlation of the columns of $\hat{\mathbf{M}}$ and $\underline{\mathbf{M}}$
  4: Create blood mask using $\hat{\mathbf{C}}_{:,3} > thr$
  5: Remove blood-positive pixels.
  6: Second stage: set $S = 2$. Re-estimate $\hat{\mathbf{M}}$ and $\hat{\mathbf{C}}$ on remaining pixels using BKSVD
  7: **return** $\hat{\mathbf{M}}$, $\hat{\mathbf{C}} = \hat{\mathbf{M}}^+\mathbf{Y}$ and the blood mask.

2.3   Databases

- **Synthetic Blood Dataset (SBD).** We use the Warwick Stain Separation Benchmark (WSSB) [1], which contains 24 H&E images of different tissues from different laboratories for which their ground truth color-vector matrix $\mathbf{M}_{GT}$ is known. WSSB images are free from artifacts, therefore we synthetically combine them with blood pixels obtained in the Stavanger University Hospital, incrementally added as columns to each image. This allows us to control the amount of blood in the image, which makes it possible to measure the effect of blood on the estimation of the color-vector matrix. The

---

[1] The code will be made available at https://github.com/vipgugr/.

amounts of blood considered are $\{0, 0.1, 0.2, \ldots, 0.9\}$ times the size of the original images, creating a dataset of 240 images.

- **TCGA Blood Dataset.** In order to test the method on real images containing blood, we used 8 breast biopsies from The Cancer Genome Atlas (TCGA). Breast biopsies often contain blood due to the biopsy procedure [6] and it is also possible to find blood vessels in the tissue. We selected 16 2000 × 2000 H&E image patches where blood was manually labeled.

3 Experiments

We have designed a set of experiments to assess our method. First, we examine its blood detection capability and then, its robustness using BCD.

3.1 Blood Detection

In this experiment, we assess the performance of the proposed method to identify blood in the image. We evaluate the most appropriate value for thresholding the third channel [13] and compare the results with other approaches for blood detection.

The concentration channels take values $\in [0, -log(1/255)]$. We have evaluated thresholds for the blood component thr in the whole range with a 0.1 step. Pixels with a value above the thr are marked as blood, and the mask is then compared with the ground truth label in the SBD and TCGA datasets. We use the Jaccard index (intersection over union) and the F1-score.

Table 1 shows the performance of the method on the SBD images for different values of the threshold. Low values of the threshold are able to correctly separate the blood in the image without including an excessive number of false positives. The best results in mean are obtained using $thr = 0.3$. Values of $th > 1$ were experimentally found to be irrelevant for blood detection.

Table 1. Jaccard index and F1-score for the blood mask obtained with the proposed method and different threshold values on SBD images.

Amount of blood	Jaccard index						F1-score					
	Threshold value						Threshold value					
	0.2	0.3	0.4	0.5	0.6	0.7	0.2	0.3	0.4	0.5	0.6	0.7
0.1	0.615	0.715	0.797	0.814	0.798	**0.861**	0.722	0.791	0.860	0.871	0.850	**0.913**
0.2	0.729	0.794	**0.874**	0.789	0.778	0.839	0.823	0.859	**0.923**	0.845	0.834	0.896
0.3	0.759	**0.804**	0.786	0.797	0.749	0.720	0.849	**0.868**	0.845	0.865	0.818	0.794
0.4	0.741	0.808	0.802	**0.858**	0.786	0.808	0.828	0.872	0.867	**0.909**	0.852	0.870
0.5	0.791	0.818	0.825	**0.830**	0.731	0.748	0.871	0.885	0.881	**0.888**	0.801	0.825
0.6	0.780	**0.865**	0.774	0.803	0.759	0.764	0.863	**0.920**	0.843	0.866	0.828	0.845
0.7	0.795	0.853	0.794	**0.863**	0.727	0.721	0.873	0.912	0.856	**0.919**	0.808	0.798
0.8	0.794	**0.839**	0.834	0.757	0.759	0.784	0.870	**0.902**	0.894	0.829	0.833	0.864
0.9	**0.833**	0.808	0.809	0.714	0.728	0.655	**0.901**	0.878	0.876	0.797	0.803	0.738
Mean	0.760	**0.811**	0.810	0.803	0.757	0.767	0.844	**0.876**	0.871	0.865	0.825	0.838

Using a value of $thr = 0.3$ we compare the proposed method with the blood detection approaches in [4,16,17] by thresholding the red RGB, the saturation HSV and the magenta CMYK channels, respectively. Results are summarized in Table 2. RGB and CMYK do not obtain a close mask, while HSV obtains a fair estimation. Our approach obtains the best Jaccard index and F1-score in most cases and also has the best mean value. The proposed method obtains the best mean value in both SBD and TCGA datasets. Figure 1 depicts an example using a TCGA image.

Table 2. Jaccard index and F1-score for different approaches to blood detection on the SBD images. The proposed method uses $thr = 0.3$.

Amount of blood	SBD images							
	Jaccard index				F1-score			
	R-RGB	S-HSV	M-CMYK	Proposed	R-RGB	S-HSV	M-CMYK	Proposed
0.1	0.0078	0.6984	0.5369	**0.7152**	0.0154	**0.8010**	0.6814	0.7911
0.2	0.0125	0.7484	0.5964	**0.7936**	0.0247	0.8447	0.7381	**0.8588**
0.3	0.0157	0.7605	0.6156	**0.8037**	0.0309	0.8569	0.7566	**0.8677**
0.4	0.0165	0.7819	0.6385	**0.8078**	0.0324	**0.8727**	0.7756	0.8720
0.5	0.0173	0.7914	0.6498	**0.8181**	0.0340	0.8800	0.7850	**0.8850**
0.6	0.0179	0.7978	0.6577	**0.8646**	0.0352	0.8848	0.7914	**0.9204**
0.7	0.0184	0.8041	0.6650	**0.8531**	0.0360	0.8893	0.7972	**0.9121**
0.8	0.0186	0.8087	0.6700	**0.8391**	0.0365	0.8925	0.8011	**0.9023**
0.9	0.0187	**0.8104**	0.6730	0.8076	0.0368	**0.8938**	0.8034	0.8777
Mean	0.0159	0.7780	0.6337	**0.8114**	0.0313	0.8684	0.7700	**0.8763**
	TCGA images							
	Jaccard index				F1-score			
Mean	0.0402	0.3299	0.1245	**0.4342**	0.0723	0.4667	0.2065	**0.5648**

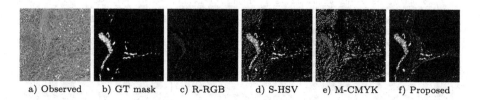

a) Observed   b) GT mask   c) R-RGB   d) S-HSV   e) M-CMYK   f) Proposed

Fig. 1. Blood detection comparison on a TCGA image. a) Observed image. b) Ground truth manual mask. c-f) Blood mask obtained by thresholding different channels and with the proposed method.

3.2 Effect of Blood in the Estimation of the Color-Vector Matrix

All the images in SBD were deconvolved and the obtained color-vector matrix $\hat{\mathbf{M}}$ for H&E was compared with the expected $\mathbf{M}_{GT}$ using the euclidean distance. Table 3 summarizes the results obtained according to the amount of blood in the image for different values of thr. When the amount of blood in the image is none or small, removing excessive pixels from the image is not beneficial, and therefore the optimal *thr* is higher. However, in concordance with the previous experiment, a value of *thr* = 0.3 obtains the best result for a realistic range of the amount of blood.

Table 3. Mean Euclidean distance between ground truth $\mathbf{M}_{GT}$ and obtained $\mathbf{M}$ color-vector matrix using the proposed method and different threshold values.

Amount of blood	Threshold value							
	0.2	0.3	0.4	0.5	0.6	0.7	0.8	0.9
0	0.1386	0.1248	0.0927	0.0826	0.0726	0.0775	0.0748	**0.0653**
0.1	0.0790	0.0852	0.0971	**0.0699**	0.0704	0.0699	0.0920	0.0893
0.2	0.0910	0.0768	0.0683	**0.0661**	0.1038	0.0994	0.1391	0.1500
0.3	0.0984	**0.0760**	0.0942	0.0826	0.1020	0.1498	0.1607	0.1720
0.4	**0.0843**	0.0966	0.0934	0.0992	0.1296	0.1489	0.1775	0.1881
0.5	0.1019	**0.0985**	0.1214	0.1264	0.1427	0.1663	0.1814	0.1903
0.6	0.1145	0.1210	**0.1001**	0.1342	0.1558	0.1724	0.1817	0.1953
0.7	0.1228	**0.1016**	0.1229	0.1294	0.1505	0.1842	0.1908	0.2034
0.8	0.1201	**0.1165**	0.1365	0.1391	0.1676	0.1862	0.1949	0.2009
0.9	0.1338	**0.1287**	0.1294	0.1588	0.1710	0.1844	0.1939	0.2110
Mean	0.1084	**0.1026**	0.1056	0.1088	0.1266	0.1439	0.1587	0.1666

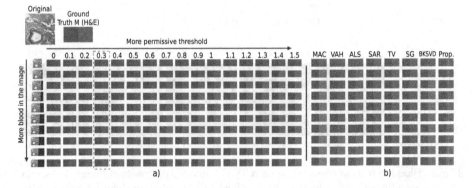

Fig. 2. Qualitative comparison of color-matrices on an SBD image. a) Obtained color-vector matrix $\mathbf{M}$ for different thresholds and amounts of blood in the image. The first row corresponds to the clean image and the amount of blood increases with the row number. The columns fix the threshold value, with zero being the most restrictive. The grey dashed box indicates the column with a closer euclidean distance to the ground truth. b) Obtained $\mathbf{M}$ for different state-of-the-art methods. (Color figure online)

A visual example is depicted in Fig. 2a. We represent the color-vector matrix **M** as a tuple where each square shows the normalized color vector for each stain, H and E, respectively. The effect of blood is most appreciated in the E channel, which becomes reddish as the blood increases, and a higher *thr* is chosen. The E channel changes to represent blood rather than eosin, and the H channel shifts to represent the mix of H&E which are closer in color between them than with blood.

Table 4 compares the results of the proposed method with *thr* = 0.3 with state-of-the-art methods in terms of euclidean distance to the ground truth color-vector matrix. Specifically, we compare our method with the methods by Macenko *et al.* [8] (MAC), the robust method by Vahadane *et al.* [19] (VAH), Alsubaie *et al.* [1] (ALS), the reference-prior based Bayesian methods using Simultaneous AutoRegressive (SAR) [5], Total Variation (TV) [13], and Super Gaussian (SG) [14], and the previous non-robust BKSVD [12]. The proposed method outperforms previous methods, showing that the current state-of-the-art for stain separation is not well adapted when there is blood on the image.

Table 4. Mean Euclidean distance between ground truth $\mathbf{M}_{GT}$ and obtained **M** color-vector matrix for different BCD methods. The proposed method uses *thr* = 0.3.

Amount of blood	Method							
	MAC	VAH	ALS	SAR	TV	SG	BKSVD	Proposed
0	0.2428	0.1019	0.3059	0.3118	0.3328	0.3119	**0.0838**	0.1248
0.1	0.3004	0.0972	0.3291	0.3118	0.3783	0.3119	0.2427	**0.0852**
0.2	0.3074	0.1116	0.3983	0.3118	0.3889	0.3119	0.2484	**0.0768**
0.3	0.3090	0.1227	0.3824	0.3118	0.3971	0.3119	0.2487	**0.0760**
0.4	0.3099	0.1340	0.4116	0.3118	0.4061	0.3119	0.2515	**0.0966**
0.5	0.3101	0.1398	0.4305	0.3118	0.4141	0.3119	0.2510	**0.0985**
0.6	0.3096	0.1893	0.4272	0.3118	0.4244	0.3119	0.2525	**0.1210**
0.7	0.3091	0.2494	0.4012	0.3118	0.4321	0.3119	0.2489	**0.1016**
0.8	0.3089	0.2710	0.5008	0.3118	0.4417	0.3119	0.2481	**0.1165**
0.9	0.3090	0.3002	0.5228	0.3118	0.4503	0.3119	0.2492	**0.1287**
Mean	0.3016	0.1717	0.4110	0.3118	0.4066	0.3119	0.2325	**0.1026**

In Fig. 2.b, we present a qualitative comparison of the color-vector matrix obtained by the different methods. Small amounts of blood affect most of the methods. VAH works relatively well when there are small amounts of blood. The Bayesian methods SAR, TV, and SG that use a reference prior are not strongly affected by blood, but the estimated color-vector matrix remains close to the prior. The proposed method solves the limitations of the base BKSVD and is able to correctly estimate the stains in the image even with the presence of a large amount of blood.

4   Conclusions

In this work, we have extended the BKSVD method for stain separation to acknowledge the presence of blood on the images, obtaining a robust estimation of the stains and making it possible to detect blood. This work proposed, for the first time, the use of BCD techniques for blood detection, connecting the fields of color-preprocessing and artifact detection, which benefits both fields. On the one hand, the presence of blood hampers the estimation of the stains in the images and the correct separation of the stains. On the other hand, considering the appearance of the stains provides a framework that facilitates the detection of blood.

The proposed approach has been tested on synthetic and real images containing blood, showing promising performance. The method is able to accurately estimate the color-vector matrix and separate the stains when there is blood in the image, and correctly identify blood pixels.

We believe that our work remarks on an important issue of stain separation techniques that are the basis for the color-preprocessing of histopathological images. In future research, we plan to extend these results and assess how blood impacts the color normalization and augmentation of histological images for cancer classification.

References

1. Alsubaie, N., et al.: Stain deconvolution using statistical analysis of multi-resolution stain colour representation. PLoS ONE **12**, e0169875 (2017)
2. Anghel, A., et al.: A high-performance system for robust stain normalization of whole-slide images in histopathology. Front. Med. **6**, 193 (2019)
3. Bukenya, F., et al.: An automated method for segmentation and quantification of blood vessels in histology images. Microvas. Res. **128**, 103928 (2020)
4. Chen, Z., et al.: Histological quantitation of brain injury using whole slide imaging: a pilot validation study in mice. PLOS ONE **9**(3), 1–10 (2014)
5. Hidalgo-Gavira, N., Mateos, J., Vega, M., Molina, R., Katsaggelos, A.K.: Variational Bayesian blind color deconvolution of histopathological images. IEEE Trans. Image Process. **29**(1), 2026–2036 (2020)
6. Kanwal, N., Pérez-Bueno, F., Schmidt, A., Molina, R., Engan, K.: The devil is in the details: Whole slide image acquisition and processing for artifacts detection, color variation, and data augmentation. A review. IEEE Access, p. 1 (2022)
7. Kim, N.T., et al.: An original approach for quantification of blood vessels on the whole tumour section. Anal. Cell. Pathol. **25**(2), 63–75 (2003)
8. Macenko, M., et al.: A method for normalizing histology slides for quantitative analysis. In: International Symposium on Biomedical Imaging (ISBI), pp. 1107–1110 (2009)
9. Morales, S., Engan, K., Naranjo, V.: Artificial intelligence in computational pathology - challenges and future directions. Digit. Sig. Process **119**, 103196 (2021)
10. Mosaliganti, K., et al.: An imaging workflow for characterizing phenotypical change in large histological mouse model datasets. J. Biomed. Inform. **41**(6), 863–873 (2008)

11. Perry, T.S.: Andrew ng x-rays the AI hype. IEEE Spectrum (2021)
12. Pérez-Bueno, F., Serra, J., Vega, M., Mateos, J., Molina, R., Katsaggelos, A.K.: Bayesian K-SVD for H&E blind color deconvolution. Applications to stain normalization, data augmentation, and cancer classification. Comput. Med. Imaging Graph. **97**, 102048 (2022)
13. Pérez-Bueno, F., López-Pérez, M., Vega, M., Mateos, J., Naranjo, V., Molina, R., et al.: A TV-based image processing framework for blind color deconvolution and classification of histological images. Digit. Signal Process. **101**, 102727 (2020)
14. Pérez-Bueno, F., Vega, M., Sales, M.A., Aneiros-Fernández, J., Naranjo, V., Molina, R., Katsaggelos, A.K.: Blind color deconvolution, normalization, and classification of histological images using general super gaussian priors and Bayesian inference. Comput. Meth. Prog. Bio. **211**, 106453 (2021)
15. Ruifrok, A.C., Johnston, D.A.: Quantification of histochemical staining by color deconvolution. Anal. Quant. Cytol. Histol. **23**, 291–299 (2001)
16. Sertel, O., et al.: Texture classification using nonlinear color quantization: application to histopathological image analysis. In: 2008 IEEE International Conference on Acoustics, Speech, and Signal Processing (ICASSP), pp. 597–600 (2008)
17. Swiderska-Chadaj, Z., et al.: Automatic quantification of vessels in hemorrhoids whole slide images. In: International Conference on Computational Problems of Electrical Engineering (CPEE), pp. 1–4 (2016)
18. Tellez, D., et al.: Quantifying the effects of data augmentation and stain color normalization in convolutional neural networks for computational pathology. Med. Image Anal. **58**, 101544 (2019)
19. Vahadane, A., et al.: Structure-preserving color normalization and sparse stain separation for histological images. IEEE Trans. Med. Imag. **35**, 1962–1971 (2016)
20. Wetteland, R., Engan, K., et al.: A multiscale approach for whole-slide image segmentation of five tissue classes in urothelial carcinoma slides. Technol. Cancer Res. Treat. 19, 153303382094678 (2020)

Can Knowledge Transfer Techniques Compensate for the Limited Myocardial Infarction Data by Leveraging Hæmodynamics? An *in silico* Study

Riccardo Tenderini[1]([✉]) [iD], Federico Betti[1], Ortal Yona Senouf[1],
Olivier Muller[2], Simone Deparis[1] [iD], Annalisa Buffa[1] [iD], and Emmanuel Abbé[1,3]

[1] École Polytechnique Fédérale de Lausanne (EPFL), Institute of Mathematics,
1015 Lausanne, Switzerland
riccardo.tenderini@epfl.ch
[2] Department of Cardiology, Lausanne University Hospital (CHUV),
1011 Lausanne, Switzerland
[3] École Polytechnique Fédérale de Lausanne (EPFL), School of Computer
and Communication Sciences, 1015 Lausanne, Switzerland

Abstract. The goal of this work is to investigate the ability of transfer learning (TL) and multitask learning (MTL) algorithms to predict tasks related to myocardial infarction (MI) in a small–data regime, leveraging a larger dataset of hæmodynamic targets. The data are generated *in silico*, by solving steady–state Navier–Stokes equations in a patient–specific bifurcation geometry. Stenoses, whose location, shape, and dimension vary among the datapoints, are artificially incorporated in the geometry to replicate coronary artery disease conditions. The model input consists of a pair of greyscale images, obtained by postprocessing the velocity field resulting from the numerical simulations. The output is a synthetic MI risk index, designed as a function of various geometrical and hæmodynamic parameters, such as the diameter stenosis and the wall shear stress (WSS) at the plaque throat. Moreover, the Fractional Flow Reserve (FFR) at each outlet branch is computed. The *ResNet18* model trained on all the available MI labels is taken as reference. We consider two scenarios. In the first one, we assume that only a fraction of MI labels is available. For TL, models pretrained on FFR data — learned on the full dataset — reach accuracies comparable to the reference. In the second scenario, instead, we suppose also the number of known FFR labels to be small. We employ MTL algorithms in order to leverage domain–specific feature sharing, and significant accuracy gains with respect to the baseline single–task learning approach are observed. Ultimately, we conclude that exploiting representations learned from hæmodynamics–related tasks improves the predictive capability of the models.

Keywords: Myocardial infarction · Fractional flow reserve ·
Hæmodynamics · Knowledge transfer · Transfer learning · Multitask
learning · Coronary angiography

© The Author(s), under exclusive license to Springer Nature Switzerland AG 2023
J. M. Juarez et al. (Eds.): AIME 2023, LNAI 13897, pp. 218–228, 2023.
https://doi.org/10.1007/978-3-031-34344-5_26

1   Introduction

Among the reasons that hinder the application of deep learning (DL) based techniques in the medical context on a daily basis, one should not forget the limited amount of data at disposal. In fact, machine learning (ML) has established itself as a pillar of a new generation of scientific development for its ability to exploit data abundance in problems that are difficult to be solved through classic algorithmic paradigms (such as, for instance, image recognition or natural language processing). However, the effective applicability of DL algorithms in small–data regimes, i.e. in frameworks where the amount of accessible data is either limited or partial or subject to a high degree of inaccuracy (because of the high cost and/or complexity of data acquisition procedures), is still questionable. In particular, this is a quite common regime in various medical applications, among which the prediction of myocardial infarction (MI) culprit lesions [10]. Our claim is that data–driven models can compensate for the limited amount of available MI data by exploiting hæmodynamics. In this project, we provide a proof of concept to demonstrate our thesis with data generated *in silico*. We show that leveraging hæmodynamics awareness, by means of knowledge transfer techniques, can improve MI risk prediction.

2   Dataset Generation

As discussed in Sect. 1, the dataset has been generated *in silico*, by solving the steady–state Navier–Stokes equations in a vessel featuring sub–critical stenotic formations.

We considered the patient–specific geometry of a femoropopliteal bypass, segmented from CT scans [11], as reference. The stenoses, whose location, shape and dimension vary among the different datapoints, were artificially incorporated in the geometry by enforcing a deformation to the vessel wall. Figure 1 (left) shows the reference geometry, with the positions of the potential stenoses, and examples of deformed geometries. Let $\widehat{\Omega} \in \mathbb{R}^3$ be the reference domain, $\partial\widehat{\Omega}$ its boundary and $\widehat{\Gamma}_w \subset \partial\widehat{\Omega}$ the vessel wall. The boundary displacement $\phi : \mathbb{R}^3 \to \mathbb{R}^3$ is defined as

$$\phi(\boldsymbol{x}) := -A\,\widehat{\phi}\left(\frac{||\boldsymbol{x} - \boldsymbol{c}||_*}{r}\right)\boldsymbol{n}(\boldsymbol{c}) \quad \text{with} \quad \widehat{\phi}(y) = \exp\left(-\frac{1}{1 - y^2}\right)\mathbb{1}_{|y|\leq 1}\,, \quad (1)$$

where $\boldsymbol{c} \in \widehat{\Gamma}_w$ is the stenosis center, $\boldsymbol{n}(\boldsymbol{c})$ is the outward unit normal vector to $\widehat{\Gamma}_w$ at $\boldsymbol{c}$, $A, r \in \mathbb{R}^+$ are, respectively, the stenosis depth (also called diameter stenosis) and length. The norm function $|| \cdot ||_* : \mathbb{R}^3 \to \mathbb{R}$ is suitably defined in order to obtain stenoses that are elongated in the main direction of the flow.

Even if hæmorheology indicates that blood is a non–Newtonian fluid, the latter can be approximated as a Newtonian one if the vessels where it flows are sufficiently large. Under this assumption, the blood flow is governed by the incompressible Navier–Stokes equations. Let $\Omega \subset \mathbb{R}^3$ be the computational domain (obtained upon deformation of the reference one $\widehat{\Omega}$) and $\partial\Omega$ its boundary. Let

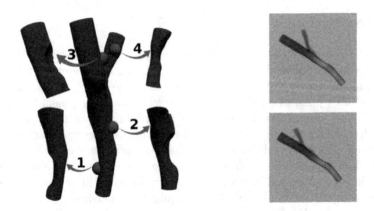

Fig. 1. (left) Reference geometry (taken from [11]) with locations of the four potential stenoses and stenoses representation. (right) Example of angiography–like snapshot.

us introduce the following partition: $\partial\Omega = \Gamma_w \cup \Gamma_{in} \cup \Gamma_{out}^1 \cup \Gamma_{out}^2$; Γ_w denotes the lateral wall, Γ_{in} the inlet, $\{\Gamma_{out}^i\}_{i=1}^2$ the outlets. The following non–linear system of partial differential equations (PDEs) is solved:

$$\begin{cases} \rho_f \left(\boldsymbol{u} \cdot \nabla\right) \boldsymbol{u} - \nabla \cdot \boldsymbol{\sigma}_f(\boldsymbol{u}, p) = 0 & \text{in } \Omega, \\ \nabla \cdot \boldsymbol{u} = 0 & \text{in } \Omega. \end{cases} \qquad (2)$$

$\boldsymbol{u} : \Omega \rightarrow \mathbb{R}^3$ and $p : \Omega \rightarrow \mathbb{R}$ denote, respectively, the fluid velocity and pressure; $\boldsymbol{\sigma}_f(\boldsymbol{u}, p) := -p\boldsymbol{I} + 2\mu_f \left((\nabla\boldsymbol{u} + (\nabla\boldsymbol{u})^T)/2\right)$ is the Cauchy stress tensor. The constants $\rho_f = 1.06 \; g \cdot cm^{-3}$ and $\mu_f = 3.5 \cdot 10^{-3} \; g \cdot cm^{-1} \cdot s^{-1}$ are, respectively, the blood density and viscosity. The first equation (momentum conservation equation) represents the generalization of Newton's second law of motion to continuums, while the second equation (continuity equation) enforces the incompressibility constraint. Concerning boundary conditions, we impose free flow at the outlets, zero velocity (no slip) on the lateral wall, and a prescribed flow rate $Q \in \mathbb{R}^+$ at the inlet. It is worth remarking that, to reduce the computational cost of the simulations, we decided to neglect temporal dynamics and to solve the Navier–Stokes equations in a steady–state formulation. For the numerical discretization of Eq. (2), we employed the finite element (FE) method with P1–P1 Lagrangian basis functions. The well–posedness of the problem has been guaranteed by means of SUPG stabilization [3]. The Newton's method has been used to handle the non–linearity, stemming from the convective term $(\boldsymbol{u} \cdot \nabla)\boldsymbol{u}$ in the momentum conservation equation. All simulations have been performed using *LifeV*, a C++ FE library with support to high–performance computing [2].

The dataset has been generated from 8'100 numerical solutions to Eq. (2), obtained for different stenoses locations (see Fig. 1), shapes and sizes and for different values of Q. An equispaced deterministic stratified sampling strategy has been chosen, so that a good coverage of the parametric space could be guaranteed. In particular, we considered 4 possible stenosis locations, 9 values for the inflow rate Q (equispaced in $[0.7, 1.5]$) and 15 values for both the relative (with respect to the local vessel diameter) stenosis depth A (equispaced in $[0.1, 0.5]$) and length r (equispaced in $[0.1, 0.3]$). Upon a postprocessing phase, the following data are generated and available for each simulation:

- The values of Q, A, r and c.
- 5 different couples of 2D greyscale images (200×200 pixels), displaying the velocity field; as a result, the dataset is made of 40'500 datapoints. Each couple is characterized by a different viewpoint (selected at random) and the two images are obtained upon a rotation of roughly $30°$ around the z–axis. Moreover, Poisson random noise has been added to increase the variability of the dataset. We remark that there may be views from which the stenoses are completely hidden.
- The value of the fractional flow reserve (FFR) at the two outlet branches. We computed the FFR as $FFR = Q_s/Q_h$, where Q_s is the flow rate (measured after the stenosis) in presence of the lesion and Q_h is the one of a healthy patient. The latter could be computed by running a simulation on the original reference geometry. We remark that this definition of the FFR differs from the one commonly used in clinical practice, which is $FFR = P_d/P_a$, being P_d the pressure distal to the lesion and P_a the pressure proximal to the lesion (often approximated with the aortic one) [1,12,16]. Nevertheless, the latter is nothing but an approximation of the former, under the assumption of neglecting the micro–vasculature pressure and of considering the micro–vasculature resistance to flow as independent from the presence of the lesion. To highlight this aspect, from now on we will use the notation Q–FFR.
- The sample average of the wall shear stress (WSS) euclidean norm at the mesh nodes in the plaque throat region, opposite to the stenosis. We denote this value by $\mathcal{A}(\|WSS\|_2)$.
- A synthetic MI risk index, designed as follows:

$$MI := \tanh\left(\sqrt{\frac{R^2}{2G}}\exp(A)\right) \qquad \in (0,1), \qquad (3)$$

where $R \in [0,1]$ is a risk factor related to the location of the lesion and decreasing from the inlet to the outlets. The term $\sqrt{(2G)^{-1}}$ — with $G := \left(1 + \frac{\mathcal{A}(\|WSS\|_2)}{\mathcal{A}(\|WSS_h\|_2)}\right)^{-1}$ — is a factor common to many mathematical models of atherosclerotic plaque growth [5,14,17]. Here WSS_h denotes the WSS of a healthy patient.

3   Deep Learning Models for MI Risk Prediction

Our goal is to demonstrate, in a controlled *in silico* environment, the benefit of exploiting hæmodynamics while training DL models for the MI risk prediction task. To this end, we consider two different scenarios. First, we assume that only a fraction of MI labels is available, whereas all the hæmodynamics–related

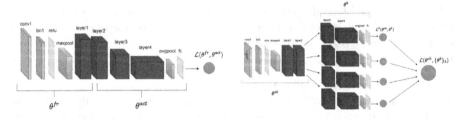

Fig. 2. Schematic diagrams of the TL (left) and MTL (right) algorithms applied to the *ResNet18* architecture.

targets are at disposal. In the second scenario, instead, we suppose that also the number of known hæmodynamics targets is small, thus working in a proper small–data regime. Two different knowledge transfer techniques, namely transfer learning (TL) and multitask learning (MTL), are employed in order to improve the models' predictive capability. For all the experiments, we chose the *ResNet18* architecture [6]. As a baseline, the model is trained on all the available MI labels.

3.1 Scenario 1 – Transfer Learning

Since many "auxiliary" data are available, we take advantage of representational knowledge transfer and, in particular, of the transfer learning (TL) paradigm [13]. The basic idea behind TL is that the predictive power of a ML model on a certain target domain $\mathcal{D}_T$ can be enhanced by transforming and transferring features learned by the same model on a source domain $\mathcal{D}_S$, under the assumption that $\mathcal{D}_T$ and $\mathcal{D}_S$ are somehow related. In particular, we focus on homogeneous TL, where the only difference between the source and the target domain lies on the label space (i.e. the output), whereas the feature space (i.e. the input) is the same. In practice, we first train the model to predict either the diameter stenosis A or the Q–FFR on the entire dataset. Incidentally, it should be mentioned that Q–FFR is not explicitly present in the expression of the MI risk (see Eq. (3)). The learned weights are then used to initialize a second model, whose task is to predict MI labels on a much smaller dataset. In order to boost and facilitate the training process, we also "freeze" some initial layers, i.e. we set a very low learning rate to their parameters. In this way, the number of parameters to be fine–tuned upon inductive bias initialization is smaller, and convergence properties can be improved. The objective function in both source and target tasks is chosen to be the Mean Squared Error (MSE) with $L2$ regularization. Optimization is performed by the *Adam* scheme [7]. A schematic representation of the TL algorithm applied to the *ResNet18* model is shown in Fig. 2 (left).

   All numerical tests have been conducted considering a 70%–10%–20% splitting of the dataset into training, validation and testing. Hence, the (maximal) number of training datapoints is 28'350, whereas 8'100 data are reserved for testing. We highlight that all 5 couples of images generated from the same numerical simulation are set to belong to the same sub–dataset. The initial learning rates have been set to $\alpha_0^{act} = 5 \cdot 10^{-4}$ and $\alpha_0^{fr} = 5 \cdot 10^{-6}$ for the active and frozen

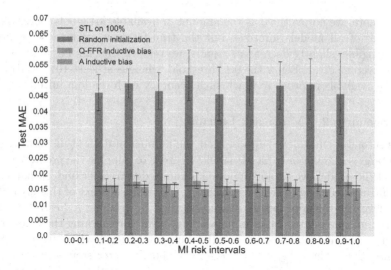

Fig. 3. Test MAE distribution on MI risk targets for TL models using Q–FFR (orange) and diameter stenosis A (green) as inductive bias. Single task learning (STL) on the full dataset (red line) is taken as reference, while STL on the available MI labels (blue) serves as baseline. Black vertical bars represent 95% confidence intervals. The tests have been conducted supposing 20% of the MI labels to be known. (Color figure online)

Table 1. 95% confidence intervals of test RMSE on MI labels for different percentages of the latter, considering random initialization or Q–FFR and A as inductive biases.

MI fraction	Random	Q–FFR bias	A bias
1%	$(2.2 \pm 0.2)\cdot 10^{-1}$	$(9.1 \pm 1.1)\cdot 10^{-2}$	$(6.6 \pm 1.0)\cdot 10^{-2}$
5%	$(1.4 \pm 0.3)\cdot 10^{-1}$	$(4.9 \pm 0.5)\cdot 10^{-2}$	$(2.9 \pm 0.3)\cdot 10^{-2}$
10%	$(7.7 \pm 1.2)\cdot 10^{-2}$	$(3.1 \pm 0.3)\cdot 10^{-2}$	$(2.2 \pm 0.2)\cdot 10^{-2}$
20%	$(5.1 \pm 1.0)\cdot 10^{-2}$	$(2.1 \pm 0.2)\cdot 10^{-2}$	$(1.8 \pm 0.2)\cdot 10^{-2}$

layers, respectively, and an exponential decay law with rate 0.9 every 5 epochs has been chosen. All models have been trained for 100 epochs. Figure 3 shows the test MAE distribution on MI labels if 20% of the latter are available, considering Q–FFR or A as inductive biases. The benefits of TL are well evident. In particular, leveraging knowledge of the Q–FFR on the entire dataset allows to drastically reduce the error and to attain performances comparable to the ones of the reference model. Moreover, it is worth remarking Q–FFR and A are somehow equally efficient as inductive biases, even though the former does not appear in the synthetic MI risk formula of Eq. (3). Table 1 reports instead the 95% confidence intervals of the test root mean squared error (RMSE) on MI labels for different percentages of the latter, considering random initialization or Q–FFR and A as inductive biases. In general, it is noteworthy how representations learned from the hæmodynamic target improve the model performances.

Furthermore, we notice that, as the amount of available MI labels increases, the accuracies of all models improve, but the transfer gain, despite being always non–negligible, slightly decreases. Finally, we remark that all TL–based models feature average errors below the 0.1 threshold, whereas at least 10% of MI labels must be available in order to reach such accuracy with random initialization.

3.2 Scenario 2 – Multitask Learning

In the second scenario, we suppose that the number of data at disposal is small for all targets and not only for MI labels. In order to improve the predictive capability of the model, we exploit functional knowledge transfer and, in particular, the multitask learning (MTL) paradigm [4,15]. In the context of this project, we try to leverage domain–specific feature sharing by training a *ResNet18* model to learn multiple related tasks in parallel. It is worth remarking that model performances are always assessed on MI labels predictions, which we refer to as the main task. The other targets (Q–FFR and diameter stenosis) are, in this context, auxiliary tasks, and their role is thus to help the model learn better (with respect to the main task) shared representations of the input data.

A schematic representation of the MTL algorithm applied to the *ResNet18* model can be found in Fig. 2 (right). Let K be the number of tasks to be learned in parallel. Let us partition the trainable parameters (at iteration t of the network optimization process) as follows: $\Theta_t = [\Theta_t^{sh}, \Theta_t^1, \cdots, \Theta_t^K]$. The parameters Θ_t^{sh} are the so–called shared parameters, since belong to the initial part of the model, common to all the tasks. The parameters $\{\Theta_t^k\}_{k=1}^K$ are instead task–specific, since they are related to the network layers that are specific to a single task. The choice of common and task–specific layers is a key hyperparameter in MTL models. In this work, we decided to split the *ResNet18* network as displayed in Fig. 2 (right). Let now $\mathcal{L}^k(\Theta_t^{sh}, \Theta_t^k)$ ($k \in \{1, \ldots, K\}$) be the loss function associated to the k–th task. The overall loss function is defined as the weighted average of the single–task losses, i.e.

$$\mathcal{L}(\Theta_t^{sh}, \{\Theta_t^k\}_{k=1}^K) = \sum_{k=1}^K w_t^k \mathcal{L}^k(\Theta_t^{sh}, \Theta_t^k) \ .$$

Ultimately, different MTL algorithms stem from different choices of the weighting parameters $\{w_t^k\}_{k=1}^K \in \mathbb{R}^+$, that are typically normalized so that $\sum_k w_t^k = K$. In this work, we considered two different weighting strategies:

- **Dynamic Weight Average (DWA)** [9]: the weights update writes as:

$$w_t^k = \frac{K \exp(\lambda_k(t-1)/T)}{\sum_{k'} \exp(\lambda_{k'}(t-1)/T)} \quad \text{with} \quad \lambda_k(t-1) = \frac{\mathcal{L}^k(\Theta_{t-1}^{sh}, \Theta_{t-1}^k)}{\mathcal{L}^k(\Theta_{t-2}^{sh}, \Theta_{t-2}^k)} \ ,$$

where $T \in \mathbb{R}^+$ controls the softness of task weighting. In a nutshell, DWA tries to balance model performances across the different tasks, by looking at the rate of change of single–task losses. Hence, it increases the weights of the tasks on which the model performed poorly and, conversely, it decreases the weights of the tasks on which the model improved the most.

- **Online Learning for Auxiliary losses (OL–AUX)** [8]: differently from DWA, this approach has been designed assuming the presence of a main task — on which model accuracy is assessed — and several auxiliary tasks. Hence, the goal is to find weights for the auxiliary tasks such that main task loss decreases the fastest. Let $\mathcal{L}_{main}$ be the main task loss and $\{\mathcal{L}_{aux}^k\}_{k=1}^{K-1}$ be the auxiliary task losses. The weights update strategy writes then as follows:

$$w_t^k = w_{t-1}^k - \frac{\alpha_t \beta}{N} \sum_{j=0}^{N-1} \left(\nabla_{\Theta_{t-j}^{sh}} \mathcal{L}_{main}(\Theta_{t-j}^{sh}, \Theta_{t-j}^k) \right)^T \left(\nabla_{\Theta_{t-j}^{sh}} \mathcal{L}_{aux}^k(\Theta_{t-j}^{sh}, \Theta_{t-j}^k) \right)$$

where $N \in \mathbb{N}$ is the horizon and $\beta \in \mathbb{R}^+$ the learning rate. In our tests, we fixed $\beta = 0.5$ and we considered different values of N; we denote the N-step OL–AUX method by OL–AUX–N.

All numerical tests have been conducted under the same conditions of Scenario 1. Figure 4 shows the test MAE distribution on MI labels for the DWA, OL–AUX–1, and OL–AUX–10 weighting strategies, considering Q–FFR as auxiliary task. Moreover, the errors obtained with the baseline single–task learning (STL) model on 20% of the MI labels and with the reference STL model on the full dataset are reported. The benefits of domain–specific feature sharing provided by MTL are evident for all the considered weighting strategies. Indeed, for all MI labels intervals, the performances are significantly improved with respect to the baseline. Furthermore, OL–AUX outperforms DWA and reaches a level of accuracy that is, for most of the intervals, comparable to the one of the reference STL model, even though being trained only on one fifth of the data. In fact, the OL–AUX method perfectly fits in our framework, since we can identify a main task (MI risk) — through which model performances are assessed — and several auxiliary tasks (Q–FFR, A). Table 2 reports the 95% confidence intervals for the test RMSE on MI labels for different percentages of the latter and different weighting strategies. If only 1% of the data is available (i.e. 285 couples of images, coming from 57 different simulations), the accuracy of all models is low and MTL does not lead to any improvement. However, if a sufficient amount of data is at disposal, then substantially better accuracies are attained by exploiting MTL, and errors are roughly halved compared to STL. Among the considered weighting strategies, the best performances are realized by OL–AUX–10.

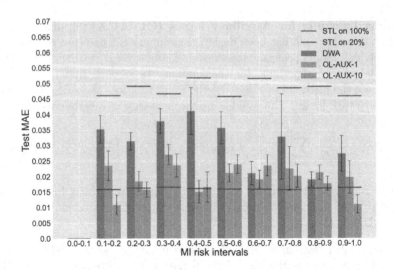

Fig. 4. Test MAE distribution on MI risk targets for different MTL weighting strategies using Q–FFR targets as auxiliary task. Single task learning (STL) on the full dataset (red line) is taken as reference, while STL on the available MI labels (blue line) serves as baseline. The tests have been conducted supposing 20% of the labels to be known. (Color figure online)

Table 2. 95% confidence intervals of the test RMSE on MI labels for different percentages of the latter, considering STL, DWA, and OL–AUX algorithms.

MI fraction	STL	DWA	OL–AUX	
			N=1	N=10
1%	$(2.2 \pm 0.2) \cdot 10^{-1}$	$(2.6 \pm 0.3) \cdot 10^{-1}$	$(2.2 \pm 0.2) \cdot 10^{-1}$	$(2.4 \pm 0.3) \cdot 10^{-1}$
5%	$(1.4 \pm 0.3) \cdot 10^{-1}$	$(1.3 \pm 0.2) \cdot 10^{-1}$	$(1.1 \pm 0.1) \cdot 10^{-1}$	$(8.9 \pm 1.0) \cdot 10^{-2}$
10%	$(7.7 \pm 1.2) \cdot 10^{-2}$	$(7.0 \pm 1.2) \cdot 10^{-2}$	$(5.7 \pm 1.3) \cdot 10^{-2}$	$(4.9 \pm 0.8) \cdot 10^{-2}$
20%	$(5.1 \pm 1.0) \cdot 10^{-2}$	$(3.1 \pm 0.5) \cdot 10^{-2}$	$(2.4 \pm 0.4) \cdot 10^{-2}$	$(2.2 \pm 0.3) \cdot 10^{-2}$

4   Conclusions

In this work, we developed DL models for MI risk prediction from angiography–like images in a small–data regime. Working on *in silico* data, generated by solving steady–state Navier–Stokes equations in geometries featuring artificial sub–critical stenotic formations, we demonstrated that hæmodynamic knowledge (Q–FFR in our case) can be leveraged in order to improve the predictive capability of the models. In particular, if many hæmodynamic targets are available, TL can be exploited and accuracy levels comparable to a reference STL model trained on all MI labels are attained. Conversely, if also the amount of hæmodynamics targets is scarce, then MTL algorithms can be employed and, for clever choices of the weighting strategy of the losses, significant performance gains can be realized. Further developments involve a more sophisticated dataset construction, a better design of the MI risk function, and a more thorough analysis of the models' sensitivity with respect to their main hyperparameters.

Acknowledgements. We thank the *Center for Intelligent Systems (CIS)* at EPFL for the support. RT, SD have received funding from the *Swiss National Science Foundation (SNSF)*, grant agreement No 200021 197021.

References

1. Achenbach, S., et al.: Performing and interpreting fractional flow reserve measurements in clinical practice: an expert consensus document. Intervent. Cardiol. Rev. **12**(2), 97 (2017)
2. Bertagna, L., Deparis, S., Formaggia, L., Forti, D., Veneziani, A.: The LifeV library: engineering mathematics beyond the proof of concept. arXiv preprint arXiv:1710.06596 (2017)
3. Brooks, A.N., Hughes, T.J.: Streamline upwind Petrov-Galerkin formulations for convection dominated flows with particular emphasis on the incompressible Navier-Stokes equations. Comput. Methods Appl. Mech. Eng. **32**(1–3), 199–259 (1982)
4. Caruana, R.: Multitask learning. Mach. Learn. **28**(1), 41–75 (1997)
5. Frei, S., Heinlein, A., Richter, T.: On temporal homogenization in the numerical simulation of atherosclerotic plaque growth. PAMM **21**(1), e202100055 (2021)
6. He, K., Zhang, X., Ren, S., Sun, J.: Deep residual learning for image recognition. In: Proceedings of the IEEE Conference on Computer Vision and Pattern Recognition, pp. 770–778 (2016)
7. Kingma, D.P., Ba, J.: Adam: a method for stochastic optimization. arXiv preprint arXiv:1412.6980 (2014)
8. Lin, X., Baweja, H., Kantor, G., Held, D.: Adaptive auxiliary task weighting for reinforcement learning. Advances in Neural Information Processing Systems (2019)
9. Liu, S., Johns, E., Davison, A.J.: End-to-end multi-task learning with attention. In: Proceedings of the IEEE/Conference on Computer Vision and Pattern Recognition (2019)
10. Mahendiran, T., et al.: Deep learning-based prediction of future myocardial infarction using invasive coronary angiography: a feasibility study. Open Heart **10**(1), e002237 (2023)
11. Marchandise, E., Crosetto, P., Geuzaine, C., Remacle, J.F., Sauvage, E.: Quality open source mesh generation for cardiovascular flow simulations. In: Ambrosi, D., Quarteroni, A., Rozza, G. (eds.) Modeling of Physiological Flows. MS&A – Modeling, Simulation and Applications, vol 5. Springer, Milano (2012). https://doi.org/10.1007/978-88-470-1935-5_13
12. Pagnoni, M., et al.: Future culprit detection based on angiography-derived FFR. Catheter. Cardiovasc. Interv. **98**(3), E388–E394 (2021)
13. Pan, S.J., Yang, Q.: A survey on transfer learning. IEEE Transactions on Knowledge and Data Engineering (2009)
14. Rodriguez, E.K., Hoger, A., McCulloch, A.D.: Stress-dependent finite growth in soft elastic tissues. J. Biomech. **27**(4), 455–467 (1994)
15. Ruder, S.: An overview of multi-task learning in deep neural networks. arXiv preprint arXiv:1706.05098 (2017)

16. Tu, S., et al.: Diagnostic accuracy of fast computational approaches to derive fractional flow reserve from diagnostic coronary angiography: the international multicenter FAVOR pilot study. Cardiovasc. Interventions **9**(19), 2024–2035 (2016)
17. Yang, Y., Jäger, W., Neuss-Radu, M., Richter, T.: Mathematical modeling and simulation of the evolution of plaques in blood vessels. J. Math. Biol. **72**(4), 973–996 (2016)

COVID-19 Diagnosis in 3D Chest CT Scans with Attention-Based Models

Kathrin Hartmann and Enrique Hortal[(⊠)] [iD]

Department of Advanced Computing Sciences, Maastricht University, Maastricht,
The Netherlands
kathrin.hartmann@student.maastrichtuniversity.nl,
enrique.hortal@maastrichtuniversity.nl

Abstract. The three-dimensional information in CT scans reveals notorious findings in the medical context, also for detecting symptoms of COVID-19 in chest CT scans. However, due to the lack of availability of large-scale datasets in 3D, the use of attention-based models in this field is proven to be difficult. With transfer learning, this work tackles this problem, investigating the performance of a pre-trained TimeSformer model, which was originally developed for video classification, on COVID-19 classification of three-dimensional chest CT scans. The attention-based model outperforms a DenseNet baseline. Furthermore, we propose three new attention schemes for TimeSformer improving the accuracy of the model by 1.5% and reducing runtime by almost 25% compared to the original attention scheme.

Keywords: Vision Transformer · Medical imaging · Attention Schemes · COVID-19 · 3D CT scan

1 Introduction

One recent research field that has rapidly evolved due to the pandemic is the identification of COVID-19 symptoms in lung images. Previous work includes approaches for the classification, detection and segmentation of COVID-19 images among others [14]. COVID-19 classification with deep learning models leads to especially good results when the models are trained on three-dimensional CT scans which are likely to reveal most information about the disease as the symptoms of COVID-19 might be present at different depth levels in the lung [14]. This leads us to hypothesize that, incorporating depth dependencies can help with the task at hand.

At the same time, Vision transformers [3], which are an adaption of the classical transformer architecture for images, are gaining popularity in computer vision. Similarly, the use of attention-based models in the medical context has grown significantly in the last couple of years [14]. With enough data available, these models have been able to outperform classical Convolutional Neural Networks in several tasks in the medical field [14]. However, due to the lack of availability of large-scale datasets, especially three-dimensional ones, research in

© The Author(s), under exclusive license to Springer Nature Switzerland AG 2023
J. M. Juarez et al. (Eds.): AIME 2023, LNAI 13897, pp. 229–238, 2023.
https://doi.org/10.1007/978-3-031-34344-5_27

COVID-19 classification has mostly been focusing on 2D CT scans using CNN-based models, as attention-based models need a large amount of data to be trained on [3].

To investigate this research gap of attention-based model classification on 3D images in the medical field, we aim to apply attention-based models on 3D chest CT images to identify lungs affected by COVID-19. With transfer learning (using models pre-trained on images from a different domain), we want to overcome the challenges posed when using attention-based models on small datasets. We aim to achieve more accurate results compared to traditional Convolutional Neural Network approaches by embedding information globally across the overall image using attention schemes. Apart from that, we want to investigate how our different, newly developed attention schemes perform compared to previously developed attention schemes both performance and time-wise (reducing the computational power required). To the best of our knowledge, 3D attention-based models have not been applied to the proposed task. The only 3D approaches in the literature are using CNN or U-Net approaches such as [6] and those using a 3D attention-based model are not intended for medical image classification but 3D segmentation such as [15].

2   Related Works

2.1   COVID-19 Datasets

A few datasets containing 3D chest CT scan images like the ones presented in [10] and [11], MIA-COV19 [9], COV19 CT DB [13] and CC-CCII [17] have been collected. However, most of them are not publicly available. On its part, COV19 CT DB contains 3D CT scans of lungs infected with COVID-19 from around 1000 patients but no other healthy or scans from lugs with other medical conditions are considered. Finally, the CC-CCII dataset contains three classes, namely Common Pneumonia, COVID-19 and Normal lung scans. This database is open access and was pre-processed and restructured in [5]. In this work, we are using and adjusting this pre-processed CC-CCII dataset to conduct our research.

2.2   Convolutional Neural Networks for COVID-19 Detection

Convolutional Neural Networks (CNNs) are very popular for image classification and are essential for computer vision in medical imaging. Naturally, CNNs have been widely used for COVID-19 classification over the last three years, also using chest CT scans. Several works state that ResNet [4] is the best-performing network when comparing performance to other nets [1, 10, 12]. Other research [5], however, shows that DenseNets are able to outperform ResNets when using 3D convolutions. In this respect, a DenseNet121 achieves the highest scores on the dataset CC-CCII [17]. In view of the above, the DenseNet121 model is utilized as the baseline model in this work.

2.3   Attention-Based Models for COVID-19 Detection

A few approaches have been developed using 3D CT scans. Two of them are the works presented in [7] and [18] which use the Swin transformer in their network to distinguish 3D CT scans based on the classes "COVID-19" and "healthy". Both models are trained on the MIA-COV19D dataset. The work in [18] uses a U-Net for lung segmentation and then classifies the segmented lung scans with a Swin transformer network. Authors in [7] propose two networks: The first one determines the importance of single slices in a scan based on symptoms shown in it via the Wilcoxon signed-rank test [16] on features extracted with a Swin transformer block. The second network is hybrid, consisting of a CNN that extracts features of each CT scan slice and two different Swin transformers: one captures within slice dependencies and the other is used to identify between slice dependencies. In this way, the full context of the 3D scan is captured. Both of the proposed models outperform a DenseNet201 that was trained for comparison. In our work, on the contrary, we are not using a Swin transformer but an original vision transformer model that was adjusted for three-dimensional, originally video, input. We are leveraging the 3D information in the CT scans by applying several attention schemes that process the CT scans in different ways to explore both spatial and temporal dependencies.

3   Methodology

We propose the first application of a pure vision transformer-based model for COVID-19 CT scan classification that is using the 3D information in the CT scans. This is done by applying a pre-trained TimeSformer model [2] on a pre-processed dataset.

3.1   Dataset

The CC-CCII dataset [5, 17] is a publicly available 3D chest CT scan dataset that we modify for our research purpose with appropriate corrections. The dataset contains three different classes: lungs diagnosed with Common Pneumonia (CP), lungs diagnosed with Novel Corona Virus (NCP), and lungs without any condition (Normal). In this study, only the first two classes, namely CP and NCP, are considered. Slices containing mistakes in the order of the lung slices were discarded. Apart from that, for consistency and to reduce the required computational power, we sample the number of slices in a scan to 32 (scans with less than 32 slices are also discarded), crop, and resize the lung slices. Our final dataset contains a total of 1874 scans of width $\times$ height $\times$ number slices = $160 \times 128 \times 32$, 824 of them in class CP and 1047 in class NCP. A part of a randomly selected sample of a lung scan from the final dataset can be seen in Fig. 1.

3.2   TimeSformer

To be able to efficiently train a model that can distinguish between diseases of 3D lung scans, we examine a domain outside of 3D medical imaging that also requires

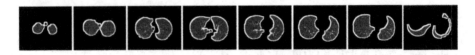

Fig. 1. The extract of a random lung scan from the utilized dataset. The scans, from left to right, correspond with lung slices from top to bottom.

3D inputs: video classification. One video can be seen as 2D images (frames) stacked together, which also corresponds to our application of 3D CT scans. To build efficient vision transformer models for video classification, authors in [2] have developed TimeSformer, a model that takes 3D inputs (videos), divides the video frames into patches and feeds these patches to a transformer network that consists of n attention blocks. Within these blocks, embeddings and attention schemes are applied to efficiently classify the videos. Our modified version of their architecture can be seen in Fig. 2.

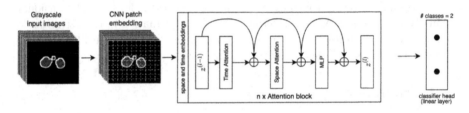

Fig. 2. Our TimeSformer setup. 3D grey-scale input images are embedded into patches and fed into a Transformer encoder. The encoder first applies spacial and time embeddings and then inputs these embeddings into n attention blocks. The architecture of the attention block depends on the attention scheme. Eventually, the outcome of the encoder is classified by a linear layer.

3.3   Original Attention Schemes

Authors in [2] compare the implementation of three different attention schemes, namely joint-space-time attention, space-only attention and divided-space-time attention. Joint-space-time-attention calculates the attention between all patches of all layers (see Fig. 3 (b)). This approach is, however, computationally intensive, as the attention between all existing patches is calculated. For this reason, other attention schemes were proposed. Space-only attention, on the contrary, calculates attention only between patches on the same layer (Fig. 3 (a)). This way of calculating attention is less computationally intensive than joint-space-time attention but does not consider dependencies across layers and therefore, ignores the "depth" information. The third attention scheme, divided-space time attention is developed to focus on both spacial and time information in the video. It calculates the attention within all patches at the same temporal position (across frames) and the attention with all patches in the current layer (spatial attention within the frame) (Fig. 3 (c)). Both of these attentions are calculated separately and then combined and fed into a multilayer perceptron. The paper also proposes temporal embedding in addition to spatial embedding to give the model more depth information.

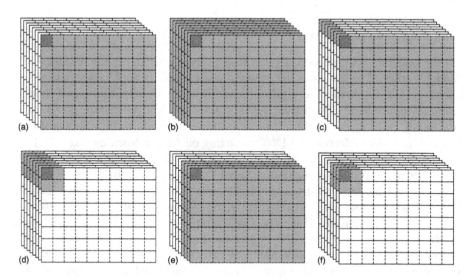

Fig. 3. The attention schemes presented in [2]: space-only attention (a), joint-space-time attention (b), divided-space-time attention (c); and the three newly proposed schemes: space-limited attention (d), time-limited attention (e) and space-and-time-limited attention (f). The patch under analysis is highlighted in orange while the patches considered in each attention scheme are represented in grey. (Color figure online)

3.4   New Attention Schemes

The attention schemes described above were developed for video classification. In our use case, we hypothesize that the parts affected by COVID-19 or Common Pneumonia may be spread in depth over a bigger area than a single patch. From the proposed schemes, divided-space-time attention only considers one patch in the time dimension while space-only attention does not consider depth information. As joint-space-time attention is highly computationally intensive, we want to evaluate other attention schemes able to 1) capture both spatial and time dependencies and 2) reduce the time computational power required. We propose three new attention calculation schemes for our use case: space-limited attention, time-limited attention and space-and-time-limited attention. Per each patch, the space-limited attention considers the total time dimension but only focuses on a subarea around the patch under analysis (see Fig. 3 (d)). To that end, non-overlapping squares of adjacent patches, for example, 4 × 4 patches or 2 × 2 patches, are utilized on each slice. Time-limited attention does the opposite: while considering the total space dimension, a limited number of adjacent slices in the time (depth in our case) direction are considered (Fig. 3 (e)). Finally, a combination of both attention schemes is proposed as space-and-time-limited attention. This attention scheme uses non-overlapping cubes of patches, smaller than the original width and height (Fig. 3 (f)).

3.5 Experimental Setup

As the baseline, a DenseNet121 [8] is trained from scratch on our dataset. This model is selected as it utilizes 3D convolutions to capture depth information and therefore, can be considered a fairer comparison to our attention-based models than 2D approaches. The DenseNet is trained with a batch size of 16 for a maximum of 50 epochs with a patience factor of 15 for early stopping and a learning rate decay factor of $0.1^{\frac{epoch}{20}}$, starting off from a learning rate of 0.1.

Due to the large amount of data necessary to train attention-based models, we utilize a pre-trained TimeSformer model made available by [2] that we fine-tune on our dataset. Pre-trained weights from training on the video dataset Kinetics-600 are used. It is worth mentioning that we modified the architecture to fit our dataset input. Thus, the CNN input layer from TimeSformer, with 3 (RGB) channels, is replaced by a 1-channel (grey-scale) input. Similarly, the output layer is modified to accommodate two classes, instead of the 600 classes on the Kinetics-600 dataset. Figure 4 shows our modified TimeSformer model. As an initial evaluation, this modified divided-space-time attention (dst) model is compared with the DenseNet model explained in Sect. 2.2.

Consecutively, the proposed space-limited (sl), time-limited (tl) and space-and-time-limited (stl) attention schemes are evaluated. For consistency, the implementation of these models is as similar as possible to the initial divided-space-time attention approach. The modified input and output layers of TimeSformer and the number of attention blocks are consistent across the three models. The sole implementation difference across the proposed models is the attention blocks. Depending on the attention scheme applied, the original dst attention blocks are replaced by the proposed ones. The same pre-trained weights from training on the video Kinetics-600 dataset, as in the previous experiment, are used. This means that the model was pre-trained with a different attention scheme (divided-space-time attention) and is now fine-tuned on one of the newly proposed ones. In total, 12 attention blocks are used for all experiments.

All the above-mentioned experiments are conducted using the dataset described in Sect. 3.1. The train-validation-test split is set fixed during pre-processing to avoid having scans of the same patient in different sets. We randomly split the patients in the dataset into training, validation and test data such that the ratio of scans in each set is 60-20-20. After pre-processing, we have a final input dataset with 1181 scans in the training dataset, 361 scans in the validation dataset and 332 scans in the test dataset. The classes are also balanced as much as possible, with 827 and 1047 CP and NCP instances respectively. All the experiments have been conducted in the same machine, using an NVIDIA® Tesla V100 32GB GPU.

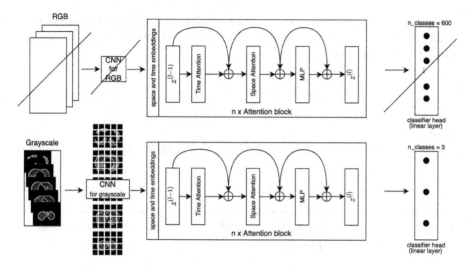

Fig. 4. Modification of TimeSformer model to make it suitable for our dataset. The 3-channel CNN input layer that divides the input images into patches is replaced by a 1-channel grey-scale input layer. The attention blocks remain unchanged, containing the calculation of time attention, space attention and a multilayer perceptron, connected with skip connections. The output layer is modified to classify two classes instead of the original 600.

3.6   Evaluation Metrics

The performance of the models is statistically analyzed by using the following evaluation metrics:

Accuracy (acc), calculated as the number of correctly classified instances divided by the total number of them:

$$acc = (\# \text{ instances correctly classified})/(\# \text{ total instances}) \qquad (1)$$

Precision (prec), calculated by dividing the number of true positive (TP) samples by the number of true positive and false positive (FP) samples:

$$prec = (TP)/(TP + FP) \qquad (2)$$

Recall (rec), calculated by dividing the number of true positive instances by the number of true positive and false negative (FN) ones:

$$rec = (TP)/(TP + FN) \qquad (3)$$

Additionally, we also calculate specificity. This metric assesses how the negative class performs and it is calculated as the number of true negatives (TN) instances divided by the number of TN and FP:

$$spec = (TN)/(TN + FP) \qquad (4)$$

Finally, the weighted average F1 score is calculated. This metric measures a combination of precision and recall for each class and weights it by the number of instances in the class as:

$$weighted\_avg\_f1 = \sum_{n=1}^{C} f1_{C_n} * W_{c_i} \tag{5}$$

where

$$W_{c_i} = (\# \text{ instances in class } c_i)/(\# \text{ total instances}) \tag{6}$$

$$f1 = 2 * (prec * rec)/(prec + rec) \tag{7}$$

Apart from these statistical metrics, the training runtime was also measured. This information is a good indicator of the computational power required to train each of the models proposed.

4   Results and Discussion

Table 1 shows the resulting metrics after running the proposed baseline, the 3D DenseNet121 (*3D DN*) and five different attention schemes on the dataset described in Sect. 3.1. We successfully fine-tuned the customized divided-space-time attention (*dst*) and space-only attention (*so*) models presented in [2] and the three newly proposed schemes, namely space-limited (*sl*), time-limited (*tl*) and space-and-time-limited attention (*stl*). The joint-space-time attention scheme from [2] could not be evaluated on this dataset due to computational power limitations. Additionally, it is worth mentioning that, also due to computational power restrictions, the attention schemes evaluated could only be run with a batch size of 16. The best results for all models were achieved when fine-tuning for 20 epochs. After exhaustive experimentation, the best-performing window size for space-limited attention was a window of 2×2 and the depth with the highest accuracy for time-limited attention was 8. Thus, these results are combined and a cube size of $2 \times 2 \times 8$ is used for training the model with space-and-time-limited attention.

Table 1. The results for the 3D DenseNet121 baseline and the fine-tuning TimeSformer models with five different attention schemes.

Attention type	acc	prec	recall	specificity	weighted avg f1	runtime (mins)
3D DN	0.777	0.818	0.818	0.713	0.777	**32**
TSf dst	0.798	0.818	0.862	0.698	0.796	75
TSf so	0.798	0.812	**0.872**	0.682	0.796	54
TSf sl	**0.813**	**0.844**	0.852	0.752	**0.813**	57
TSf tl	0.633	0.758	0.586	0.705	0.637	68
TSf stl	0.783	0.839	0.798	**0.760**	0.784	55

Among all the attention-based models evaluated, the proposed *TSf sl* (TimeSformer with space-limit attention) scheme outperforms the rest, achieving around 4% improvement in accuracy over the 3D DenseNet121 [8] baseline model (from 0.777 to 0.813) and a 1.5% improvement over the original schemes, namely *dst* and *so*. This model also surpasses the original models in precision (higher than 3% improvement), specificity (between 5.4 and 7%) and the weighted average F1 (1.7%). Additionally, this model is able to reduce the training runtime by almost 25% compared to the original attention scheme divided-space-time attention (from 75 to 57 min in our setup).

5   Conclusion

In this work, we proposed three newly developed attention schemes in addition to the attention schemes developed for TimeSformer [2]. These schemes are proposed with the aim of reducing the computationally intensive training process while maintaining or even improving the performance of our classification models. Our results indicate that our space-limited attention scheme yields better results compared to all other schemes and baseline for distinguishing between scans from lugs affected by COVID-19 and Common Pneumonia. This finding corroborates our hypothesis that capturing the time (depth in our case) dependencies play an important role in the detection of lung diseases.

However, as future work, it would be advisable for a more in-depth evaluation of the proposed attention schemes. These newly proposed attention schemes should be further investigated by validating more combinations of patch shapes, both in the spatial and temporal dimensions. Nevertheless, it is worth stressing that, even though the use of bigger patches could help identify the time and space dependencies more accurately, it will come at the expense of a higher computational power. Furthermore, with more computational power, it would be also possible to design more comparable experiments and train DenseNets and TimeSformer on the same batch size as the original works. Moreover, more fine-tuning of the models could be done to boost their performance. With more resources, it would also be possible to evaluate how the joint-space-time attention scheme performs compared to the proposed attention approaches. To conclude, another very interesting experiment we are planning to conduct is the evaluation of the proposed models on different 3D lung CT scan datasets and more classes to get further insights into how generally valid the results achieved in this work are. Finally, for the medical field, more research in the visualization of attention and the explanation of models for this application is of interest.

References

1. Ardakani, A.A., Kanafi, A.R., Acharya, U.R., Khadem, N., Mohammadi, A.: Application of deep learning technique to manage covid- 19 in routine clinical practice using CT images: results of 10 convolutional neural networks. Computers in biology and medicine 121 (2020)
2. Bertasius, G., Wang, H., Torresani, L.: Is space-time attention all you need for video understanding? In: ICML, vol. 2, p. 4 (2021)
3. Dosovitskiy, A., Beyer, L., Kolesnikov, A., et al.: An image is worth 16x16 words: transformers for image recognition at scale. arXiv preprint arXiv:2010.11929 (2020)
4. He, K., Zhang, X., Ren, S., Sun, J.: Deep residual learning for image recognition. In: Proceedings of the IEEE Conference on Computer Vision and Pattern Recognition, pp. 770–778 (2016)
5. He, X., et al.: Benchmarking deep learning models and automated model design for covid-19 detection with chest CT scans. MedRxiv (2021)
6. Hatamizadeh, A., Tang, Y., Nath, V., et al.: Unetr: transformers for 3D medical image segmentation. In: Proceedings of the IEEE/CVF Winter Conference on Applications of Computer Vision, pp. 574–584 (2022)
7. Hsu, C.-C., Chen, G.-L., Wu, M.-H.: Visual transformer with statistical test for covid-19 classification. arXiv preprint arXiv:2107.05334 (2021)
8. Huang, G., Liu, Z., Van Der Maaten, L., Weinberger, K.Q.: Densely connected convolutional networks. In: Proceedings of the IEEE Conference on Computer Vision and Pattern Recognition, pp. 4700–4708 (2017)
9. Kollias, D., Arsenos, A., Soukissian, L., Kollias, S.: Mia-cov19d: Covid-19 detection through 3-D chest CT image analysis. In: Proceedings of the IEEE/CVF International Conference on Computer Vision, pp. 537–544 (2021)
10. Li, L., Qin, L., Xu, Z., et al.: Using artificial intelligence to detect COVID-19 and community-acquired pneumonia based on pulmonary CT: evaluation of the diagnostic accuracy. Radiology **296**(2), 65–71 (2020)
11. Mishra, A.K., Das, S.K., Roy, P.: Bandyopadhyay, S.: Identifying covid19 from chest CT images: a deep convolutional neural networks based approach. J. Healthcare Eng. (2020)
12. Pham, T.D.: A comprehensive study on classification of COVID-19 on computed tomography with pretrained convolutional neural networks. Sci. Rep. **10**(1), 1–8 (2020)
13. Shakouri, S., et al.: Covid19-CT-dataset: an open-access chest CT image repository of 1000+ patients with confirmed covid-19 diagnosis. BMC Res Notes (2021)
14. Shamshad, F., et al.: Transformers in medical imaging: a survey. arXiv preprint arXiv:2201.09873 (2022)
15. Shin, Y., Eo, T., Rha, H., et al.: Digestive Organ Recognition in Video Capsule Endoscopy Based on Temporal Segmentation Network. In: Wang, L., Dou, Q., Fletcher, P.T., Speidel, S., Li, S. (eds.) MICCAI 2022. LNCS, vol. 13437, pp. 136–146. Springer, Cham (2022). https://doi.org/10.1007/978-3-031-16449-1_14
16. Woolson, R.F.: Wilcoxon signed-rank test. Wiley encyclopedia of clinical trials, 1–3 (2007)
17. Zhang, K., Liu, X., Shen, J., et al.: Clinically applicable AI system for accurate diagnosis, quantitative measurements, and prognosis of covid-19 pneumonia using computed tomography. Cell **181**(6), 1423–1433 (2020)
18. Zhang, L., Wen, Y.: A transformer-based framework for automatic covid19 diagnosis in chest CTs. In: Proceedings of the IEEE/CVF International Conference on Computer Vision, pp. 513–518 (2021)

Generalized Deep Learning-Based Proximal Gradient Descent for MR Reconstruction

Guanxiong Luo[1,2], Mengmeng Kuang[1], and Peng Cao[1(✉)]

[1] The University of Hong Kong, HKSAR, Pok Fu Lam, China
caopeng1@hku.hk
[2] University Medical Center Göttingen, Göttingen, Germany

Abstract. The data consistency for the physical forward model is crucial in inverse problems, especially in MR imaging reconstruction. The standard way is to unroll an iterative algorithm into a neural network with a forward model embedded. The forward model always changes in clinical practice, so the learning component's entanglement with the forward model makes the reconstruction hard to generalize. The deep learning-based proximal gradient descent was proposed and use a network as regularization term that is independent of the forward model, which makes it more generalizable for different MR acquisition settings. This one-time pre-trained regularization is applied to different MR acquisition settings and was compared to conventional ℓ_1 regularization showing ~3 dB improvement in the peak signal-to-noise ratio. We also demonstrated the flexibility of the proposed method in choosing different undersampling patterns.

Keywords: Magnetic Resonance Imaging · Image reconstruction · Deep Learning · Proximal gradient descent · Learned regularization term

1 Introduction

Before employing deep learning in accelerated MRI reconstruction, conventional methods for parallel MR imaging are based on the numerical pseudo-inversion of ill-posed MRI encoding matrix, which could be prone to reconstruction error at poor conditioning [4,6,7]. The encoding matrix comprises the k-space under-sampling scheme, coil sensitivities, Fourier transform. The traditional reconstruction involves some gradient descent methods for minimizing the cost function of the k-space fidelity and the regularization term [6,9]. There is a tradeoff between the image artifact level and the under-sampling rate as limited by the encoding capacity of coil sensitivities. Nevertheless, the parallel imaging technique robustly provides the acceleration at factor 2–4 [4,7]. The compressed sensing technique exploits the sparsity property of MR images in a specific transform domain, such as the wavelet domain, in combination with the incoherent under-sampling in k-space, which enables even larger acceleration factor.

With the fast growth of machine learning, the supervised learning have been applied to MRI reconstruction [1,8,11]. Those methods MRI encoding matrices were fully

---

P. Cao—Department of Diagnostic Radiology, LKS Faculty of Medicine, Pokfulam Road, HK, SAR, China.

J. M. Juarez et al. (Eds.): AIME 2023, LNAI 13897, pp. 239–244, 2023.
https://doi.org/10.1007/978-3-031-34344-5_28

included in the neural network models. These models were trained with predetermined encoding matrices and corresponding under-sampling artifacts. After training, imaging configurations, including the k-space under-sampling schemes and coil sensitivities, associated encoding matrices, must also be unchanged or changed only within pre-determined sampling patterns, during the validation and application, which could be cumbersome or to some extent impractical for the potential clinical use.

To tackle this design challenge, we unroll proximal gradient descent steps into a network and call it Proximator-Net. Inspired by [3], the proposed method was adapted from proximal gradient descent. This study's objective was to develop a flexible and practical deep learning-based MRI reconstruction method and implement and validate the proposed method in an experimental setting regarding changeable k-space under-sampling schemes.

2   Method - Proximal Network as a Regularization Term

The image reconstruction for MR k-space data acquired with different k-space trajectories can be formulated as an inverse problem for the corresponding forward model. We followed the MRI reconstruction problem formulation used in l_1-ESPIRiT [9]. Let a forward operator $\mathcal{A}$ to map the MR image $\mathbf{x}$ to the sampled k-space data $\mathbf{y}$. The operator $\mathcal{A}$ consists of a Fourier transform operator $\mathcal{F}$, coil sensitivity $\mathcal{S}$, and a k-space sampling mask $\mathcal{P}$, i.e., $\mathcal{A} = \mathcal{PFS}$ [9]. The well-known solution to the inverse problem can be formulated as an optimization problem with a regularization term [9] as

$$\bar{\mathbf{x}} = \arg \min_x \|\mathcal{A}\mathbf{x} - \mathbf{y}\|_2^2 + \lambda \phi(\mathbf{x}), \tag{1}$$

where $\phi(\mathbf{x})$ is a regularization term that introduces the prior knowledge, and λ the regularization parameter. The proximal operator for $\phi(\mathbf{x})$ in the proposed approach, $\mathbf{prox}_\phi(\mathbf{x})$, is defined as:

$$\hat{\mathbf{x}} = \mathbf{prox}_{\phi,1/\lambda}(\mathbf{v}) := \arg \min_{\mathbf{x}} \|\mathbf{x} - \mathbf{v}\|_2^2 + \lambda \phi(\mathbf{x}) \tag{2}$$

$$\hat{\mathbf{x}} \in \chi \sim P_{model} \approx P_{data} \tag{3}$$

where $\hat{\mathbf{x}}$ is the proximate value at $\mathbf{v}$, χ is the sub-space that contains MR images (including the ground truth), P_{data} is the probability distribution of the observed MR images, and P_{model} is learned the probability distribution. Taking the sparse constraint in compressed sensing as an analogy, $\phi(\mathbf{x})$ is a ℓ_1-norm of coefficients from the wavelet transform, the Equation (2) can be approximated by the shrinkage method (FISTA). Back to the definition of the proximal operator, we treat Equation (2) as another optimization problem solved iteratively by the proximal gradient descent method, which is expressed as follows

$$\mathbf{g}^{(t+1)} = \mathbf{x}^{(t)} + \alpha^{(t)}(\mathbf{v} - \mathbf{x}^{(t)}) \tag{4a}$$

$$\mathbf{x}^{(t+1)} = \mathbf{Net}_\phi(\mathbf{g}^{(t+1)}) \tag{4b}$$

then, we unroll above iterative steps as a Proximator-Net shown in Fig. 1.

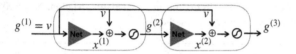

Fig. 1. The unrolled proximate gradient descent as neural network, the $\mathbf{Net}_\phi$ are the red triangles whose detailed structure is detailed in material.

At the iteration t, $\mathbf{x}^{(t+1)}$ is supposed be closer to P_{data} than $\mathbf{x}^{(t)}$. In order to achieve this, images corrupted by Gaussian noises are used as input. The corresponding noise-free images are used as labels. The ℓ_2-loss between outputs and labels is used. To solve Equation (3) and make the prediction close to the data distribution, the ℓ_2 regularization of the gradient of output with respect to input is used, as proposed in [2]. Therefore the final training loss is

$$\mathbb{E}\left[\left\|r_\theta(\tilde{\mathbf{x}}) - \mathbf{x}\right\|_2^2 + \sigma^2 \left\|\frac{\partial r_\theta(\mathbf{x})}{\partial \mathbf{x}}\right\|_2^2\right], \tag{5}$$

where $\tilde{\mathbf{x}}$ is the image perturbed by the Gaussian noise with σ and $r_\theta(\mathbf{x})$ is the neural network parameterized by θ. Proximator-Net is recast as a neural network-based proximal operator for the proposed regularization term, and the proposed approach for reconstruction is

$$\mathbf{m}^{(t)} = \mathbf{x}^{(t)} + \alpha^{(t)} \mathcal{A}^H (\mathbf{y} - \mathcal{A}\mathbf{x}^{(t)}) \tag{6a}$$

$$\mathbf{x}^{(t+1)} = (1 - \lambda) \cdot \mathbf{x}^{(t)} + \lambda \cdot r_\theta(\mathbf{m}^{(t)}) \tag{6b}$$

where λ is the parameter for the learned regularization term.

3   Experiments and Results

Dataset and Pre-processing: The dataset we used in this work is from Ref. [5]. Training images were reconstructed from 8 channels k-space data without undersampling. Then, these image data-sets after coil combination were scaled to a magnitude range of $[-1, 1]$ and resized to an image size of 256×256. In the end, 900 images were used for training, and 300 images were used for testing. We set the level of Gaussian noise with $\sigma = 0.03$, $\mu = 0$. The training was performed with Tensorflow on 4 GTX 2080Ti GPUs and took about 2 h for 500 epochs. Real and imaginary parts of all 2D images were separated into two channels when inputted into the neural network.

Table 1. Comparison of PSNRs (in dB, mean $\pm$ standard deviation, N = 300) and SSIMs(%) between ℓ_1-ESPIRiT and the proposed method.

Sampling rate	l_1-ESPIRiT		Ours	
	PSNR(dB)	SSIM(%)	PSNR(dB)	SSIM(%)
1D mask				
20%	31.63±5.34	68.28±14.48	35.12±3.41	91.34±5.32
30%	34.25±5.66	76.31±12.54	37.96±3.58	93.87±4.37
2D mask				
20%	34.73±5.84	71.48±16.06	37.90±2.86	93.55±4.09
30%	37.55±5.34	80.95±11.84	40.12±2.98	95.09±3.41

Reconstruction and Comparisons: The proposed method in Eq. 6 was implemented with Python. We performed l_1-ESPIRIT (regularization parameter=0.005) reconstruction with the BART toolbox [9, 10]. The coil sensitivity map was estimated from a 20×20 calibration region in the central k-space using ESPIRiT. To compare the images reconstructed using different methods, we calculated the peak signal-to-noise ratio (PSNR) and structural similarity index measure (SSIM) with respect to ground truth. Also, we validated the proposed method in different acquisition settings such as acceleration along one dimension, two dimensions, or radial acquisition. The metrics of different reconstructions were shown in Table 1. Generally, the result demonstrated that the proposed method could restore the high-quality MRI with both high acceleration factor (i.e., 20% samples) and high PSNR (i.e., >35 dB). Figure 2 shows the comparison between l_1-ESPIRIT and the proposed method.

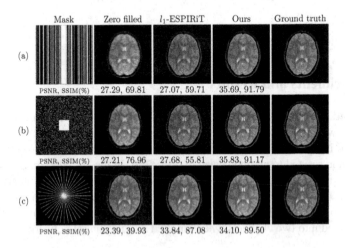

Fig. 2. Row (a) shows the proposed method can remove the aliasing artifacts along phase encoding direction in the case of using 30% k-space. Row (b) shows the proposed method can eliminate the burring fog on the image in the case of using 20% k-space. Row (c) shows the two methods' performances were close. With 40 radial spokes acquired in k-space.

Smoothing Effect of the Learned Regularization Term: The smoothing effect of the proposed regularization term was shown in Fig. 3. The parameter λ was tunable in Equation (6b). When λ was set to be zero, it was an iterative parallel imaging (i.e., SENSE) reconstruction [7]. With the increase of λ, the artifacts and noises disappeared, and also the image appeared more smooth with some details eclipsed. We noticed that setting λ around 0.1 provided better PSNR and SSIM. We plotted SSIM and PSNR over iterations to monitor the quality of reconstruction, using varied tuning parameter λ.

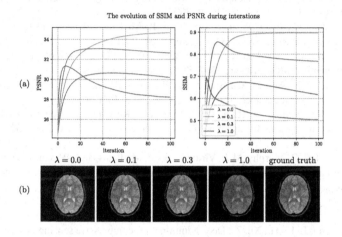

Fig. 3. (a) The curve of SSIM and PSNR over iterative steps. (b) Images reconstruction from 40 radial k-space spokes with different λ.

4   Discussion and Conclusion

In this study, a once pre-trained neural network - Proximator-Net - is used as a regularization in iterative algorithm for MRI reconstruction and can be applied to different reconstruction tasks, taking advantage of the separation of the learned information from the fidelity enforcement. That makes it different to previous methods [1,8,11]. In this initial experiment, we focused on demonstrating the utility of the proposed method in classic compressed sensing and radial k-space acquisition, and we used the brain MRI data to evaluate the method. Like conventional iterative reconstruction algorithms, k-space fidelity in the proposed hybrid approach was enforced by the least-square term and implemented outside the neural network, allowing the high flexibility to change k-space under-sampling schemes and RF coil settings. For quantitative comparison, our methods achieved 3dB higher PSNR in the tested acquisition settings compared with l_1-ESPIRiT.

References

1. Aggarwal, H.K., Mani, M.P., Jacob, M.: Modl: model-based deep learning architecture for inverse problems. IEEE Trans. Med. Imaging **38**(2), 394–405 (2018)
2. Alain, G., Bengio, Y.: What regularized auto-encoders learn from the data-generating distribution. J. Mach. Learn. Res. **15**(1), 3563–3593 (2014)
3. Chen, D., Davies, M.E., Golbabaee, M.: Compressive mr fingerprinting reconstruction with neural proximal gradient iterations (2020)
4. Griswold, M.A., et al.: Generalized autocalibrating partially parallel acquisitions (grappa). Magn. Reson. Med. **47**(6), 1202–1210 (2002)
5. Luo, G., Zhao, N., Jiang, W., Hui, E.S., Cao, P.: MRI reconstruction using deep bayesian estimation. Magn. Reson. Med. **84**(4), 2246–2261 (2020)
6. Lustig, M., Donoho, D., Pauly, J.M.: Sparse mri: the application of compressed sensing for rapid mr imaging. Magn. Reson. Med. **58**(6), 1182–1195 (2007)
7. Pruessmann, K.P., Weiger, M., Börnert, P., Boesiger, P.: Advances in sensitivity encoding with arbitrary k-space trajectories. Magn. Reson. Med. **46**(4), 638–651 (2001)
8. Schlemper, J., Caballero, J., Hajnal, J.V., Price, A.N., Rueckert, D.: A deep cascade of convolutional neural networks for dynamic mr image reconstruction. IEEE T.M.I. **37**(2), 491–503 (2017)
9. Uecker, M., Lai, P., Murphy, M.J., Virtue, P., Elad, M., Pauly, J.M., Vasanawala, S.S., Lustig, M.: Espirit an eigenvalue approach to autocalibrating parallel mri: where sense meets grappa. Magn. Reson. Med. **71**(3), 990–1001 (2014)
10. Uecker, M., et al.: Berkeley advanced reconstruction toolbox. In: Proc. Intl. Soc. Mag. Reson. Med., vol. 23, p. 2486 (2015)
11. Yang, Y., Sun, J., Li, H., Xu, Z.: Deep admm-net for compressive sensing mri. In: Lee, D.D., Sugiyama, M., Luxburg, U.V., Guyon, I., Garnett, R. (eds.) Advances in Neural Information Processing Systems 29, pp. 10–18. Curran Associates, Inc. (2016)

Crowdsourcing Segmentation of Histopathological Images Using Annotations Provided by Medical Students

Miguel López-Pérez[1]([✉]) [iD], Pablo Morales-Álvarez[2] [iD], Lee A. D. Cooper[3,4] [iD],
Rafael Molina[1] [iD], and Aggelos K. Katsaggelos[3,5] [iD]

[1] Department of Computer Science and Artificial Intelligence, University of Granada,
18010 Granada, Spain
{mlopez,rms}@decsai.ugr.es
[2] Department of Statistics and Operations Research, University of Granada,
18010 Granada, Spain
pablomorales@ugr.es
[3] Center for Computational Imaging and Signal Analytics, Northwestern University,
Chicago, IL 60611, USA
{lee.cooper,aggk}@northwestern.edu
[4] Department of Pathology at Northwestern University, Chicago, IL 60611, USA
[5] Department of Electrical and Computer Engineering at Northwestern University,
Evanston, IL 60208, USA

Abstract. Segmentation of histopathological images is an essential task
for cancer diagnosis and prognosis in computational pathology. Unfortunately, Machine Learning (ML) techniques need large labeled datasets to
train accurate segmentation algorithms that generalize well. A possible
solution to alleviate this burden is crowdsourcing, which distributes the
effort of labeling among a group of (non-expert) individuals. The bias
and noise from less experienced annotators may hamper the performance
of machine learning techniques. So far, crowdsourcing approaches in ML
leveraging these noisy labels achieve promising results in classification.
However, crowdsourcing segmentation is still a challenge in histopathological images. This paper presents a novel crowdsourcing approach to
the segmentation of Triple Negative Breast Cancer images. Our method
is based on the UNet architecture incorporating a pre-trained ResNet-34
as a backbone. The noisy behavior of the annotators is modeled with a
coupled network. Our methodology is validated on a real-world dataset
annotated by medical students, where five classes are distinguished. The
results show that our method with crowd labels achieves a high level of
accuracy in segmentation (DICE: 0.7578), outperforming the well-known
STAPLE (DICE: 0.7039) and close to the segmentation model using
expert labels (DICE: 0.7723). In conclusion, the initial results of our
work suggest that crowdsourcing is a feasible approach to segmentation
in histopathological images https://github.com/wizmik12/CRowd_Seg.

Supported by FEDER/Junta de Andalucia under Project P20_00286, FEDER/Junta
de Andalucía and Universidad de Granada under Project B-TIC-324-UGR20, and
"Contrato puente" of University of Granada.

J. M. Juarez et al. (Eds.): AIME 2023, LNAI 13897, pp. 245–249, 2023.
https://doi.org/10.1007/978-3-031-34344-5_29

Keywords: Histopathological images · Segmentation · Crowdsourcing

1 Introduction

The analysis of tissue biopsies is the gold standard for cancer diagnosis and prognosis. In computational pathology, one essential task is tissue segmentation, i.e., identifying and outlining structures of interest in a histopathological image. Machine Learning (ML) techniques have achieved excellent results in this task, but they need large labeled datasets to generalize well. This need for data is the main challenge. Crowdsourcing is a feasible strategy to obtain larger labeled datasets. It distributes the effort among several participants, who may have different expertise and biases. Therefore, crowdsourcing labels usually suffer from noise and subjectivity, which may hamper the performance of ML methods. Several works have studied how to leverage these noisy labels with different strategies [1,2]. The most effective approach to crowdsourcing classification is to simultaneously estimate the reliability of annotators and the underlying classifier, as shown in studies on histopathological images [3–5]. Recently, this approach has been adapted to crowdsourcing segmentation but it has only been applied to computed tomography imaging [8]. Crowdsourcing segmentation remains a challenge in computational pathology, where most methods aggregate the labels in a previous step, e.g., the well-known STAPLE algorithm [7].

This paper proposes a crowdsourcing segmentation method for Triple Negative Breast Cancer images labeled by 20 medical students following the methodology presented in [8]. Our method consists of two coupled networks, one for segmentation and another for estimating the annotators' reliability. Figure 1 shows the pipeline of our method. We validate our model in the experimental section against several baselines. To the best of our knowledge, this paper presents the first results in crowdsourcing segmentation in computational pathology, jointly estimating the segmentation network and the annotators' reliability during the learning process.

2 Proposed Methodology

We address the problem of semantic segmentation using a noisy labeled dataset $\mathcal{D} = \{(\mathbf{x}_n, \tilde{\mathbf{y}}_n^{(r)})\}_{n=1,...,N}^{r \in S(\mathbf{x}_n)}$, where $S(\mathbf{x}_n)$ denotes the set of annotators from $\{1, ..., R\}$ that segment the n-th patch $\mathbf{x}_n$. Each pixel belongs to one of the C classes. Our goal is to learn and predict on new images the latent real segmentation $\mathbf{y}$. To tackle this problem, we build two coupled networks: a segmentation network and an annotator network. The segmentation network estimates the real segmentation by computing the probability (pixel-wise) to belong to each class $\hat{p}_\theta(\mathbf{x}_n) \in [0,1]^{W \times H \times C}$. For this task, we use a UNet architecture because of its excellence in biomedical segmentation tasks [6]. Since computational pathology poses a challenging problem in the biomedical field, we enhance the training using a pre-trained ResNet-34 on ImageNet as the backbone. The annotator network gives us the pixel-wise reliability matrix for the r-th annotator, $\hat{\mathbf{A}}_\phi^{(r)}(\mathbf{x}) \in [0,1]^{W \times H \times C \times C}$. For each pixel, we have a $C \times C$ matrix. The

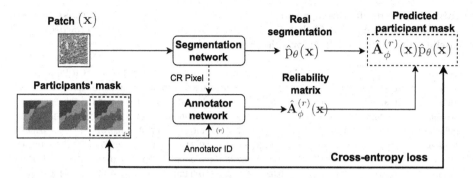

Fig. 1. Proposed crowdsourcing model. It couples two networks to predict the noisy observed masks (segmentation and annotator networks). We consider two different models for the annotator network: i) in *CR Global*, the behavior only depends on the annotator ID, $\hat{\mathbf{A}}_\phi^r(\mathbf{x}) := \hat{\mathbf{A}}_\phi^r$; ii) in *CR Pixel*, the behavior additionally depends on the patch $\mathbf{x}$. For prediction, we take the output of the segmentation network, $\hat{p}_\theta(\cdot)$.

(i, j)-th element of this matrix describes the probability that the annotator label that pixel with the class i while the real class is j. We use two different models for the annotator network. The first one, *CR Global*, assumes that the reliability matrix of each annotator does not depend on the features of the images. We use a fully connected layer on the one-hot encoding ID of the annotators followed by a SoftPlus activation to ensure positivity. The second one, *CR Pixel*, estimates a different reliability matrix for each annotator and pixel. The inputs to this model are the ID and the last feature map of the segmentation network. Note that the latter depends on the features of the images and it concatenates 2D convolutions with 2D instance normalization and PReLU activation. The final activation is a Softplus. In [8], they built different branches for each annotator. In contrast, our annotator network has only one branch. We compute an embedding for the ID of the annotators, which makes the model cheaper regarding the memory cost. We also assume independence among annotators. Finally, the noisy masks of the participants are estimated pixel-wise by $\hat{p}_{\theta,\phi}^{(r)}(\mathbf{x}) := \hat{\mathbf{A}}_\phi^{(r)}(\mathbf{x})\hat{p}_\theta(\mathbf{x})$. Notice that here we perform pointwise matrix multiplications in the spatial dimensions W and H. We train the model and find the parametrization of the two networks $\{\phi, \theta\}$ by minimizing the cross-entropy between our estimation and the real noisy mask of the annotators. Since many products of $\hat{\mathbf{A}}_\phi^{(r)}(\mathbf{x})$ and $\hat{p}_\theta(\mathbf{x})$ produce the same $\hat{p}_{\theta,\phi}^{(r)}(\mathbf{x})$, we add $-\mathrm{tr}(\hat{\mathbf{A}}_\phi^{(r)})$ as a regularizer to the cross-entropy to warm up the annotator network. This regularizer favors annotators to be reliable.

3   Experiments

3.1   Experimental Setup

Data and Preprocessing. We use a real-world dataset of Triple Negative Breast Cancer with crowdsourcing masks [1]. This dataset is composed of 151

Table 1. Dice results (average and standard deviation of three different runs) for proposed crowdsourcing (CR Global and CR Pixel) segmentation models, crowdsourcing baselines (MV and STAPLE), and the expert masks.

Dice	MV	STAPLE	Expert	CR Global	CR Pixel
Others	0.8113 ± 0.0189	0.8257 ± 0.0066	0.8309 ± 0.0110	**0.8327 ± 0.0140**	0.8145 ± 0.0344
Tumor	0.8252 ± 0.0078	0.8089 ± 0.0087	0.8322 ± 0.0066	**0.8364 ± 0.0053**	0.8323 ± 0.0123
Stroma	0.7462 ± 0.0220	0.7512 ± 0.0012	**0.7731 ± 0.0061**	0.7434 ± 0.0128	0.729 ± 0.0407
Inflammation	0.6650 ± 0.0137	0.6224 ± 0.0617	**0.7434 ± 0.0410**	0.7146 ± 0.0701	0.7285 ± 0.0073
Necrosis	0.6297 ± 0.0798	0.5111 ± 0.0643	**0.6819 ± 0.2106**	0.6621 ± 0.1254	0.6176 ± 0.171
Micro	0.7687 ± 0.0097	0.7549 ± 0.0091	**0.7953 ± 0.0099**	0.7812 ± 0.0115	0.7721 ± 0.0201
Macro	0.7355 ± 0.0179	0.7039 ± 0.0207	**0.7723 ± 0.0430**	0.7578 ± 0.0207	0.7444 ± 0.0371

Whole Slide Images (WSIs). The 161 Regions of Interest (ROIs) were extracted from these images and 20 medical students were asked to annotate them. Ten ROIs were annotated by everyone, the rest by only one participant. Every ROI has a curated (expert) annotation provided by a senior pathologist. We extracted 512×512 patches (training: $10,173$; validation: $1,264$; test: 399). For our segmentation problem, we considered five classes: tumor, stroma, inflammation, necrosis, and others.

Model Hyperparameters. We utilized data augmentation (blur, rotation, saturation, ...). We warmed up the annotator network using the trace regularizer and a learning rate of 10^{-3} during five epochs. After the fifth epoch, we remove the regularizer and set the learning rate to 10^{-4}. We fine-tuned the segmentation network with a learning rate of 10^{-4} for the whole training process. We trained the networks for 20 epochs, a mini-batch size of 16, and made use of the Adam optimizer. The methods were run three times using different seeds to ensure robustness.

Baseline Methods. The hyperparameters, architecture, and training procedure were the same as for the segmentation network of our crowdsourcing model. We name the expert model the one trained using the curated labels. We also used two different methods that combine the multiple noisy labels: Majority Voting (MV) and the well-known STAPLE [7].

Evaluation Metric. We used the DICE coefficient as our figure of merit. DICE measures the pixel-wise agreement between a predicted segmentation and its corresponding ground truth. In our experiments, we calculated the DICE between the real segmentation provided by senior pathologists and the methods' prediction for the following cases: per-class, macro (averaging across classes), and micro (weighted class average).

3.2   Results

Table 1 presents the main numerical results of the three runs. We can see that our crowdsourcing models, CR Global (DICE macro: 0.7578) and CR Pixel (DICE macro: 0.7444), perform slightly worse than the expert model (DICE Macro:

0.7723) and outperform MV (DICE Macro: 0.7355) and STAPLE (DICE Macro: 0.7039). CR Global also performed better than the expert model on certain classes and better than CR Pixel in most cases. We believe that this may be due to the higher complexity of CR Pixel, which may need a wider number and variety of patches and annotations to outperform the rest.

4   Conclusion

To the best of our knowledge, this is the first work addressing crowdsourcing segmentation in computational pathology, jointly estimating the underlying segmentation network and the annotators' reliability during the learning process. Our results indicate that the proposed methodology performs better than widely used approaches, and is competitive with expert annotations. We have also found a decrease in the performance using the flexible pixel-wise crowdsourcing model with respect to the global one. We may avoid this effect by collecting more images and annotations. Future work will aim to improve the annotator network by considering correlations among annotators. Future work will also provide additional insights into annotators' behavior.

References

1. Amgad, M., et al.: Structured crowdsourcing enables convolutional segmentation of histology images. Bioinformatics **35**(18), 3461–3467 (2019)
2. Grote, A., Schaadt, N.S., Forestier, G., Wemmert, C., Feuerhake, F.: Crowdsourcing of histological image labeling and object delineation by medical students. IEEE Trans. Med. Imaging **38**(5), 1284–1294 (2018)
3. López-Pérez, M., et al.: Learning from crowds in digital pathology using scalable variational Gaussian processes. Sci. Rep. **11**(1), 1–9 (2021)
4. López-Pérez, M., Morales-Álvarez, P., Cooper, L.A.D., Molina, R., Katsaggelos, A.K.: Deep Gaussian processes for classification with multiple noisy annotators. Application to breast cancer tissue classification. IEEE Access **11**, 6922–6934 (2023)
5. Nir, G., et al.: Automatic grading of prostate cancer in digitized histopathology images: learning from multiple experts. Med. Image Anal. **50**, 167–180 (2018)
6. Ronneberger, O., Fischer, P., Brox, T.: U-net: convolutional networks for biomedical image segmentation. In: Navab, N., Hornegger, J., Wells, W.M., Frangi, A.F. (eds.) MICCAI 2015. LNCS, vol. 9351, pp. 234–241. Springer, Cham (2015). https://doi.org/10.1007/978-3-319-24574-4_28
7. Warfield, S.K., Zou, K.H., Wells, W.M.: Simultaneous truth and performance level estimation (staple): an algorithm for the validation of image segmentation. IEEE Trans. Med. Imaging **23**(7), 903–921 (2004)
8. Zhang, L., et al.: Disentangling human error from ground truth in segmentation of medical images. In: NeurIPS, vol. 33, pp. 15750–15762 (2020)

Automatic Sleep Stage Classification on EEG Signals Using Time-Frequency Representation

Paul Dequidt[1]([✉])[ID], Mathieu Seraphim[1][ID], Alexis Lechervy[1][ID],
Ivan Igor Gaez[2][ID], Luc Brun[1][ID], and Olivier Etard[2][ID]

[1] GREYC, UMR CNRS 6072, Normandie Univ, CNRS, UNICAEN, ENSICAEN,
Caen, France
paul.dequidt@unicaen.fr
[2] COMETE, UMR-S 1075, Normandie Univ, CNRS, UNICAEN, Caen, France

Abstract. Sleep stage scoring based on electroencephalogram (EEG) signals is a repetitive task required for basic and clinical sleep studies. Sleep stages are defined on 30 s EEG-epochs from brainwave patterns present in specific frequency bands. Time-frequency representations such as spectrograms can be used as input for deep learning methods. In this paper we compare different spectrograms, encoding multiple EEG channels, as input for a deep network devoted to the recognition of image's visual patterns. We further investigate how contextual input enhance the classification by using EEG-epoch sequences of increasing lengths. We also propose a common evaluation framework to allow a fair comparison between state-of-art methods. Evaluations performed on a standard dataset using this unified protocol show that our method outperforms four state-of-art methods.

Keywords: Sleep scoring · time-frequency representation · computer vision · EEG · signal processing

1 Introduction

Sleep is an important physiological process which can be monitored through polysomnography (PSG). A PSG involves multiple signals, such as electro-encephalogram (EEG), electro-oculogram (EOG) or electro-myogram (EMG). The American Academy of Sleep Medicine (AASM) edited guidelines [3] to classify sleep into different stages based on a 30 s time-frame called an epoch (EEG-epoch). The actual AASM standard identifies 5 stages: wakefulness (W), rapid eye movement (REM) and Non-REM sleep (N1, N2, N3). Sleep studies often

This study is co-funded by the Normandy County Council and the European Union (PredicAlert European Project - FEDER fund). Part of this work was performed using computing resources of CRIANN (Normandy, France). This work was performed using HPC resources from GENCI-IDRIS (Grant 2022-102446).

J. M. Juarez et al. (Eds.): AIME 2023, LNAI 13897, pp. 250–259, 2023.
https://doi.org/10.1007/978-3-031-34344-5_30

Table 1. Frequency bands for brain activity

	Delta	Theta	Alpha	Beta$_{low}$	Beta$_{high}$	Gamma
Frequencies (Hz)	[0.5; 4[	[4; 8[	[8; 12[	[12; 22[	[22; 30[	[30; 45[

need to score each EEG-epoch, which is a tedious task for a human expert. Therefore, sleep scoring could benefit from automation, especially in the case of whole night recordings. Frontal EEG are often artefacted by ocular movements similar to the EOG signal. Moreover, the placement of multiple sensors for multi-modal analysis complicates the acquisition stage while the contribution of multimodality to the stage analysis is not clearly established.

Therefore, we want to investigate if we can simplify the acquisition by only using multiple EEG channels. The AASM groups different frequencies into frequency bands as detailed in Table 1. Each band corresponds to specific graphic elements in EEG signal used to analyze the brain activity and for sleep scoring. Representation of EEG signals as time-frequency images, or spectrograms, can show the variations in brain activity during sleep depth [12]. One important advantage of such representations is that it enables the use of deep networks devoted to pattern recognition and classification problems on images. Such networks have been subject to intensive investigations for more than 10 years.

2   State of the Art

Sleep scoring methods can be analyzed through their input modalities, the computed features on each EEG-epoch, the way they take into account contextual information, or the type of method used in order to obtain the final classification.

Many authors use EEG, EOG and EMG as multimodal inputs for their networks [4,6,7,9,10,18], resulting in a scoring aligned with the AASM standard. A classical heuristic [5,16,19,20] consists in substracting the EOG signal to EEG acquisitions. The resulting signal provides interesting classification results but lacks interpretability for experts.

The number of EEG channels varies according to the methods and the datasets on which they are applied. For example, Qu et al. [13] only use one EEG signal while Jia et al. [7] use up to 20 EEG signals. Using more electrodes increases the number of input signals and should improve the scoring. However, increasing the number of electrodes increases the complexity of acquisition and analysis. In addition, electrodes become closer together on the skull as the number of electrodes increases, which enhances interference between signals from nearby electrodes. The optimal number of electrodes is still an open question.

The benefits of spectral analysis is recognized by sleep researchers as early as the 1980s [15]. However, spectral estimation based on Fourier transform assumes the signal is infinite, periodic and stationary. Therefore, used on EEG signals, which are finite, aperiodic and non-stationary, the spectrogram can be artefacted. Multitaper convolutions, or Tapers, have been proven to reduce this bias [12].

Tapers-based spectrograms have been used by Vilamala et al. [21] in conjunction with the VGG-16 convolutional neural network (CNN), on a 5 EEG-epoch sequence on a single EEG signal. VGG-16 takes (3,224,224) RGB images as input. Vilamala et al. transform the spectrogram with a colormap; then use a VGG-16 pretrained on the ImageNet dataset, but with parameters in the fully connected layers randomized, and all weights unfreezed in their best performing experiments.

Manual sleep scoring often involves some form of contextual input. Many studies test the impact of using a temporal sequence of EEG-epochs as input of their network, instead of a single EEG epoch [5,10,11,16]. The length of sequence differs greatly for one author to another. For example, Dong et al. [5] tried sequences from 1 to 6 EEG-epochs and found their best results between 2 and 4 EEG-epochs, while Phan et al. [10] found that sequences greater than 10 EEG-epochs had minimal impact on classification. Therefore, networks can often be divided into two sections: one that extracts features within each EEG-epoch (intra-epoch), and another that compares the features from neighboring EEG-epochs to get a better classification (inter-epoch).

The intra-epoch features can be handcrafted. For example, Dong et al. [5] and Sun et al. [18] extract handcrafted features from the power spectral density (PSD). Features can also be learned using CNNs [5,9,13,16,18] or recurrent neural networks (RNNs) [10].

Inter-epoch networks are mainly based on RNN structures, mostly Long-Short-Term-Memory (LSTM) layers, both bi-directional [16,18,19] and not [5, 20]. Phan et al. [10] also used RNNs, with bi-directional Gated Recurrent Units (GRU) at both the intra-epoch and inter-epoch levels. Contextual information within the inter-epoch section may also be analysed using non-recurrent networks. For example, Qu et al. [13] used a Transformer-like attention network to extract inter-epoch features, Jia et al. [7] combines a temporal convolution with an attention mechanism and Dong et al. [5] only used a softmax layer.

3   Method

3.1   Dataset Preprocessing

Our network takes spectrograms extracted from C EEG signals, spatially diverse, allowing us to include rich spatial information. We used two time-frequency representations of our signals: FFT and Tapers. We computed spectrograms using the Fast Fourier Transform (FFT) and the parameters provided by Phan et al. [10]: a 2-s Hamming window with 50% overlap. This gives us an image where 1 pixel encodes 0.5 Hz. Unlike Vilamala et al. [21], we cut the frequency axis 45 Hz included, as higher frequencies do not carry relevant brainwave information. We then convert the amplitude spectrum to a logarithmic scale as done by Vilamala et al. After computing the spectrograms, we divided them into 30 s EEG-epochs. With C the number of EEG channels used as input, the resulting image for 1 EEG-epoch is a $(C,30,90)$ tensor. For each EEG-epoch, we also computed the $C \times C$ electrodes covariance matrices, as a way to convey spatial

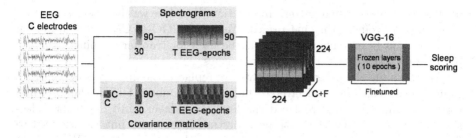

Fig. 1. The preprocessing and classification pipeline. The EEG signals are transformed into C spectrograms and F covariance matrices. Contextual information is added (here, $T = 7$ and vertical lines have been added for visualization only). Spectrograms and covariance matrices are concatenated into a $(C + F, 224, 244)$ tensor, used as input for a finetuned VGG-16 network.

co-variation information, with F being the number of covariance matrices. These matrices have been computed on the native EEG-epoch signal, and on a filtered version for each frequency band described in Table 1, resulting in $F = 7$. With $C = 8$, these F (8,8) matrices have been reshaped through repetition of rows and concatenated into a $(F, 30, 90)$ tensor, then concatenated as supplementary channels to the $(C, 30, 90)$ spectrograms. The resulting $(C + F, 30, 90)$ tensor is then zero-padded to fit inside a $(C + F, 224, 224)$ shape, which is our adapted VGG-16 input layer, as shown in Fig. 1.

To get the same spectrogram frequency sampling when computing the FFT and multitapers spectrograms, we used the heuristic described in Prerau et al. [12] where the number of tapers L is $L = \lfloor 2W \rfloor - 1$, with W being the half-time bandwith product and $\lfloor x \rfloor$ the floor function. To get comparable spectra between FFT and Tapers, we used $L = 3$.

3.2 Contextual Information

In order to investigate the impact of contextual information for scoring, we used different EEG sequence lengths as input for the VGG input space, with T being the length of the sequence. Following the AASM guidelines, information above 3 to 5 EEG-epochs should not be relevant for scoring. Therefore, we tested $T = 1,3,5,7$ EEG-epochs fitted inside the $(C + F,224,224)$ input tensor. In these samples, the EEG-epoch to classify has been centered in the input tensor, and we added past EEG-epochs on the left of the central epoch and future EEG-epochs on the right. We added future EEG-epochs as human experts also use them for manual scoring, especially during state transitions.

3.3 Finetuning a Deep Network to Multi-electrode Data

Manual sleep scoring is a visual process, therefore we use computer vision techniques to extract information from EEG signals. Extensive research has been done in the field of visual pattern recognition, therefore we chose a classification network that is efficient in that regard. We used a VGG-16 CNN [17] with batch normalization layers. This network can be used for transfer learning, as it has been trained on ImageNet [14] natural images.

After loading the pretrained weights, we replaced the first layer by a $C + F$-channel deep convolutional layer to match the number of EEG channels and covariance matrices, and replaced the final dense layer to fit our 5-class classification problem. As VGG has been pretrained on color images, we need to finetune the first and last layer before training the whole network. We initially froze the weights of the network except the first and last layers, for 10 epochs, then unfroze the whole VGG for the rest of the training. We used checkpoints after each training epoch and early stopping to save the best network based on validation MF1 score.

4   Experiments

4.1   Dataset Used

This study is focused on healthy subjects described by multiple EEG signals. We therefore use the Montreal Archive of Sleep Studies (MASS) dataset [8] which fits these criteria while remaining easily accessible.

The MASS dataset is divided in 5 studies (SS1 to SS5). The only study which involves healthy subjects, scored on 30-s epochs is the SS3 subset. This subset gathers one whole-night recording of 62 subjects using 20 EEG channels, which allows comparison between different brain region signals. On each EEG channel, 60 Hz notch filter has been applied, as well as a low and high-pass filter with a cutoff frequency of 0.30 Hz 100 Hz respectively. The different sleep stages are not equally distributed during the night. Consequently, this dataset is imbalanced, with around 50.2% of epochs being in the N2 sleep stage, as shown in Table 2. To correct the class imbalance, we repeat samples during training so that each class has the same number of samples. For validation and testing, we used unbalanced data as these subsets are supposed to represent real case studies, meaning unbalanced classes.

As stated in Sect. 3.1, we have elected to study 8 EEG signals in particular, recorded from a variety of locations on the skull: F3, F4, C3, C4, T3, T4, O1 and O2. We will refer to each electrode couple by their location (F for {F3;F4}). Unless stated otherwise in Sect. 4.5 of the ablation study, each result is presented with C=8 electrodes (FCTO).

Table 2. Class imbalance on the MASS SS3 dataset

	Awake	REM	N1	N2	N3	Total
Number of EEG-epochs	6.442	10.581	4.839	29.802	7.653	59.317
Percentage	10.86%	17.84%	8.16	50.24%	12.90%	100%

4.2   Folds and Metrics

We divided the SS3 dataset using the 31 folds proposed by Seo et al. [16], and available on their Github repository. Each fold has 50 training subjects, 10 validation subjects and 2 test subjects. Since there are 62 subjects, the set of test folds covers the whole datatest set without overlap. We used a Tree-structured Parzen Estimator (TPE) approach [2] as implemented in Optuna [1] to determine the best hyperparameters for the first fold. Namely, the learning rate, momentum, weight decay and learning rate decay. Due to time constraints, we applied the TPE only on the first fold, and used the resulting fine-tuned set of hyperparameters for training and validation on the 31 folds. The resulting model for each fold is then evaluated on the test set of the fold, leading to one value of each metric per fold. We monitored the main metrics used in the literature: macro-averaged F1 (MF1), macro-averaged accuracy (MaccroAccuracy) and Cohen's kappa. The MASS dataset has a strong class imbalance, therefore overall accuracy and F1 score become biased metrics. Consequently, MF1 and Macro-accuracy are more indicative of relative per-class accuracy. In our results, we present the mean and standard deviation of each metric based on the 31 fold-based predictions.

4.3   Time-Frequency Representations

We first test the impact of the spectral representation by testing FFT against Tapers. We insure that both representations have the same spectral resolution by using $L = 3$ as number of Tapers, which gives a frequency resolution of 0.5 Hz, identical to a 2-s window FFT. The comparison of FFT and Tapers scoring is provided in Table 3. Tapers gets better results on all three statistics while the gap between both representations remains within the standard deviation. We therefore use Tapers spectrograms in the following tests.

Table 3. Performances reached for FFT and Tapers

	MF1	MacroAccuracy	Kappa
FFT	77.79 ± 3.80	80.68 ± 3.84	0.759 ± 0.051
Tapers	**78.53 ± 3.77**	**81.06 ± 3.39**	**0.766 ± 0.057**

4.4   Increasing Context Information

Using Tapers as time-frequency representation, we tested the effect of contextual inputs by gradually increasing the length of the sequence T. Results are shown in Table 4. We observe an improvement of all the metrics and a decreasing standard deviation as the length of the EEG sequences grows. Reaching better performances from 1 to 3 EEG-epochs underlines the importance of contextual information, which is congruent with the AASM standard. Our best performances is reached for 7 epochs, which does not align with Dong et al. [5], as their best results were reached between 2 and 4 EEG-epochs. These results suggest that even longer sequences can enhance classification. Especially, reducing the standard deviation seems to be relevant when using the fold provided by Seo et al. [16]. Consequently, we used $T=7$ for the remaining tests.

Table 4. Results from 1 to 7 EEG epochs in the VGG space

T EEG epochs	MF1	MacroAccuracy	Kappa
1	78.53 ± 3.77	81.06 ± 3.39	0.766 ± 0.057
3	80.00 ± 3.65	81.83 ± 3.70	0.782 ± 0.052
5	80.02 ± 3.80	81.89 ± 3.40	0.7869 ± 0.049
7	**81.79 ± 2.95**	**82.96 ± 2.88**	**0.809 ± 0.038**

4.5   Removing Spatial Information

We want to study the effect of reducing the number of EEG channels. We tested 5 decreasing sets of electrodes, from only left electrodes {F3, C3, T3, O1} (FCTO_left) or right {F4, C4, T4, O2} (FCTO_right), then tested on FCO_right, FO_right and F_right to see the influence of the number of electrodes. Comparing (FCTO_left and FCTO_right), we observe a slight decrease in performance compared to the full FCTO set, while still giving good performances. The right side performs better, so we successively removed the T, C and O electrodes. We observe a rise of the standard deviation, but also good performances for the FCO_right subset. This aligns with the AASM standard, which recommends using at least FCO from one side for human expert scoring. These results also suggest that the temporal (T) electrode does not seem relevant for this task, as removing it seems to give slightly better results.

Table 5. Reducing the number of electrodes

Electrodes	MF1	MaccroAccuracy	Kappa
FCTO_left	79.82 ± 3.81	81.38 ± 3.58	0.788 ± 0.046
FCTO_right	80.71 ± 3.01	82.07 ± 3.25	0.798 ± 0.039
FCO_right	**81.49 ± 3.21**	**83.11 ± 3.37**	**0.802 ± 0.043**
FO_right	80.60 ± 3.44	82.67 ± 3.58	0.790 ± 0.049
F_right	78.77 ± 3.62	81.01 ± 3.19	0.776 ± 0.049

4.6   Comparison with State-of-the-Art Methods

In order to compare our results to the state of the art, we ran 3 recent methods [7, 16,19]. Jia et al. [7] is a method which gives good results on MASS, Supratak et al. [19] is often cited as the state-of-art baseline, and Seo et al. [16] uses the folds we based our method on. All methods used the MASS dataset and have their code publicly available on Github. To get a fair comparison, we ran their code using the same folds and metrics that we used, as presented in Sect. 4.2.

All three studies originally used 31 folds on MASS, but both Supratak et al. [19] and Jia et al. [7] have a different fold composition than Seo et al. [16]. Their fold involves 60 subjects per fold as a training set, and the remaining 2 as a validation set. To the best of our knowledge, their published results have been obtained on their validation set only. This is not best practice but can be understood, as neither their code nor their papers show signs of hyperparameter optimization. In order to allow a robust comparison between methods, we retrained Supratak et al. and Jia et al. on Seo et al.'s folds.

All three studies computed their metrics by concatenating the predictions of each fold into a single array of predictions, and comparing those predictions with a similarly obtained array of targets. This way, they compute their metrics on all EEG-epochs of all 62 subjects without omission or repetition. However, this technique is debatable, as it groups together results obtained from 31 different networks (one per fold) without taking into account that different folds have different number of EEG-epochs during prediction. Therefore, computing their metrics on a concatenated prediction array creates an implicit weighting relative to each fold's prediction set size. Moreover, they used overall accuracy instead of Maccro-Accuracy, and none of them give the standard deviation. Consequently, we computed the Macro-Accuracy, MF1 and Kappa score for each fold, then computed the averaged and standard deviation.

These differences may explain the discrepancy between the results published and the results we obtained by running their code using our folds and methodology, as seen in Table 6. We also tested Vilamala et al. [21] on the MASS dataset, as they are also using Tapers and transfer-learning on VGG-16. While training Vilamala's method, we froze the convolution layers beforehand as it gave this method better results.

With this shared protocol, our method outperforms all four methods, as shown in Table 6.

Table 6. Comparison with SOA methods

Methods	MF1	MaccroAccuracy	Kappa
Seo [16]	77.36 ± 4.76	77.17 ± 4.08	0.774 ± 0.052
Supratak [19]	79.67 ± 4.49	79.99 ± 4.26	0.792 ± 0.047
Jia [7]	76.03 ± 4.01	76.35 ± 4.53	0.751 ± 0.056
Vilamala [21]	72.87 ± 5.72	73.24 ± 5.78	0.666 ± 0.072
Our method	**81.79 ± 2.95**	**82.96 ± 2.88**	**0.809 ± 0.038**

5   Discussion

Our results suggest that automatic sleep scoring could benefit from using multitapers spectrograms as time-frequency representation. Using a high number of epochs as contextual input gave higher results, and seems to reduce the standard deviation. Our results are congruent with the fact that some state transitions can be influenced by the previous epochs in the AASM standard, and unlike Dong et al. [5], we still got improvement with a sequence length greater than 6 EEG-epochs. We showed in Table 5 than halving the number of electrode could maintain strong classification results, thus questioning on the redundancy between left and right side. However, left-side EEGs are often artefacted by the cardiac activity, which may explain why we reach slightly better performances when using only right-side electrodes compared to left-side electrodes. Using multiple EEG, we got better results than Supratak et al. [19] which used one mixed signal resulting from the difference between one EOG and one EEG channel, and better result than Jia et al. [7] which use EEG, EOG and EMG. This underline that EEG alone can give robust results for classification, thus leading to simpler acquisition protocols.

6   Conclusion

In this paper we compared two types of time-frequency spectrograms for sleep scoring. Using a fine-tuned deep visual network, we outperforms state-of-the-art results. We did an ablation study to determine the number of contextual EEG-epochs needed, and study how the number of electrodes could impact classification results. Our results seems relevant with the AASM standard and EEG expertise regarding the number of EEG-epochs and when comparing left and right side for electrodes. Our results suggest that acquisition protocol could be reduced to a lesser number of modalities and sparser electrodes. Finally, we propose a common methodology for training and method comparison, using the same folds as Seo et al. [16], which allows hyperparameter research.

References

1. Akiba, T., Sano, S., Yanase, T., Ohta, T., Koyama, M.: Optuna: a next-generation hyperparameter optimization framework. In: Proceedings of the 25th ACM SIGKDD International Conference on Knowledge Discovery & Data Mining, pp. 2623–2631 (2019)
2. Bergstra, J., Bardenet, R., Bengio, Y., Kégl, B.: Algorithms for hyper-parameter optimization. In: Advances in Neural Information Processing Systems, vol. 24 (2011)
3. Berry, R.B., et al.: AASM scoring manual updates for 2017 (version 2.4) (2017)
4. Chambon, S., Galtier, M., Arnal, P.J., Wainrib, G., Gramfort, A.: A deep learning architecture for temporal sleep stage classification using multivariate and multimodal time series. IEEE Trans. Neural Syst. Rehabil. Eng. **26**(4), 17683810 (2018)

5. Dong, H., Supratak, A., Pan, W., Wu, C., Matthews, P.M., Guo, Y.: Mixed neural network approach for temporal sleep stage classification. IEEE Trans. Neural Syst. Rehabil. Eng. **26**(2), 324–333 (2018)

6. Jia, Z., et al.: Multi-view spatial-temporal graph convolutional networks with domain generalization for sleep stage classification. IEEE Trans. Neural Syst. Rehabil. Eng. **29**, 1977–1986 (2021)

7. Jia, Z., et al.: Graphsleepnet: adaptive spatial-temporal graph convolutional networks for sleep stage classification. In: IJCAI, pp. 1324–1330 (2020)

8. O'reilly, C., Gosselin, N., Carrier, J., Nielsen, T.: Montreal archive of sleep studies: an open-access resource for instrument benchmarking and exploratory research. J. Sleep Res. **23**(6), 628–635 (2014)

9. Phan, H., Andreotti, F., Cooray, N., Chén, O., de Vos, M.: Joint classification and prediction CNN framework for automatic sleep stage classification. IEEE Trans. Biomed. Eng. **66**, 1285–1296 (2019)

10. Phan, H., Andreotti, F., Cooray, N., Chén, O.Y., De Vos, M.: Seqsleepnet: end-to-end hierarchical recurrent neural network for sequence-to-sequence automatic sleep staging. IEEE Trans. Neural Syst. Rehabil. Eng. **27**(3), 400–410 (2019)

11. Phan, H., Mikkelsen, K., Chén, O.Y., Koch, P., Mertins, A., De Vos, M.: Sleep-transformer: automatic sleep staging with interpretability and uncertainty quantification. IEEE Trans. Biomed. Eng. **69**(8), 2456–2467 (2022)

12. Prerau, M.J., Brown, R.E., Bianchi, M.T., Ellenbogen, J.M., Purdon, P.L.: Sleep neurophysiological dynamics through the lens of multitaper spectral analysis. Physiology **32**(1), 60–92 (2017)

13. Qu, W., et al.: A residual based attention model for EEG based sleep staging. IEEE J. Biomed. Health Inform. **24**(10), 2833–2843 (2020)

14. Russakovsky, O., et al.: Imagenet large scale visual recognition challenge. Int. J. Comput. Vision **115**(3), 211–252 (2015)

15. Salinsky, M., Goins, S., Sutula, T., Roscoe, D., Weber, S.: Comparison of sleep staging by polygraph and color density spectral array. Sleep **11**(2), 131–8 (1988)

16. Seo, H., Back, S., Lee, S., Park, D., Kim, T., Lee, K.: Intra- and inter-epoch temporal context network (iitnet) using sub-epoch features for automatic sleep scoring on raw single-channel eeg. Biomed. Signal Process. Control **61**, 102037 (2020)

17. Simonyan, K., Zisserman, A.: Very deep convolutional networks for large-scale image recognition. arXiv preprint arXiv:1409.1556 (2014)

18. Sun, C., Chen, C., Li, W., Fan, J., Chen, W.: A hierarchical neural network for sleep stage classification based on comprehensive feature learning and multi-flow sequence learning. IEEE J. Biomed. Health Inform. **24**(5), 1351–1366 (2020)

19. Supratak, A., Dong, H., Wu, C., Guo, Y.: Deepsleepnet: a model for automatic sleep stage scoring based on raw single-channel EEG. IEEE Trans. Neural Syst. Rehabil. Eng. **25**(11), 1998–2008 (2017)

20. Supratak, A., Guo, Y.: Tinysleepnet: an efficient deep learning model for sleep stage scoring based on raw single-channel EEG. In: 2020 42nd Annual International Conference of the IEEE Engineering in Medicine Biology Society (EMBC), pp. 641–644 (2020)

21. Vilamala, A., Madsen, K.H., Hansen, L.K.: Deep convolutional neural networks for interpretable analysis of EEG sleep stage scoring. In: 2017 IEEE 27th International Workshop on Machine Learning for Signal Processing (MLSP), pp. 1–6. IEEE (2017)

Learning EKG Diagnostic Models with Hierarchical Class Label Dependencies

Junheng Wang[✉][iD] and Milos Hauskrecht[iD]

Department of Computer Science, University of Pittsburgh, Pittsburgh, PA, USA
{juw100,milos}@pitt.edu

Abstract. Electrocardiogram (EKG/ECG) is a key diagnostic tool to assess patient's cardiac condition and is widely used in clinical applications such as patient monitoring, surgery support, and heart medicine research. With recent advances in machine learning (ML) technology there has been a growing interest in the development of models supporting automatic EKG interpretation and diagnosis based on past EKG data. The problem can be modeled as multi-label classification (MLC), where the objective is to learn a function that maps each EKG reading to a vector of diagnostic class labels reflecting the underlying patient condition at different levels of abstraction. In this paper, we propose and investigate an ML model that considers class-label dependency embedded in the hierarchical organization of EKG diagnoses to improve the EKG classification performance. Our model first transforms the EKG signals into a low-dimensional vector, and after that uses the vector to predict different class labels with the help of the conditional tree structured Bayesian network (CTBN) that is able to capture hierarchical dependencies among class variables. We evaluate our model on the publicly available PTB-XL dataset. Our experiments demonstrate that modeling of hierarchical dependencies among class variables improves the diagnostic model performance under multiple classification performance metrics as compared to classification models that predict each class label independently.

Keywords: Electrocardiogram · Machine Learning · Bayesian Network

1 Introduction

Electrocardiogram (EKG/ECG) is a key diagnostic tool to assess patient's cardiac condition and is widely used in patient monitoring, surgery support, and heart medicine research. Until recent years, most EKG processing depends largely on domain knowledge from experts and requires signal filtering and enhancing. Thanks to advances in machine learning (ML) methodologies and the increasing quantity and quality of EKG data, recent years has seen increased

J. M. Juarez et al. (Eds.): AIME 2023, LNAI 13897, pp. 260–270, 2023.
https://doi.org/10.1007/978-3-031-34344-5_31

interest in the development of data driven solutions that can automatically interpret EKG signals and use it for the diagnosis of the underlying patient conditions. Such conditions are often labelled using standardized EKG vocabulary that is organized into a diagnostic class hierarchy [20]. For example, the ILMI (inferolateral myocardial infarction) and IPMI (inferoposterior myocardial infarction) labels are aggregated in the IMI (inferior myocardial infarction) class at a lower level and the MI (myocardial infarction) class at a higher level of the hierarchy. On the other hand, the ASMI (anteroseptal myocardial infarction) is a member of AMI (anterior myocardial infarction) class which in turns also belongs to the same MI (myocardial infarction) class.

The problem of assigning class labels to EKG signals can be cast as a multi-label classification (MLC) problem. Unlike the multi-class classification problem where each instance belongs to exactly one class, in MLC each instance can have multiple class labels. In general, class labels that are assigned to individual EKGs can be at the same or the different level of abstraction. For example, a specific EKG can be assigned ILMI (inferolateral myocardial infarction), IMI (inferior myocardial infarction) as well MI (myocardial infarction) labels. The majority of the previous works on EKG classification does not consider hierarchical dependencies among class labels [1,14,19] and hence may result in inconsistent predictions at different levels. For example, the model may predict a class to be true while its parent class being false, or a class to be true while all its children classes are false.

In this paper, we propose and investigate an ML model that can perform MLC of EKG signals based on the hierarchical organization of EKG diagnoses and their corresponding class label dependencies. The proposed model starts from EKG signals that are transformed via ML architectures into a lower dimensional vector representation of EKG, and after that, it relies on a hierarchical organization of classes to make class label predictions. The hierarchical class dependencies our model relies on are implemented in a conditional tree structured Bayesian network (CTBN) [2] where the tree structure encodes the class hierarchy. We use multiple logistic regression models as classifiers of the CTBN, where each logistic regression model comes with its own set of trainable parameters. The trained CTBN model can make multi-label predictions in time linear in the number of classes, by computing the most probable assignment of all class variables using the tree structure of the CTBN.

We evaluate our CTBN model on the PTB-XL [20] EKG dataset that annotates each EKG using a mix of class labels at the different levels of abstraction. We show that by explicitly including the label dependencies we can improve the EKG classification performance in terms of both the exact match accuracy (EMA) and conditional log likelihood loss (CLL-loss), two criteria commonly to evaluate the MLC predictions. To prove the robustness of the CTBN model for the MLC problem we test its benefits by combining it with (five) different EKG signal transformations solutions: multilayer perceptron (MLP), recurrent neural network (RNN), long short-term memory (LSTM), gated recurrent unit (GRU), and fully convolutional network (FCN).

2   Related Work

In this section we briefly review the work related to our methodology, in particular, the work on ML models for EKG annotation, and multi-label and hierarchical classification tasks.

EKG Labeling. Many efforts have been made to organize an extensive amount of EKG data into datasets [10,11,20] that consist of EKG waveform and diagnostic labels. Early research on EKG labeling required domain knowledge from experts, complex preprocessing on EKG signals [3,5,23] and hand-crafted features. More recent research efforts have explored modern ML solutions based on neural networks [1,22] and deep learning [4,14,15] models to implement EKG processing and suitable EKG low-dimensional representations. Although this research does not share a common set of prediction targets, it leads to many promising results that demonstrate the potential of ML in automatic EKG labeling and diagnosis.

Multi-label Classification. Specifically to the PTB-XL dataset, [17] examines several ML models including a five-layer one-dimensional convolutional neural network, a channel-wise SincNet-based [12] network architecture, and a convolutional neural network with hand picked entropy features calculated for every channel, and reports results on pair-wise classification tasks and 5- and 20-class MLC. [18] uses both the entropy features from [17] and features generated from R-wave detection methods, and performs MLC using aggregated models with different combinations of features. [19] provides a comprehensive collection of ML methods including Wavelet with shallow neural network [16], LSTM [7], FCN [21], ResNet [6], Inception [8] models and their variants. These models are benchmarked in various MLC tasks on different targets such as diagnosis, forms, and rhythms. Note that all the works mentioned above perform MLC without class-label hierarchy, and no dependency is considered when training and evaluating the models.

Hierarchical Classification. The ML approach of hierarchical classification can be applied to the medical field by learning a collection of diagnostic classification models explicitly related via hierarchy. Malakouti and Hauskrecht [9] proposes a set of predictive models for multiple diagnostic categories organized in a hierarchy, and uses the hierarchy to guide the transfer of model parameters. The algorithm uses a two passes approach: the first pass follows the hierarchy in top-down fashion where the model parameters are transferred from higher-level diagnostic categories to lower-level ones; the second pass transfers the information bottom-up by adapting model parameters from lower level to their immediate parents. Their results shows improved performance when compared to independently learned models, especially for diagnosis with low priors and well-defined parent categories. [2] introduces conditional tree-structured Bayesian network (CTBN), a probabilistic approach that models conditional dependencies between classes in an effective yet computationally efficient way. Parameters of the CTBN model are captured in probabilistic prediction functions, and the structure is learned automatically from the data by maximizing the conditional log likelihood. The model makes predictions by finding the MAP assignment of class variables, with complexity linear to the number of class variables due to its tree structure. This dependency structure of class variables produces reliable probabilistic estimates and allows better performance of CTBN when comparing to other probabilistic methods.

3 Methodology

Our proposed model consists of: i) a model for generating low-dimensional summaries of the EKG inputs, and ii) an MLC model based on the CTBN supporting MLC tasks with label dependencies. The dependencies reflect the hierarchical organization of EKG classes at different levels of abstraction.

3.1 Low-dimensional EKG Representation

The EKG is defined by a complex high-frequency time-series signal. Hence, the key challenge is to summarize this signal more compactly so that it can be linked to different MLC frameworks working with vector-based inputs. In this work, we consider modern neural network architectures to perform this step and generate low-dimensional representation of the EKG signals. Briefly, given an input instance X, formed by time-series of measurements, the model defines a function g that maps X to a lower-dimensional vector space $X' = g(X)$. In the case of EKG the input signal X is a tensor of shape (c, l), where c is the number of EKG channels and l is the length of the signal. The output is a k-dimensional real-valued vector $X' = (x_1, ..., x_k)$ that reduces the temporal dimension of the original EKG signal and aims to capture the key information needed to support the classification task.

The transformation of the EKG signal to a specific low-dimensional representation can be defined and learned with the help of different ML architectures. Here we consider: multilayer perceptron (MLP), fully convolutional network (FCN), and recurrent neural network (RNN). All of them have been applied to time series classification and prediction tasks, some specifically to EKG classification [15, 19] and have shown decent performance.

3.2 Multi-label Classification Model

The low-dimensional vector-based representations X' of EKG support MLC tasks. Briefly, we are interested in learning a model $f : X' \rightarrow Y$ where X' is a low-dimensional representation of EKG, and Y is a binary vector of m class labels. One way to define and train an MLC model $f : X' \rightarrow Y$ is to express it in terms of conditional probability $P(Y|X')$. The best assignment of labels to the input vector X' is then obtained by calculating $Y^* = \arg\max_Y P(Y|X')$.

There are different ways to represent and train $P(Y|X')$. One solution is to rely on a set of independent classification models, one model per class variable, to define $P(Y|X') = \prod_i P(Y_i|X')$. However, using an independent set of classifiers to support the MLC may fail to represent the dependencies among classes. This is important especially for EKG classification where typical multi-label annotation combines classes at the different level of abstraction. To alleviate the problem, one can resort to different MLC methods such as those defined by classifier chains [13]. Briefly, a classifier chain model decomposes the conditional probability $P(Y|X')$ over class labels into a product of conditional probabilities over components of Y using the chain rule: $P(Y|X') = \prod_i P(Y_i|pa(Y_i), X')$ where class label Y_i depends on a subset of so called parent class variables $pa(Y_i)$ that Y_i depends on.

In general, selecting and optimizing the parent subsets defining the chain classifier is a hard task. In this work we resort to **tree structured label dependency model** where each individual label may depend on at most one other class label. The model we adopt is the Conditional Tree Structure Bayesian Network (CTBN) model proposed by Batal et al. [2]. It consists of a set of binary probabilistic classifiers for the transformed EKG vector X' and the value of at most one other class label. The classifiers are organized in a tree structure where the parent class denotes the class the variable depends on. Figure 1 illustrates a CTBN model with three binary class variables.

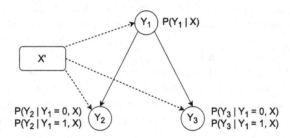

Fig. 1. An example of a CTBN with three binary class variables. The dash lines indicate the input and the solid lines model the hierarchy. Class variable Y_1 can be modeled with only one classifier, whereas Y_2 and Y_3 each requires two classifiers.

The advantages of the CTBN model are: (i) the tree structure can fit the hierarchy of class labels often used to annotate EKG, see Sect. 4.1, and (ii) the optimal assignment of the classes for the transformed input X' can be found efficiently in the time linear in the number of classes [2]. Following Batal et al. [2] we use logistic regression classifiers to define the individual classification models in the CTBN model, that is: $P(Y_i = 1|pa(Y_i) = j, X' = \mathbf{x}') = \sigma(\mathbf{w}_{i,j}^T \mathbf{x}')$ where σ is the logistic function and $\mathbf{w}_{i,j}$ are learnable weight vectors parameterizing individual models.

3.3   Training Models

The model that consists of the low-dimensional transformation g of the EKG signal and the CTBN implementing the MLC with dependencies among labels can be trained jointly using standard neural network optimization frameworks. Briefly, the CTBN part concurrently optimizes the weights of multiple logistic regression models implementing the different classifiers, where tree-based dependencies among the class variables are used to automatically select (via masking) all logistic regression models responsible for the specific training instance. We use average binary cross-entropy (BCE) loss and the AdamW optimizer with weight decay to optimize the models.

3.4 Making Predictions

The trained model allows us to transfer the original EKG signal x to its low-dimensional vector representation x', which in turn let us estimate $P(Y_i = 1|pa(Y_i), x')$ for all class labels. To make a prediction we want to identify the best possible assignment of class labels to all class variable, or in other words, the assignment that maximizes $y* = \arg\max_y P(Y = y|X' = x')$.

For the CTBN models, this optimization can be carried out efficiently across the tree-structure [2]. More specifically, we use a variant of the max-sum/max-product algorithm that runs in two passes with complexity linear to the number of class variables. In the first pass, information is sent upward from the leaves to the root, where each node compute the product of its local conditional probability and all probabilities sent from its children, and maximize and send the result over its value to its parent node. In the second pass, information is sent downward from the root to the leaves, where each node find its optimal assignment base on the assignment of its parent (if any) and its local conditional probability, and propagates the optimal assignment to its leaves.

4 Experiments

4.1 Data

We evaluate the proposed model on the EKG diagnostic task with 2 hierarchy levels using the publicly available PTB-XL dataset. The dataset consists of 21837 10-s long 12-lead EKGs from 18885 patients. This data is evenly distributed in terms of gender, with age covering a wide range of 0 to 95 years old. The EKG waveform is collected at 500 Hz sampling rate with 16 bit precision, and is down-sampled 100 Hz frequency. Each EKG instance is annotated by up to two cardiologists and labelled with 71 different EKG statements, using the SCP-ECG standard that covers diagnostic, form, and rhythm statements. We use the 44 diagnostic interpretations, and map them into 5 superclasses and 23 subclasses, as shown in Fig. 2. Each EKG signal can be assigned multiple class labels, even at the same hierarchy level. In addition to EKG readings, PTB-XL dataset provides extensive metadata on demographics, but we do not use them in our classification models.

4.2   Methods

Our experiments compare the performance of the CTBN model to the baseline multi-label classifier that relies on a set of independent classification models where each class is predicted from a low-dimensional EKG summary vector x'. We compare these two models on five different EKG transformation approaches: MLP, RNN, LSTM, GRU, and FCN. For the MLP approach, we flatten the EKG signal input over the temporal dimension, and apply 5 fully connected layers with activation, using a hidden size of 4096 perceptrons per layer. For the recurrent approaches (RNN, LSTM, GRU), we use one unidirectional recurrent layer with hidden dimension 128. For the FCN approach, we use 5 convolution layers with activation and max pooling, followed by an adaptive average pooling layer to reduce the temporal dimension.

All models are built to make joint predictions on 28 class labels that consist of all 5 superclass and 23 subclass labels. Since superclass labels are binary and each subclass variable has exactly one superclass, two logistic regressions are needed to cover the predictions of each subclass in CTBN. However, since no EKG instance can belong to a subclass without belonging to its parent superclass label we can simplify the CTBN by considering just one logistic regression model for each subclass variable, that is, predicting probability conditioned on its parent class being true. This reduces the total number of logistic regression classifiers used in our CTBN model to 28, which is the same number of as used for the baseline MLC model that predicts each class independently.

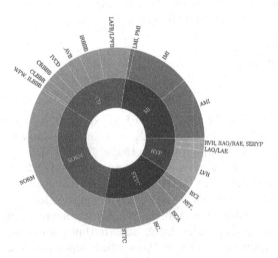

Fig. 2. A hierarchy of diagnostic classes describing the EKG readings in the PTB-XL dataset [20]. Diagnostic superclasses correspond to labels in the inner circle and subclasses to labels in the outer circle.

4.3   Metrics

We consider three different metrics to evaluate the models: exact match accuracy (EMA), conditional log likelihood loss (CLL-loss), and macro F1. Briefly, the exact match accuracy (EMA) computes the percentage of the instances whose predicted class vectors are exactly the same as their true class vector, i.e. all class variables are correctly predicted. The conditional log likelihood loss (CLL-loss) is defined for probabilistic methods as: $CLL\text{-}loss = -\sum_{k=1}^{n} \log P(\mathbf{y}^{(k)}|\mathbf{x}^{(k)})$. The CLL-loss for a test instance $\mathbf{x}^{(k)}$ is small if the predicted probability of the true class vector $\mathbf{y}^{(k)}$ is close to 1, and the CLL-loss is large if the predicted probability of $\mathbf{y}^{(k)}$ is close to 0. Finally, the macro F1 score is the unweighted arithmetic mean of all the per-class F1 scores, which for each individual class variable is calculated as the harmonic mean of its precision and recall.

EMA evaluates the success of the models in finding the conditional joint distribution for all class variables $P(\mathbf{Y}|\mathbf{X})$ and thus is appropriate in our MLC setting. CLL-loss evaluates how much probability mass the model assigns to the true class vector and is a useful measurement for probabilistic methods. For two models that both misclassify an instance according to EMA, CLL-loss will still favor the one that assigns higher probability to the correct output. We report the F1 score of the models, but note that it is not a very suitable metric for the MLC problem, since it is calculated separately for each class variable and then aggregated, and thus does not consider the dependencies between classes.

4.4   Results

All experimental results are obtained using 10-fold cross validation that is kept the same for all evaluated models. All the splits are obtained via stratified sampling while respecting patient assignments such that all records of the same patient are assigned to the same fold. As recommended in the PTB-XL dataset, we use folds 1–8 as training set, and folds 9 and 10 that underwent at least one human evaluation with high label quality as validation set and test set.

Table 1. Performance of the CTBN model vs non-CTBN baseline for five different EKG transformations on the PTB-XL dataset. The MLC problem is defined on the superclass and subclass labels.

Transformation	Baseline (non-CTBN)			CTBN		
	EMA	CLL	macro F1	EMA	CLL	macro F1
MLP	0.368	13.0	0.168	0.385	10.4	0.153
RNN	0.237	6.7	0.045	0.332	4.2	0.043
LSTM	0.367	5.1	0.188	0.419	3.3	0.248
GRU	0.393	4.6	0.248	0.473	3.0	0.301
FCN	0.448	4.0	0.337	0.498	2.6	0.329

Table 1 compares the performance of our CTBN model to the non-CTBN baseline defined by a set independent logistic regression models, one model per class variable. For all five EKG transformations representing three types of neural networks (fully connected, recurrent, convolutional), the CTBN model outperform its non-CTBN counterpart in terms of EMA and CLL-loss. We observe bigger improvements in EMA when using more complex EKG transformations, like the recurrent and convolutional neural networks. The results on F1 score are not consistent, as we see improvements when using LSTM and GRU, but not on MLP or FCN. This can be explained by the property of F1 score that it does not capture label dependencies and thus shows no advantage of CTBN.

Table 2. Performance of the CTBN model vs non-CTBN baseline evaluated with pair-wise statistical significance testing using 95% confidence interval.

	95% CI of $\Delta metric$ (lower, mean, upper)	
Transformation	ΔEMA (higher the better)	ΔCLL (lower the better)
MLP	(0.002, 0.016, 0.031)	(−3.497, −2.834, −2.154)
RNN	(0.070, 0.094, 0.117)	(−2.584, −2.488, −2.395)
LSTM	(0.043, 0.056, 0.075)	(−1.884, −1.761, −1.640)
GRU	(0.061, 0.080, 0.102)	(−1.801, −1.672, −1.540)
FCN	(0.031, 0.050, 0.067)	(−1.504, −1.402, −1.306)

Table 2 evaluates the performance of our CTBN model and the non-CTBN baseline using pair-wise statistical significance testing. We generate (with replacement) 1000 random bootstrap samples of sample size 1024 from the test set, and for each sample we evaluate our CTBN models and non-CTBN baselines using the desired metrics. We define $\Delta metric = metric_{ctbn} - metric_{baseline}$ and report the mean, upper bound, and lower bound of the 95% confidence interval of all 1000 samples. We conjecture that our CTBN models are statistically significantly better than corresponding baselines since the model consistently outperforms the baseline within an acceptable confidence interval, i.e. when ΔEMA is positive and ΔCLL is negative.

5   Conclusion

In this paper, we propose an ML model that improves EKG classification by leveraging the hierarchical class label dependencies. The model generates low-dimensional summaries of EKG instances with ML methods, and performs MLC using CTBN [2] that captures the dependencies between class variables. Our model uses logistic regression as probabilistic classifiers for the CTBN model, and can perform exact inference with complexity linear in the number of class variables. Our experimental evaluation on the PTB-XL [20] dataset shows that our approach outperforms the same ML architectures that do not incorporate class dependencies, and produces more reliable probabilistic estimates.

Acknowledgement. The work presented in this paper was supported in part by NIH grants R01EB032752 and R01DK131586. The content of this paper is solely the responsibility of the authors and does not necessarily represent the official views of NIH.

References

1. Aziz, S., Ahmed, S., Alouini, M.S.: ECG-based machine-learning algorithms for heartbeat classification. Sci. Rep. **11**, 18738 (2021)
2. Batal, I., Hong, C., Hauskrecht, M.: An efficient probabilistic framework for multi-dimensional classification. In: Proceedings of the 22nd ACM International Conference on Conference on Information & Knowledge Management - CIKM 2013 (2013)
3. Bulusu, S.C., Faezipour, M., Ng, V., Nourani, M., Tamil, L.S., Banerjee, S.: Transient st-segment episode detection for ECG beat classification. In: 2011 IEEE/NIH Life Science Systems and Applications Workshop (LiSSA) (2011)
4. Darmawahyuni, A., et al.: Deep learning-based electrocardiogram rhythm and beat features for heart abnormality classification. PeerJ. Comput. Sci. **8**, e825 (2022)
5. Di Marco, L.Y., Chiari, L.: A wavelet-based ECG delineation algorithm for 32-bit integer online processing. Biomed. Eng. Online **10**, 23 (2011)
6. He, K., Zhang, X., Ren, S., Sun, J.: Deep residual learning for image recognition. In: 2016 IEEE Conference on Computer Vision and Pattern Recognition (CVPR)
7. Hochreiter, S., Schmidhuber, J.: Long short-term memory. Neural Comput. **9**(8), 1735–1780 (1997)
8. Ismail Fawaz, H., et al.: InceptionTime: finding alexnet for Time Series classification. Data Min. Knowl. Disc. **34**(6), 1936–1962 (2020)
9. Malakouti, S., Hauskrecht, M.: Hierarchical adaptive multi-task learning framework for patient diagnoses and diagnostic category classification. In: Proceedings, IEEE International Conference on Bioinformatics and Biomedicine, 2019, 19323030 (2019)
10. Moody, B., Moody, G., Villarroel, M., Clifford, G.D., Silva, I.: MIMIC-III Waveform Database (version 1.0). PhysioNet (2020)
11. Moody, G.B., Mark, R.G.: The impact of the MIT-BiH Arrhythmia Database. IEEE Eng. Med. Biol. Mag. **20**(3), 45–50 (2001)
12. Ravanelli, M., Bengio, Y.: Speaker recognition from raw waveform with SincNet. In: 2018 IEEE Spoken Language Technology Workshop (SLT) (2018)
13. Read, J., Pfahringer, B., Holmes, G., et al.: Classifier chains for multi-label classification. Mach. Learn. **85**, 333–359 (2011)
14. Ribeiro, A.H., Ribeiro, M.H., Paixão, G.M.M., et al.: Automatic diagnosis of the 12-lead ECG using a deep neural network. Nat. Commun. **11**, 1760 (2020)
15. Roy, Y., Banville, H., Albuquerque, I., Gramfort, A., Falk, T.H., Faubert, J.: Deep learning-based Electroencephalography Analysis: a systematic review. J. Neural Eng. **16**(5), 051001 (2019)
16. Sharma, L.D., Sunkaria, R.K.: Inferior myocardial infarction detection using stationary wavelet transform and machine learning approach. SIViP **12**(2), 199–206 (2017). https://doi.org/10.1007/s11760-017-1146-z
17. Śmigiel, S., Pałczyński, K., Ledziński, D.: ECG signal classification using Deep Learning techniques based on the PTB-XL dataset. Entropy **23**(9), 1121 (2021)
18. Śmigiel, S., Pałczyński, K., Ledziński, D.: Deep learning techniques in the classification of ECG signals using R-peak detection based on the PTB-XL dataset. Sensors **21**(24), 8174 (2021)

19. Strodthoff, N., Wagner, P., Schaeffter, T., Samek, W.: Deep learning for ECG analysis: Benchmarks and insights from PTB-XL. IEEE J. Biomed. Health Inform. **25**(5), 1519–1528 (2021)
20. Wagner, P., Strodthoff, N., Bousseljot, R., Samek, W., Schaeffter, T.: PTB-XL, a large publicly available electrocardiography dataset (version 1.0.1). PhysioNet (2020)
21. Wang, Z., Yan, W., Oates, T.: Time Series classification from scratch with Deep Neural Networks: a strong baseline. In: 2017 International Joint Conference on Neural Networks (IJCNN) (2017)
22. Westhuizen, J.V., Lasenby, J.: Techniques for visualizing LSTMs applied to electrocardiograms. arXiv:1705.08153: Machine Learning (2017)
23. Zidelmal, Z., Amirou, A., Adnane, M., Belouchrani, A.: QRS detection based on wavelet coefficients. Comput. Methods Programs Biomed. **107**(3), 490–496 (2012)

Discriminant Audio Properties in Deep Learning Based Respiratory Insufficiency Detection in Brazilian Portuguese

Marcelo Matheus Gauy[1]([✉]) [ID], Larissa Cristina Berti[2], Arnaldo Cândido Jr[3] [ID],
Augusto Camargo Neto[1], Alfredo Goldman[1] [ID], Anna Sara Shafferman Levin[1],
Marcus Martins[1], Beatriz Raposo de Medeiros[1] [ID], Marcelo Queiroz[1],
Ester Cerdeira Sabino[1] [ID], Flaviane Romani Fernandes Svartman[1] [ID],
and Marcelo Finger[1] [ID]

[1] Universidade de São Paulo, Butanta, São Paulo, SP, Brazil
marcelo.gauy@usp.br
[2] Universidade Estadual Paulista, Marília, SP, Brazil
[3] Universidade Estadual Paulista, São José do Rio Preto, SP, Brazil

Abstract. This work investigates Artificial Intelligence (AI) systems
that detect respiratory insufficiency (RI) by analyzing speech audios,
thus treating speech as a RI biomarker. Previous works [2,6] collected
RI data (*P1*) from COVID-19 patients during the first phase of the pandemic and trained modern AI models, such as CNNs and Transformers,
which achieved 96.5% accuracy, showing the feasibility of RI detection
via AI. Here, we collect RI patient data (*P2*) with several causes besides
COVID-19, aiming at extending AI-based RI detection. We also collected
control data from hospital patients without RI. We show that the considered models, when trained on P1, do not generalize to P2, indicating
that COVID-19 RI has features that may not be found in all RI types.

Keywords: Respiratory Insufficiency · Transformers · PANNs

1 Introduction

Respiratory insufficiency (RI) is a condition that often requires hospitalization,
and which may have several causes, including asthma, heart diseases, lung diseases and several types of viruses, including COVID-19. This work is part of the
SPIRA project [1], which aims to provide cheap AI tools (cellphone app) for
the triage of patients by classifying their speech as RI-positive (requiring medical evaluation). Previous works [2,6] focused on COVID-19 RI. Here, we extend
them to more general RI causes.

Partly supported by FAPESP grants 2020/16543-7 and 2020/06443-5, and by Coordenação de Aperfeiçoamento de Pessoal de Nível Superior - Brasil (CAPES) - Finance
Code 001. Carried out at the Center for Artificial Intelligence (C4AI-USP), supported
by FAPESP grant 2019/07665-4 and by the IBM Corporation.

J. M. Juarez et al. (Eds.): AIME 2023, LNAI 13897, pp. 271–275, 2023.
https://doi.org/10.1007/978-3-031-34344-5_32

We view *speech as a biomarker*, meaning that one can detect RI through speech [2,6]. In [2], we recorded sentences from patients and a Convolutional Neural Network (CNN) was trained to achieve 87.0% accuracy for RI detection. Transformers-based networks (MFCC-gram Transformers) achieved 96.5% accuracy on the same test set [6]. Here, we study multiple models in the general RI case. For that, we provide new RI data, with 26 RI patient audios with many causes and 116 (non-RI) control audios. We call the data from [2] P1 and the new data P2.

Transformers [6] trained on P1 data using MFCC-grams obtain 38.8 accuracy (0.367 F1-score) when tested on P2 data. Pretrained audio neural networks (PANNs) [11] confirm this result, with CNN6, CNN10 and CNN14 trained on P1 data are comparable to [6] on P1 test set, but achieve less than 36% accuracy (less than 0.34 F1-score) on P2 data[1]. We provide some hypotheses for this difference in Sect. 4.

2 Related Work

Transformers were proposed to deal with text [3,15]. Later, researchers succeeded in using Transformers in computer vision [10] and audio tasks [9,12]. Transformers benefit from two training phases: **pretraining** and **finetuning**. The former involves self-supervised training on (a lot of) unlabeled data using synthetic tasks [3]. The latter involves training a model extension using labeled data for the target task. One may obtain good performance after finetuning with little labeled data [3]. PANNs were proposed in [11]. There, multiple PANNs were pretrained on AudioSet [8], a 5000-hour dataset of Youtube audios with 527 classes. These pretrained models were finetuned for several tasks such as audio set tagging [11], speech emotion recognition [7] and COVID-19 detection [14].

3 Methodology

General RI Dataset. During the pandemic, we collected patient audios in COVID-19 wards. Healthy controls were collected over the internet. This data was used for COVID-19 RI detection [1,2,4–6]. Now, we collect RI data with several causes from 4 hospitals: Beneficência Portuguesa (*BP*), Hospital da Unimar (*HU*), Santa Casa de Marília (*SC*) and CEES-Marília (*CM*). We collect three utterances: 1) a sentence[2] that induces pauses, as in P1. 2) A nursery rhyme with predetermined pauses, as in P1. 3) The sustained vowel 'a'. We expect the utterances to induce more pauses, occurring in unnatural places [4], in RI patients. As all data was collected in similar environments, adding ward noise as in [2] is no longer required and results will not be affected by bias from the collection procedure. As a downside, controls have a health issue. Specifically,

---

[1] Initial tests attain above 95% accuracy (above 0.93 F1-score) when training and testing on P2 data in all 4 networks. So P2 is not harder, it is only different.

[2] "O amor ao próximo ajuda a enfrentar essa fase com a força que a gente precisa".

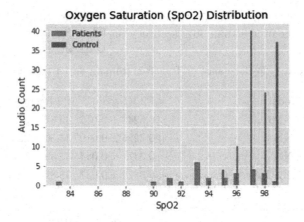

Fig. 1. SpO2 distribution in P2. Patient SpO2 mean is 94.31. For controls it is 97.66.

long COVID cases were not part of the 116 controls, as we believe they could present biases [13]. Moreover, the fewer number of RI patients relative to controls (outside the pandemic) means we should use F1-score. In P1, an RI patient was selected if his oxygen saturation level (**SpO2**) was below 92%. In P2, RI was diagnosed by physicians. As other factors may influence the diagnosis, RI patients often have SpO2 above 92%. Figure 1 shows SpO2 levels of patients and controls. We have 24 RI patients and 118 controls. However, 2 controls had SpO2 below 92%. As that fits the criteria for RI, we reclassified those 2 controls. Lastly, we only use the first utterance as in [2,6]. We have 14 RI men and 12 RI women and a mean audio duration (MAD) of 8.14 s. Also, controls comprise 36 men and 80 women and a MAD of 7.41 s.

Preprocessing. We break the audios in 4 s chunks, with 1 s steps [2,5,6]. This data augmentation prevents the audio lengths from biasing the results. For the MFCC-gram Transformers, the audios are resampled at 16 kHz[3]. We extract 128 MFCCs as in [5,6]. For the PANNs, we do the processing steps from [11].

4 Results and Discussion

Table 1 shows that P2 is substantially different from P1. We take the pretrained MFCC-gram Transformers from [5], and fine-tune it 5 times, with a learning rate of 10^{-4}, batch size 16 and 20 epochs, on P1 training set of [2], to obtain models with above 95% accuracy on P1 test set of [2]. The best model on P1 validation set of [2] after each epoch is saved. These 5 models attain an average accuracy of only 38.8% on P2. Additionally, we do the same with CNN6, CNN10 and CNN14. We take them from [11], and fine-tune[4] them 5 times each, thus

---

[3] Performance difference by resampling the audios is minimal.

[4] Again, we use 20 epochs, batch size 16, learning rate 10^{-4} and best models are saved.

obtaining models with above 95% accuracy on P1. These 5 models of the 3 CNNs attain an average of less than 36% accuracy on P2.

Table 1. Performance on P2, after training on P1 training set.

Model	P2 F1-score	P2 Accuracy
CNN6	0.3243 ± 0.052	32.67 ± 5.34
CNN10	0.3226 ± 0.019	33.56 ± 2.20
CNN14	0.3371 ± 0.044	35.39 ± 5.35
MFCC-gram Transformers	0.3674 ± 0.037	38.82 ± 4.93

Figure 2 shows the error distribution on P2 for MFCC-gram Transformers according to the hospital[5] the data was collected[6]. The left side shows true positives (TP) and false negatives (FN) of P2 RI patients. Almost all from 'BP' and the 2 'O' files (not diagnosed with RI but low SpO2) were TP. Most of the 'HU' as well as almost all from 'SC' were FN. We can see two reasons for the discrepancy: 1) COVID-19 patients are more numerous in 'BP' than 'HU' or 'SC'; 2) RI patients from 'BP' are more severe cases than 'HU' or 'SC'. As P1 was collected during the pandemic, it is filled with severe cases. The right side shows true negatives (TN) and false positives (FP) of P2 controls. 'CM' were mostly TN while 'O' were mostly FP. It is possible that certain comorbidities (more common in 'O' than 'CM') led the model to errors as it only trained on severe RI patients contrasted with healthy controls.

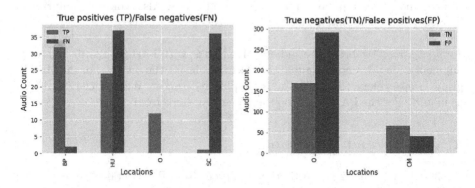

Fig. 2. RI patient audio count according to the hospital the data was collected.

Thus, our results suggest that it is possible to identify the RI cause via AI, as different forms of RI have distinct audio features that are learned by the models. However, this task will require considerably more data on each RI cause.

---

[5] 'O' (Other) and 'CM' represent controls. The other hospitals refer only to patients.
[6] Other angles do not add much. Using the PANNs yields similar results.

5 Conclusion and Future Work

We presented new RI data expanding on P1 [2]. RI in P2 data has many causes such as asthma, heart diseases, lung diseases, unlike P1 (COVID-19 only). Our results suggest P1 and P2 have relevant differences as AI models trained on P1 data perform poorly on P2 data. Thus some audio properties of COVID-19 RI are distinct from general RI causes, which should be identifiable.

Future work involves the expansion of P2 data so we may train models that detect RI as well as its cause. This would benefit more complex models as currently CNN6 and CNN10 outperform CNN14 and MFCC-gram Transformers.

References

1. Aluísio, S.M., Camargo Neto, A.C.d, et al.: Detecting respiratory insufficiency via voice analysis: the SPIRA project. In: Practical Machine Learning for Developing Countries at ICLR 2022. Proceeding. ICLR (2022)
2. Casanova, E., Gris, L., et al.: Deep learning against COVID-19: respiratory insufficiency detection in Brazilian Portuguese speech. In: Findings of the Association for Computational Linguistics: ACL-IJCNLP 2021, pp. 625–633. ACL, August 2021
3. Devlin, J., Chang, M.W., et al.: BERT: pre-training of deep bidirectional transformers for language understanding. arXiv preprint arXiv:1810.04805 (2018)
4. Fernandes-Svartman, F., Berti, L., et al.: Temporal prosodic cues for COVID-19 in Brazilian Portuguese speakers. In: Proceedings of Speech Prosody 2022, pp. 210–214 (2022)
5. Gauy, M., Finger, M.: Acoustic models for Brazilian Portuguese speech based on neural transformers (2023, submitted for publication)
6. Gauy, M.M., Finger, M.: Audio MFCC-gram transformers for respiratory insufficiency detection in COVID-19. In: STIL 2021, November 2021
7. Gauy, M.M., Finger, M.: Pretrained audio neural networks for speech emotion recognition in Portuguese. In: Automatic Speech Recognition for Spontaneous and Prepared Speech Speech Emotion Recognition in Portuguese. CEUR-WS (2022)
8. Gemmeke, J.F., Ellis, D.P., et al.: Audio set: an ontology and human-labeled dataset for audio events. In: International Conference on Acoustics, Speech and Signal Processing (ICASSP), pp. 776–780. IEEE (2017)
9. Gong, Y., Lai, C.I., et al.: SSAST: self-supervised audio spectrogram transformer. In: Proceedings of the AAAI Conference on Artificial Intelligence, vol. 36, pp. 10699–10709 (2022)
10. Khan, S., Naseer, M., et al.: Transformers in vision: a survey. ACM Comput. Surv. **54**(10s) (2022)
11. Kong, Q., Cao, Y., et al.: PANNs: large-scale pretrained audio neural networks for audio pattern recognition. IEEE/ACM Trans. Audio Speech Lang. Process. **28**, 2880–2894 (2020)
12. Liu, A.T., Yang, S.W, et al.: Mockingjay: unsupervised speech representation learning with deep bidirectional transformer encoders. In: International Conference on Acoustics, Speech and Signal Processing (ICASSP), pp. 6419–6423. IEEE (2020)
13. Robotti, C., Costantini, G., et al.: Machine learning-based voice assessment for the detection of positive and recovered COVID-19 patients. J. Voice (2021)
14. da Silva, D.P.P., Casanova, E., et al.: Interpretability analysis of deep models for COVID-19 detection. arXiv preprint arXiv:2211.14372 (2022)
15. Vaswani, A., Shazeer, N., et al.: Attention is all you need. Adv. Neural. Inf. Process. Syst. **30**, 5998–6008 (2017)

ECGAN: Self-supervised Generative Adversarial Network for Electrocardiography

Lorenzo Simone$^{(\boxtimes)}$ and Davide Bacciu

Department of Computer Science, University of Pisa, Pisa 56122, Italy
lorenzo.simone@di.unipi.it

Abstract. High-quality synthetic data can support the development of effective predictive models for biomedical tasks, especially in rare diseases or when subject to compelling privacy constraints. These limitations, for instance, negatively impact open access to electrocardiography datasets about arrhythmias. This work introduces a self-supervised approach to the generation of synthetic electrocardiography time series which is shown to promote morphological plausibility. Our model (ECGAN) allows conditioning the generative process for specific rhythm abnormalities, enhancing synchronization and diversity across samples with respect to literature models. A dedicated sample quality assessment framework is also defined, leveraging arrhythmia classifiers. The empirical results highlight a substantial improvement against state-of-the-art generative models for sequences and audio synthesis.

Keywords: generative deep learning · self-supervised · generative adversarial networks · electrocardiography

1 Introduction

Nowadays, the usage of electronic health records and the digital health (e-health) market have grown significantly due to technological advancements. Across the field, there is an upward trend in collecting heterogeneous data via diagnostic tests, clinical trials and wearable devices, supported by demographic insights. The recording of heart's electrical activity (ECG), represents the most common non-invasive diagnostic tool for the early detection of life threatening conditions. Machine learning can play a key role in assisting clinicians by efficiently monitoring and stratifying the risk of patients [3]. Although fully supervised models could seem a flawless end-to-end solution for clinical research, they come with various drawbacks. For instance, such models are typically trained leveraging clinical datasets often failing in adequately covering rare occurrences.

Deep generative models (DGMs), by approximating real conditional distributions, are usually capable of guiding the generative process towards a specific class of samples. For this reason, they represent a relevant tool for reducing

J. M. Juarez et al. (Eds.): AIME 2023, LNAI 13897, pp. 276–280, 2023.
https://doi.org/10.1007/978-3-031-34344-5_33

unbalance from rare diseases via data augmentation or patient-specific synthesis [5]. privacy and data anonymization.

In this paper, we propose a novel architecture referred to as ECGAN, combining two fields of research: self-supervised learning (SSL) and deep generative models for time series. The role of self-supervised learning in this research stands in exploiting the underlying time series dynamics via recurrent autoencoders. The features learned through a preliminary reconstruction task are transferred via weight sharing to the generator and a latent space projection. The proposed model is empirically compared with state of the art generative models for time series and audio synthesis [4,6,8], yielding competitive results.

The main contributions of this work are: (1) We introduce a deep generative model, specifically designed for electrocardiography, intersecting between self-supervised learning and the generative adversarial framework. (2) We propose a parsimonious transfer learning framework requiring lesser resources than its generative counterparts. (3) We evaluate different methods for quantitatively assessing synthetic data by inspecting latent projections of an ECG arrhythmia classifier $\mathcal{C}$.

2   The ECGAN Model

ECGAN Architecture. The architecture comprises a latent space projection learnable through self-supervision and a generative-adversarial network for the sequential domain. The model is composed of: a recurrent encoding $E_\phi : \mathcal{X} \to \mathcal{H}$, which produces a latent representation $\mathcal{H} \in \mathbb{R}^{h \times n'}$ through H and a recurrent decoder, sharing its weights with the generator $G(\mathcal{H}; \theta_g)$. We outlined each component in addition to their interaction paths and losses in Fig. 1.

There are two distinctive workflows possible, both traversing the latent space. The first path starts from the ECG input and learns to produce faithful reconstructions without involving adversarial components. Alternatively, the generative process starts by sampling through the latent feature projection and reconstructing a synthetic beat via G, which is finally evaluated by D. The non-deterministic behavior of the approach derives from learning the generator's implicit probability distribution sampling directly from the latent space instead of using random white noise as in the original setting. The weights of the projection map are learned during the pre-training task and transferred for the downstream generative task.

Sequence generation is conditioned at global level using embedding layers. During training and sampling, the hidden state of the generator is initialized with a non-linear embedding of the corresponding ECG label. Likewise, the discriminator's output is perturbed by linearly combining its output and the associated embedding. Hence, they retain an initial conditional representation about the generated/classified sequence. This allows to both condition the generative process and influence the prediction based on prior class information.

Adversarial Training. The adoption of combining WGAN principles introduced by [1] and the proposed novelties heavily mitigated mode collapse scenar-

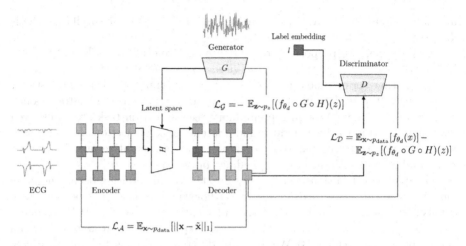

Fig. 1. Visual summary of the workflow of ECGAN. The main components (trapezoids) are correlated with their objective functions (red arrows). The sequential ECG input is either fed into the SSL block for preliminary reconstruction or to the discriminator D for adversarial training. (Color figure online)

ios. Thus, the model minimizes the Wasserstein distance via the Kantorovich-Rubinstein duality

$$W(p_{data}, \tilde{p}) = \frac{1}{K} \sup_{\|f\|_L \leq K} \mathbb{E}_{x \sim p_{data}}[f(x)] - \mathbb{E}_{x \sim \tilde{p}}[f(x)]. \tag{1}$$

Differently from literature, we sample a noise vector $\mathbf{z} \sim p_z$ processing it through the pre-trained projection map H, which is subsequently fed to the generator. As a result, the objective for the generator is

$$\mathcal{L}_G = -\mathbb{E}_{z \sim p_z}\left[(f_{\theta_d} \circ G \circ H)(z)\right]. \tag{2}$$

Meanwhile, the discriminator is a real-valued mapping $f(\mathbf{x}; \theta_d) \colon \mathbb{R}^n \to \mathbb{R}$ using a linear activation function. Its K-Lipschitz continuity constraints in (1) are enforced by gradient clipping and the resulting objective is:

$$\mathcal{L}_D = \mathbb{E}_{x \sim p_{data}}\left[f_{\theta_d}(x)\right] - \mathbb{E}_{z \sim p_z}\left[(f_{\theta_d} \circ G \circ H)(z)\right]. \tag{3}$$

In other words, the discriminator should learn to estimate the distance between real patterns and synthesized ECGs, overcoming the prior advantage of the generator. Differently, the latter focuses exclusively on minimizing the Wasserstein distance between p_{data} and $\tilde{p}$ in (1).

3   Experimental Analysis

To demonstrate the capability of the proposed model to synthesize realistic ECGs, we employed two different open datasets: MIT-BIH Arrhythmia Database

Lead II CHF

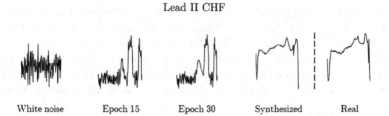

White noise   Epoch 15   Epoch 30   Synthesized   Real

Fig. 2. Sampling progression over training epochs from ECGAN. The process starts with noise vectors and ends with a fully trained (SSL and adversarial) model sampling from CHF instances on the forth column.

and BIDMC Congestive Heart Failure Database [2,7]. Each dataset is repeatedly split five times through a randomized hold-out technique, which assigns an equally partitioned 25% dedicated to hyperparameters tuning and model assessment of the classifier C. The remaining portion is entirely dedicated for the generative task (3998 and 11576 samples respectively).

The experiments compare ECGAN with state of the art generative models for signals, including WaveGAN and SpecGAN [4]. These models were originally designed for audio synthesis, hence we performed a Short-time Fourier transform (STFT) to obtain spectrograms. We report the results over five different runs of the sampling procedure in Table 1. The proposed model consistently outperforms models selected from literature in both tasks considering the proposed metrics.

Table 1. Summary of the metrics evaluated for each generative model on both datasets. We report: the inception score (IS), the Frechèt Inception Distance (FID), the linear maximum-mean-discrepancy (MMD) and the Wasserstein distance (W).

Model	FID$(p_{data}, \tilde{p}) \downarrow$	IS$(\tilde{p}) \uparrow$	MMD$(p_{data}, \tilde{p}) \downarrow$	W$(p_{data}, \tilde{p}) \downarrow$
BIDMC Dataset				
ECGAN	**233.86 ± 17.24**	**1.97 ± 0.09**	**31.71 ± 4.43**	**0.52 ± 0.01**
TimeGAN	328.03 ± 2.13	1.60 ± 0.01	144.90 ± 3.51	1.03 ± 0.00
WaveGAN	839.42 ± 7.26	1.32 ± 0.01	332.59 ± 8.93	0.81 ± 0.02
C-RnnGAN	917.22 ± 63.54	1.35 ± 0.03	788.67 ± 27.85	2.07 ± 0.03
SpecGAN	942.10 ± 8.25	1.26 ± 0.02	349.88 ± 6.56	0.79 ± 0.00
MITBIH Dataset				
ECGAN	**45.49 ± 1.82**	**1.41 ± 0.03**	**17.41 ± 1.62**	0.40 ± 0.01
C-RNNGAN	91.35 ± 3.00	1.40 ± 0.01	27.57 ± 1.15	0.52 ± 0.01
TimeGAN	113.81 ± 2.37	1.11 ± 0.01	60.51 ± 2.23	0.68 ± 0.01
WaveGAN	151.22 ± 0.94	1.03 ± 0.00	64.56 ± 0.84	**0.32 ± 0.00**
SpecGAN	151.21 ± 0.95	1.01 ± 0.00	64.97 ± 0.58	0.35 ± 0.01

A contextual estimate of the impact of the SSL module and the subsequent adversarial phase over different training epochs is shown in Fig. 2. We show, from left to right, the progression of a fixed random vector from the latent space through the SSL process. The fourth column represents the synthesized ECG pattern after the adversarial training. Lastly, the fifth column belongs to the best match, according to the DTW distance, between the generated and real pattern.

4 Discussion

We have introduced ECGAN, a generative model for ECG data leveraging self-supervised learning principles. Throughout the experiments, ECGAN has been shown to be able to produce morphologically plausible ECGs, including specific pattern abnormalities guided by input-conditioning. Our self-supervised generative adversarial framework encourages the sampling process to synthesize patterns coherently to real samples' dynamics. Anyway, a proper trade-off between preliminary and adversarial phases by a rigorous hyperparameter selection is needed.

Ultimately, we obtained competitive results for the expected properties over the monitored metrics. The synthesized patterns are also suitable for data augmentation procedures as suggested by the assessment of their functionality. We believe that our contribution could pave the way to several other biomedical applications.

References

1. Arjovsky, M., Chintala, S., Bottou, L.: Wasserstein generative adversarial networks. In: Proceedings of the 34th International Conference on Machine Learning, pp. 214–223. JMLR.org (2017)
2. Baim, D., et al.: Survival of patients with severe congestive heart failure treated with oral milrinone. J. Am. College Cardiol. **7**(3), 661–670 (1986). https://doi.org/10.1016/S0735-1097(86)80478-8
3. Beaulieu-Jones, B.K., et al.: Machine learning for patient risk stratification: standing on, or looking over, the shoulders of clinicians? NPJ Digital Med. **4**(1), 1–6 (2021)
4. Donahue, C., McAuley, J., Puckette, M.: Adversarial audio synthesis. In: International Conference on Learning Representations (2019)
5. Golany, T., Radinsky, K.: PGANs: personalized generative adversarial networks for ECG synthesis to improve patient-specific deep ECG classification. In: Proceedings of the AAAI Conference on Artificial Intelligence vol. 33, no. 01, pp. 557–564 (Jul 2019). https://doi.org/10.1609/aaai.v33i01.3301557
6. Mogren, O.: C-RNN-GAN: a continuous recurrent neural network with adversarial training. In: Constructive Machine Learning Workshop (CML) at NIPS (2016)
7. Moody, G.B., Mark, R.G.: The impact of the mit-bih arrhythmia database. IEEE Eng. Med. Biol. Mag. **20**(3), 45–50 (2001)
8. Yoon, J., Jarrett, D., Van der Schaar, M.: Time-series generative adversarial networks. In: Advances in Neural Information Processing Systems, vol. 32 (2019)

Data Analysis and Statistical Models

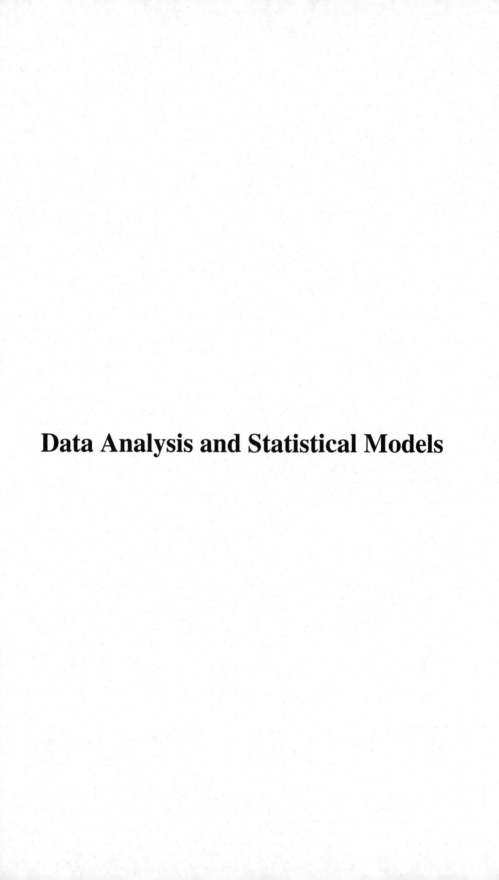

Nation-Wide ePrescription Data Reveals Landscape of Physicians and Their Drug Prescribing Patterns in Slovenia

Pavlin G. Poličar[1]([✉]), Dalibor Stanimirović[2], and Blaž Zupan[1]

[1] Faculty of Computer and Information Science, University of Ljubljana,
1000 Ljubljana, Slovenia
pavlin.policar@fri.uni-lj.si
[2] Faculty of Public Administration, University of Ljubljana, 1000 Ljubljana, Slovenia

Abstract. Throughout biomedicine, researchers aim to characterize entities of interest, infer landscapes of cells, tissues, diseases, treatments, and drugs, and reason on their relations. We here report on a data-driven approach to construct the landscape of all the physicians in Slovenia and uncover patterns of their drug prescriptions. To characterize physicians, we use the data on their ePrescriptions as provided by the Slovenian National Institute of Public Health. The data from the entire year of 2018 includes 10,766 physicians and 23,380 drugs. We describe physicians with vectors of drug prescription frequency and use the t-SNE dimensionality reduction technique to create a visual map of practicing physicians. We develop an embedding annotation technique that describes each visually-discernible cluster in the visualization with enriched top-level Anatomical Therapeutic Chemical classification terms. Our analysis shows that distinct groups of physicians correspond to different specializations, including dermatology, gynecology, and psychiatry. The visualization also reveals potential overlaps of drug prescribing patterns, indicating possible trends of physicians practicing multiple disciplines. Our approach provides a helpful visual representation of the landscape of the country's physicians, reveals their prescription domains, and provides an instrument to inform and support healthcare managers and policymakers in reviewing the country's public health status and resource allocation.

Keywords: drug prescription · drug prescribing patterns · physicians · two-dimensional embedding · explainable point-based visualisations · ATC classification

1 Introduction

The practice of medicine is a broad field comprising various medical specialties, each focusing on specific areas of the human body and its functions. One way to characterize a physician's area of interest and operation is by analyzing the drugs they prescribe to their patients. Different medical specialists will typically

J. M. Juarez et al. (Eds.): AIME 2023, LNAI 13897, pp. 283–292, 2023.
https://doi.org/10.1007/978-3-031-34344-5_34

prescribe different medications based on their expertise and the types of patients they are treating. For example, a cardiologist will typically prescribe medications for hypertension and heart disease, while a neurologist will prescribe medicines for neurological disorders such as seizures and Parkinson's.

The literature includes only a limited number of reports in that aim to characterize physicians based on their drug-prescribing patterns. Akhlaghi et al. [1] inspect data from Iran's second most populous province and identify important pairs of co-prescribed drugs. They use the occurrence of these pairs as features in a random-forest model trained to predict physician specialty. Shirazi et al. [10] use an unsupervised community-detection approach to identify nine major groups of physician specialties from a bi-partite *physician-drug* graph. In a different targeted study, Garg et al. [4] focus on identifying family physicians using machine learning methods. However, their research does not include prescription data but uses physician-specific attributes, such as sex, age, and various certifications.

In contrast to previous work, we do not have access to information about pyhsicians' actual specialties. We instead leverage an unsupervised machine learning approach to uncover and characterize groups of physicians with similar drug-prescribing patterns. Our study focuses on all physicians from Slovenia. To the best of our knowledge, this is the first study to identify physician specialties based on drug prescribing patterns encompassing the entire country's data. To this end, we develop a straightforward statistical approach based on p-value hypothesis testing that identifies characteristic top-level Anatomical Therapeutic Chemical (ATC) classification terms for visually-discernible clusters in a given two-dimensional embedding.

We start our report with a description of the data, a two-dimensional embedding approach, and a method to explain visually discernable clusters. We provide the results in annotated two-dimensional maps of physicians. In the discussion, we show that the map visually and interpretably reveals the landscape of Slovenian physicians and their drug-prescribing patterns.

2   Data and Methods

We here introduce the ePrescription data used throughout the study and provide an overview of the data construction and filtering approaches. We then describe our proposed method to identify overrepresented top-level ATC classification terms in visually discernible clusters in a t-SNE two-dimensional embedding.

2.1   Data

We use anonymized data containing all ePrescription records from Slovenia prescribed in 2018. We obtained the data from the Slovenian National Institute of Public Health. We have to note that the data, albeit anonymized, is not public and, at this stage, cannot be openly shared, subject to restrictive Slovenian law. Slovenia has a centralized healthcare system and national eHealth solution, making ePrescription records representative of the whole population of Slovenia. We chose 2018 to inspect physicians' landscape in the year before the pandemic era. In the early stages of the COVID-19 pandemic, many non-essential medical procedures were temporarily halted in Slovenia. These interruptions may have made interpreting the final embedding more difficult.

To construct the data matrix, we considered each physician and counted the number of times they prescribed each drug in 2018. To avoid drug brand preferences, we map each of the 23,380 drugs to its corresponding ATC classification term. We characterize each physician through a vector of prescription counts. The proposed procedure created a data matrix comprising 10,766 unique physicians and 920 unique level-5 ATC classification terms. To make our analysis more robust, we removed physicians that prescribed fewer than 25 drugs throughout the year, leaving us with 7,290 physicians. Metadata, such as patient age and sex, were excluded from the main analysis but also proves helpful, as discussed in Sect. 3.

2.2 Methods

Our approach aims to identify and annotate clusters in a given two-dimensional embedding. We here use t-SNE embeddings [7] because as it prioritizes cluster identification, though we could conduct a similar analysis on any other dimensionality reduction approach that yields two-dimensional maps of physicians. As is standard in dimensionality reduction of high-dimensional data, we first extract the top fifty principal components from the data [5]. We perform t-SNE dimensionality reduction with the openTSNE library v0.6.2 [8] using cosine distances, a perplexity value of 50, an exaggeration factor of 1.5, and degrees of freedom set to 0.8. We found that these parameter settings produce clear, well-separable clusters.

The first step of our pipeline is to identify visually discernible clusters in the two-dimensional embedding. As such, the embedding method of choice should produce discrete, well-separated clusters. On the resulting map, we then use the DBSCAN clustering algorithm [3] and manually tune the hyperparameters to achieve a visually sensible clustering. We found that setting DBSCAN's parameter values epsilon to 1.4 and setting the minimum number of samples to 25 identified clusters that best coincided with visually discernable groups of points in the embedding. It is worth noting that this aspect of the procedure could be automated, for instance, by maximizing the silhouette score [9] or by using DBSCANs internal parameter selection procedure [3].

To identify drugs prescribed more frequently in the identified clusters, we use a two-sample t-test and perform multiple hypothesis correction using the false discovery rate (FDR) with a threshold of $\alpha < 0.01$. To avoid selecting common drugs widely prescribed by all physicians or drugs where only minor differences in prescription frequency occur, we remove any drugs with a log-fold-change smaller than 0.25. The log-fold-change is the ratio of change between two values on a logarithmic scale and is used to compare changes over multiple orders of magnitude. As we are interested primarily in common prescription patterns in various physician specializations, we also require drugs to be prescribed by at least 25% of all physicians within the cluster. This produces a list of overrepresented level-5 ATC classification terms in each cluster. We then use these terms in the next step for identifying overrepresented, top-level ATC classification terms.

Finally, to perform enrichment of top-level ATC terms, we use the hypergeometric test and perform multiple hypothesis corrections using FDR with a threshold of $\alpha < 0.01$.

3   Results and Discussion

Figure 1 shows the resulting annotated map of physicians from Slovenia. With the proposed methods, we could determine the specific drug-prescribing patterns for most of the clusters in the physician map and abstract these patterns and their related specializations. We also find that some common physician specialties are missing from our annotations, including rheumatologists, oncologists, and radiologists. This result does not imply that there are no such specialists in Slovenia, but rather that their drug prescribing patterns likely coincide with other specializations and are probably contained within other clusters. In a related study, Shirazi et al. [10] apply community detection to ePrescription data from Iran and identify clusters of different specializations. However, they find that many identified clusters contain multiple specializations. Among these are groups of neurologists and cardiologists, internists, general practitioners, and dermatologists.

Our procedure only identifies overrepresented ATC classification terms. It is still up to us to manually determine which combination of these terms corresponds to each specialization. In many cases, the specialization of each cluster is fairly obvious, as the drugs prescribed by physicians in each cluster predominantly come from a single, top-level ATC classification group. We observed such results with gynecologists, neurologists, dermatologists, ophthalmologists, and endocrinologists.

Sometimes the top-level ATC classification terms alone are insufficient, so we examine overrepresented ATC classification terms further down the ontology by repeating the same enrichment scoring procedure. For instance, both cardiologists and hematologists predominantly prescribe drugs from the B (blood and blood-forming organs) ATC classification group making them difficult to distinguish from one another. However, upon closer inspection, we find that the top cluster tends to prescribe ACE inhibitors and diuretics more often than the lower cluster – drugs that are often used to treat cardiovascular conditions, indicating that the top cluster corresponds to cardiologists and the lower to hematologists.

We are able to confidently assign a specialization to most clusters, with the exception of the lower left cluster, which we guess corresponds to general practitioners. There are also two central clusters, colored green and light green, with no enriched top-level ATC classification terms. Inspecting lower-level ATC classification terms yielded no satisfactory conclusion. Therefore, we opt to leave these clusters unlabelled.

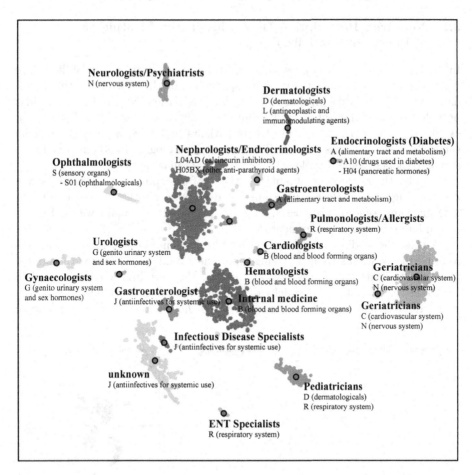

Fig. 1. Annotated t-SNE map of Slovenia phyicians characterized by their drug prescriptions. Due to the positional invariance of nonlinear embedding methods, we omit axis labels and scales. Each point on the plot represents a physician. Points are color-coded according to their cluster assignments. Physicians identified as outliers by the DBSCAN clustering algorithm are colored light gray. We indicate the center of each cluster by a red point, accompanied by the determined cluster specialization. We list the automatically-inferred, top-level ATC codes characterizing each cluster below the specialization. It is worth noting that two of the center-most clusters do not contain an annotation. Upon further investigation, we found that there were no overrepresented top-level ATC terms for these clusters, and thus, we could not assign them a distinct specialization. We hypothesize that these physicians are general practitioners. (Color figure online)

3.1   Non-drug Prescription Data Aids in the Identification of Physician Specializations

In certain instances, drug-prescribing patterns alone may not be sufficient to unambiguously determine a physician's specialization. In such cases, other meta-data available in ePrescription records, such as patient age and sex, can also provide valuable information in determining a physician's area of expertise. For example, a pediatrician typically treats patients under 18, while a geriatrician generally treats patients over 65. Similarly, a gynecologist predominantly treats female patients, while a urologist typically treats male patients.

We plot the percentage of female patients of each physician in the physician map on Fig. 2.a. While we were able to determine clusters corresponding to urologists and gynecologists based on prescription patterns alone, we use Fig. 2.a to validate our cluster assignments.

We plot the median patient age of each physician in Fig. 2.b. The red dots correspond to physicians treating younger patients. While physicians with younger patients are scattered across several clusters, they are highly concentrated in the lower-right cluster. We conclude that this cluster corresponds to pediatricians. Interestingly, we also observe a reddish hue in points corresponding to

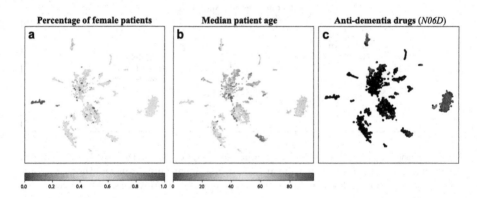

Fig. 2. Non-drug prescription data helps to gain additional insight into physician specializations or further validates the cluster annotations. For instance, panel (a) shows the proportion of female patients treated by each physician. This additional information confirms the cluster assignments of gynecologists and urologists, both very gender-specific specializations. Panel (b) depicts the median patient age treated by each physician. The cluster with the lowest median age, indicated in red, corresponds to pediatricians, who primarily treat patients under 18. We would also be interested in identifying geriatricians who treat older patients, but this plot reveals several candidate clusters. To identify geriatricians, we plot the frequency of the prescription of anti-dementia drugs in (c), as dementia is a prevalent disease in the elderly population. The plot reveals two potential clusters for anti-dementia drugs. From our annotation procedure, we have already identified the top cluster to correspond to neurologists/psychiatrists. Using extra information from meta-data gives us a high level of confidence that the right cluster corresponds to geriatricians.

gynecologists, indicating that the median age of gynecologist patients appears to be around 40 years old. Krause *et al.* [6] report similar findings in the German healthcare system, where they note that visits to gynecologists drop off with age.

Figure 2.b reveals several clusters containing physicians with older patients. We also investigate anti-dementia drug prescriptions to identify which corresponds to geriatricians. The incidence of dementia increases exponentially with age [2]. Therefore we would expect geriatricians to prescribe more anti-dementia drugs than their peers. Figure 2.c shows the frequency of the prescription of all drugs corresponding to anti-dementia drugs, with the ATC designation *N06D*. Figure 2.c reveals that physicians most frequently prescribe anti-dementia drugs from two distinct clusters, one at the top and one to the right of the embedding. Using our ATC-term enrichment approach, we determined that the top cluster corresponds to neurologists/psychiatrists. Based on this information, we infer that the cluster to the right corresponds to geriatricians. This finding is already hinted at in Fig. 2.b, as the top cluster contains a mix of red and blue dots. In contrast, the cluster on the right contains mostly physicians colored with a blueish hue, corresponding to physicians treating older patients.

3.2   Investigating Previously-Identified, Specialization-Specific Drugs

In a related study, Akhlaghi *et al.* [1] develop a predictive model for predicting physician specialization in Iran based on their drug-prescribing patterns. They identify several frequently-prescribed, specialization-specific, co-prescribed drug pairs that were found to be essential for distinguishing between different physician specializations. Unlike in our study, the authors had access to physicians' true specializations. We can then investigate whether the identified drug pairs also appear in Slovenia and evaluate their discriminatory power between different physician specializations.

We select three reliably-identified specialization clusters from our results and inspect the occurrence and discriminatory power of the drug pairs identified in Iran. We show three such combinations in Fig. 3. The plots show that while some of the identified drug pairs in Iran are, in fact, highly prescribed in their target specializations, they are not altogether discriminatory. For instance, the Clotrimazole-Fluconazole drug pair is characteristic of gynecologists in Iran. Inspecting these individually in the top two panels of Fig. 3, we can see that they are, in fact, overrepresented in the gynecology cluster. However, both of these drugs, especially Fluconazole, also appear in several other physician clusters throughout the embedding. To inspect the pairing of these two drugs, we calculate the product of the prescription counts of each of the two drugs. We show the results in the bottom panel of Fig. 3. While many gynecologists prescribe the drug pair Clotrimazole and Fluconazole, it is also prescribed by other specialists, including geriatricians, pediatricians, dermatologists, and ENT specialists. Neurologists/psychiatrists behave similarly. The Acetaminophen-Ketotifen drug pair identified for pediatricians appears to have less discriminatory power.

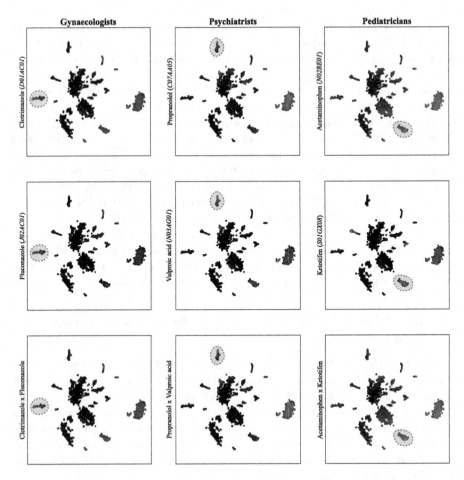

Fig. 3. We plot the prescription frequency of specilization-specific drug pairs identified by Akhlaghi *et al.* [1] for each physician. The first and second rows show the prescription frequencies of the individual drugs, while the third row plots the product of their respective frequency values. The color intensity indicates the log-transformed frequency of each drug/drug pair. The light-blue areas indicate the regions where each previously-identified physician specialty actually occurs.

This lack of specificity could result from several factors or a combination thereof. Firstly, the discrepancies between our and Iran's study could indicate potential cultural and regional differences in prescribing patterns between physicians in different countries or systemic differences in their healthcare medical and education systems. Direct comparisons between countries can be misleading, as the general population's demographic makeup and dietary habits play an essential role in the overall drug prescribing trends. Secondly, the drug pairs identified by Akhlaghi *et al.* [1] were used for prediction in a highly non-linear random-forest model and not individually, as was shown in our case. Perhaps the top-rated drug pairs are only discriminatory when used in conjunction with other drug pairs. Lastly, it may be difficult to identify drugs that are truly specific to each specialization alone. Some specializations, e.g., gynecologists or psychiatrists, are easily identifiable since most prescribed drugs originate from the corresponding ATC classification group. However, as seen in Figs. 2.c and 3, even drugs that are relatively specific to these specialists often appear in at least one other cluster as well, e.g., anti-dementia drugs are prescribed by both neurologists and geriatricians. Using their approach, the authors were able to achieve 74% accuracy when predicting physician specialization, indicating that this may not be an altogether trivial task.

4   Conclusion and Future Directions

In this study, we developed an unsupervised computational approach to infer annotated maps of physicians based on their drug prescribing characteristics. The annotations for clusters on the map use an external database with ATC classification terms. We found that the inferred physician map includes highly discernable and interpretable clusters. Unlike previous studies, we did have access to physicians' true specializations. While we used our approach to the map of physicians, we could similarly apply our method to other data modalities or clusters, e.g., patients, diseases, drugs, or institutions.

The results of this study raise several interesting points. Firstly, it is interesting that our embedding and clustering procedure uncovered such distinct groups of physicians. Using our enrichment scheme, we were able to associate these with different physician specializations. Additionally, we found that we could not unambiguously classify certain physician clusters without additional metadata, emphasizing the need to incorporate different data modalities and ontologies into similar analyses. This also highlights the benefits of constructing data maps from one data source and providing explanations from other data sources or ontologies. Interestingly, certain physician specializations were missing from our final embedding. This indicates that similar drugs and drug-prescribing patterns likely appear between different specializations and are encompassed within a single cluster. We also compare the different drug-prescribing patterns between Iranian and Slovenian physicians. Our preliminary results indicate potential differences between the drug-prescribing patterns between the two countries. This could be a result of a variety of factors, including regional, cultural, or educational factors. The differences between drug-prescribing patterns in different countries are poorly understood, and we have here provided an example of this discordance. While our results are interesting, their benefits and real-world impact are in the hands of domain experts and policymakers. In our case, we are continuing collaboration with the Slovenian National Institute of Public Health to see how the knowledge gained from our study can lead to a better understanding of the needs of the Slovenian healthcare system and provide them with the tools for a more evidence-based decision-making approach.

Acknowledgements. This work was supported by the Slovenian Research Agency Program Grant P2-0209. We would also like to thank the Slovenian National Institute of Public Health for their constructive cooperation.

References

1. Akhlaghi, M., Tabesh, H., Mahaki, B., Malekpour, M.R., Ghasemi, E., Mansourian, M.: Predicting the physician's specialty using a medical prescription database. Computat. Math. Methods Med. (2022)
2. Birkenhäger, W.H., Forette, F., Seux, M.L., Wang, J.G., Staessen, J.A.: Blood pressure, cognitive functions, and prevention of dementias in older patients with hypertension. Arch. Intern. Med. **161**(2), 152–156 (2001)
3. Ester, M., Kriegel, H.P., Sander, J., Xu, X., et al.: A density-based algorithm for discovering clusters in large spatial databases with noise. In: KDD, vol. 96, pp. 226–231 (1996)
4. Garg, A., Savage, D.W., Choudhury, S., Mago, V.: Predicting family physicians based on their practice using machine learning. In: 2021 IEEE International Conference on Big Data (Big Data), pp. 4069–4077. IEEE (2021)
5. Kobak, D., Berens, P.: The art of using t-SNE for single-cell transcriptomics. Nat. Commun. **10**(1), 5416 (2019)
6. Krause, L., Dini, L., Prütz, F.: Gynaecology and general practitioner services utilisation by women in the age group 50 years and older. J. Health Monit. **5**(2), 15 (2020)
7. Van der Maaten, L., Hinton, G.: Visualizing data using t-SNE. J. Mach. Learn. Res. **9**(11) (2008)
8. Poličar, P.G., Stražar, M., Zupan, B.: openTSNE: a modular Python library for t-SNE dimensionality reduction and embedding. BioRxiv p. 731877 (2019)
9. Rousseeuw, P.J.: Silhouettes: a graphical aid to the interpretation and validation of cluster analysis. J. Comput. Appl. Math. **20**, 53–65 (1987)
10. Shirazi, S., Albadvi, A., Akhondzadeh, E., Farzadfar, F., Teimourpour, B.: A new application of community detection for identifying the real specialty of physicians. Int. J. Med. Inform. **140**, 104161 (2020)

Machine Learning Based Prediction of Incident Cases of Crohn's Disease Using Electronic Health Records from a Large Integrated Health System

Julian Hugo[1] , Susanne Ibing[1,2(✉)] , Florian Borchert[1] ,
Jan Philipp Sachs[1,2] , Judy Cho[3] , Ryan C. Ungaro[4] ,
and Erwin P. Böttinger[1,2]

[1] Digital Health Center, Hasso Plattner Institute, University of Potsdam,
Potsdam, Germany
Susanne.Ibing@hpi.de
[2] Hasso Plattner Institute for Digital Health at Mount Sinai,
Icahn School of Medicine at Mount Sinai, New York, NY, USA
[3] The Charles Bronfman Institute for Personalized Medicine,
Icahn School of Medicine at Mount Sinai, New York, NY, USA
[4] The Henry D. Janowitz Division of Gastroenterology,
Icahn School of Medicine at Mount Sinai, New York, NY, USA

Abstract. Early diagnosis and treatment of Crohn's Disease (CD) is associated with decreased risk of surgery and complications. However, diagnostic delay is common in clinical practice. In order to better understand CD risk factors and disease indicators, we identified incident CD patients and controls within the Mount Sinai Data Warehouse (MSDW) and developed machine learning (ML) models for disease prediction.

CD incident cases were defined based on CD diagnosis codes, medication prescriptions, healthcare utilization before first CD diagnosis, and clinical text, using structured Electronic Health Records (EHR) and clinical notes from MSDW. Cases were matched to controls based on sex, age and healthcare utilization. Thus, we identified 249 incident CD cases and 1,242 matched controls in MSDW. We excluded data from 180 days before first CD diagnosis for cohort characterization and predictive modeling. Clinical text was encoded by term frequency-inverse document frequency and structured EHR features were aggregated. We compared three ML models: Logistic Regression, Random Forest, and XGBoost.

Gastrointestinal symptoms, for instance anal fistula and irritable bowel syndrome, are significantly overrepresented in cases at least 180 days before the first CD code (prevalence of 33% in cases compared to 12% in controls). XGBoost is the best performing model to predict CD with an AUROC of 0.72 based on structured EHR data only. Features with highest predictive importance from structured EHR include anemia lab values and race (white). The results suggest that ML algorithms could enable earlier diagnosis of CD and reduce the diagnostic delay.

Keywords: Crohn disease · Diagnostic delay · Electronic health records

J. Hugo and S. Ibing—These authors contributed equally to this work.

J. M. Juarez et al. (Eds.): AIME 2023, LNAI 13897, pp. 293–302, 2023.
https://doi.org/10.1007/978-3-031-34344-5_35

1   Introduction

Inflammatory bowel disease (IBD) with its main entities Crohn's disease (CD) and ulcerative colitis (UC) comprises a group of chronic immune-mediated diseases of the gastrointestinal (GI) tract with relapsing disease course [18,19].

Diagnostic delay, the time between initial manifestation of symptoms and clinical diagnosis of a disease, is a common problem in IBD.

According to a recent meta-analysis by Jayasooriya et al., the median diagnostic delay in CD is 8.0 months (6.2 months in high-income countries), compared to a significantly shorter time period of 3.7 months in UC. Diagnostic delay in CD increases the risk of complications, such as major surgery, strictures, and penetrating disease [12]. Danese et al. [8] described the development and validation of a 'Red Flags Index', a diagnostic tool comprised of 21 symptoms and signs suggestive of CD that, according to the authors, cannot be applied to general CD screening, however, possibly can serve as support tool to prioritize patients for fecal calprotectin (FC) screening [9]. Across individual studies, the identified risk factors varied and did not result in consistent patient features predictive of prolonged time to diagnosis [12]. There is a need to reduce diagnostic delay by early identification of patients presenting characteristics common to CD and early initiation of CD-specific diagnostic pathways.

In recent years, clinical predictive model (CPM) trained on patients' electronic health records (EHR) have gained interest for prognostic or diagnostic tasks to identify new predictors and build clinical decision support systems [10]. In the context of IBD, to our knowledge CPMs have only been described for prognostic tasks (e.g., to predict disease complications or therapy response) [16]. In this work, we describe the extraction of an EHR-based CD incident cohort and matched controls from the Mount Sinai Health System (MSHS) and subsequent prediction of CD diagnosis using features derived from structured EHR and clinical notes.

2   Methods

2.1   Data and Study Population

The data used in this study stems from the Mount Sinai Data Warehouse (MSDW) which contains structured EHR data and unstructured clinical notes for approximately 10.5 million patients in the Observational Medical Outcomes Partnership (OMOP) Common Data Model (CDM). We included data from November 1st, 2011, to December 31st, 2021, in our analyses.

2.2   Phenotyping Algorithm

We applied EHR-based phenotyping to identify CD incident cases, the date of their first CD diagnosis, and matched controls from MSDW (Fig. 1). Our phenotyping algorithm was developed by iterative investigations of the raw data contained in primary clinical information systems of randomly selected cases and controls.

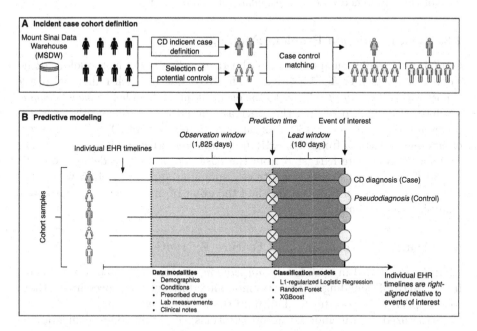

Fig. 1. Setup of Crohn's disease incident case cohort definition (A) and predictive modeling (B). Figure partially adapted from Lauritsen et al. [14]

We defined CD cases as patients with at least one IBD-specific medication prescription and two or more CD diagnosis codes coded on different days [11]. To select cases that had their incident diagnosis (index date) within MSHS (i.e., are not referral cases), healthcare encounters with non-CD diagnoses were required in the first and second year prior to the index date, and the first coded CD diagnosis within the context of an on-site encounter (excluding telehealth encounters). The former filtering step additionally ensured data availability within the observation time frame. We excluded patients with evidence of an existing IBD condition prior to the index date (i.e., specific medication, history of IBD in structured or text data). To exclude known CD patients based on unstructured clinical notes, we devised a list of strings indicative of prevalent IBD. To reduce the probability of including CD patients into the control group we excluded individuals that had an IBD condition or IBD-specific medication coded, or the presence of an IBD condition could be identified from clinical notes at any time in their EHR.

Cases and controls were exactly matched in a 1:5 ratio on year of birth in five-year bins, sex (female/male), and healthcare utilization metrics as described by Castro et al. [6]. Healthcare utilization, i.e., first and last recorded database entry and the total number of entries, was estimated from records of medication prescriptions, observation entries, diagnoses, or procedures. These features were transformed into a uniform distribution by quantile transformation and grouped in eight bins for matching [6]. Each control was then assigned the exact index date of their matched case as a *pseudodiagnosis* date [4].

2.3 Risk Prediction Problem Framing

Our CD risk prediction approach is outlined according to the methodology delineated in Lauritsen et al. [14] (Fig. 1). The EHR data of each individual of the cohort was *right-aligned* relative to the event of interest (index or *pseudodiagnosis* date). Taking into account the previously described CD diagnostic delay, the *prediction time* was set at 180 days (approximately 6 months) before the event of interest. Features from the EHR used for predictive modeling and cohort characterization were extracted from the *observation window* defined as 2,005 to 180 days before the event of interest, spanning approximately 5 years. The EHR data between *prediction time* and the event of interest was excluded from analysis (*lead window*).

2.4 Data Pre-processing and Feature Extraction

EHR data included in the *observation window* of each individual was extracted from MSDW. Depending on the data modality, different pre-processing methods were applied to aggregate the data from the *observation window*.

Structured data included drug prescriptions, condition codes, demographics and measurements. Drug prescriptions and condition codes were processed to Boolean features, indicating whether a drug or condition was recorded at any time in the *observation window* of the individual. Numeric measurements were aggregated by calculating the median, maximum, and minimum value measured during the *observation window*, additionally the absolute count per measurement type was used as feature. Missing values were imputed by median imputation. Any condition or drug coded in less than 0.1% or measurements coded in less than 5% of individuals were removed. The age in years of each individual was calculated at *prediction time*. Sex, race, smoking status were extracted from the corresponding data tables and represented as categorical features (i.e., female/male, white/non-white/unknown and never smoker/smoker/ex-smoker/unknown). The unstructured clinical notes were cleaned by deleting duplicate notes and texts shorter than three words and aggregated per individual. Notes were encoded by term frequency-inverse document frequency (TF-IDF) with stop-word removal.

We used structured features only or in combination with text features as inputs. Boruta feature selection was applied on each dataset [13].

2.5 Predictive Modeling and Evaluation

We applied three different classification models: XGBoost [7], Random Forest (RF) [5] and Logistic Regression (with L1- or L2-regularization). The data was split stratified by class into a training set (70%) for model building and a test set (30%) for performance evaluation. The training set was used for the selection of model hyperparameters by 5-fold cross validation using Bayesian hyperparameter optimization [3]. Model performance was compared based on the area under the receiver operating characteristic (AUROC), area under the precision recall curve (AUPRC), F1, and accuracy. For model explainability, we used the SHapley Additive exPlanations (SHAP) method [15].

3 Results

To extract CD incident cases within the MSHS, multiple criteria including coded conditions, prescribed medication, and healthcare utilization had to be fulfilled. Using the phenotyping algorithm we identified 7,582 likely CD cases with at least three CD diagnosis codes on different days and prescription of IBD-specific medication. To exclude potential referral cases this number was reduced to 249 incident cases by filtering based on MSHS utilization prior to the index date and due to the requirement of the first coded CD diagnosis being recorded at an on-site visit. Cases were matched to controls based on age, sex and healthcare utilization, if available in a 1:5 ratio (Table 1). For the case cohort, we observe two peaks in the age distribution of the first CD diagnosis, in their second and fourth decade of life, consistent with known epidemiological patterns [18].

Table 1. Demographic and smoking information of the 249 CD incident cases and 1,242 controls included in the study. Hypothesis testing: Kruskal-Wallis or chi-squared test with Bonferroni correction

	Controls	Cases	p-value
n	1,242	249	
Age at prediction in years median (Q1,Q3)	38.0 (25.0,60.0)	38.0 (25.0,60.0)	0.958
Sex = female (%)	59.9	59.8	1.000
Race (%)			<0.001
White	45.4	70.7	
Black or African American	15.0	10.8	
Asian	4.0	0.3	
Native Hawaiian or Other Pacific Islander	0.3	0.0	
Other	10.4	23.0	
Unknown	12.4	5.2	
Smoking (%)			0.561
Smoker	5.6	7.6	
Ex-Smoker	14.5	15.3	
Never	49.4	46.2	
Unknown	30.5	30.9	

We compared condition prevalence in cases and controls. GI conditions, such as anal fistula or abnormal stool findings, were significantly overrepresented in cases prior to the first coded CD diagnosis (Table 2). No conditions were underrepresented in cases compared to controls. Blackwell et al. developed an IBD symptoms list, which groups GI symptoms into three categories: rectal bleeding, diarrhea, and abdominal and perianal pain [4]. The prevalence of all symptom groups were significantly overrepresented in cases. In total, 33% of cases had a coded IBD symptom in comparison with 12% of the control cohort 180 days before their first coded CD diagnosis. For the 144 cases with GI symptom coded any time before first CD diagnosis, the mean time span between these two codes, a potential estimation of the diagnostic delay, was 23.5 months (standard deviation (SD) = 28.8, median = 11.7).

Table 2. Conditions with significant overrepresentation in the CD incident case cohort in comparison to controls. Hypothesis testing: Fisher's Exact Test with false discovery rate (FDR) adjustment (q-value). OR, odds ratio.

Conditions	Prevalence (%)			
	Case	Control	OR	q-value
Anal fistula	2.81	0.08	35.90	0.010
Stool finding	2.81	0.16	17.93	0.022
Hemorrhage of rectum and anus	3.21	0.24	13.71	0.017
Rectal hemorrhage	5.22	0.40	13.63	<0.001
Anal fissure	4.02	0.40	10.35	0.010
Irritable bowel syndrome	6.02	0.64	9.89	<0.001
Generalized abdominal pain	4.82	0.81	6.24	0.017
Diarrhea	11.65	2.33	5.51	<0.001
Nonspecific abdominal pain	7.23	1.77	4.32	0.010
Abdominal pain	6.83	1.69	4.26	0.014
Gastroesophageal reflux disease	11.65	4.67	2.69	0.022
Grouped symptoms [4]	**Case**	**Control**	**OR**	**q-value**
Rectal bleeding	10.84	1.69	7.07	<0.001
Diarrhea	14.06	3.86	4.07	<0.001
Abdominal and perianal pain	20.88	8.21	2.95	<0.001
Any gastrointestinal symptom	32.93	12.08	3.57	<0.001

Using EHR data from the *observation window* of each individual, we built machine learning (ML) models to predict the risk of a CD diagnosis code 180 days after the *prediction time* (Table 3). In total 1,637 features from structured EHR were used as model input (901 conditions, 660 drugs, 64 measurement, and 12 demographic features). From 22,204 clinical notes (mean count per individual: cases 12.9, controls 15.3), depending on the optimal TF-IDF hyperparameter combination, between 12,358 and 1,355,059 text features were included. The EHRs of only 68% of controls and 73% of cases include at least one clinical note in MSDW2 during the *observation window*. Boruta Feature selection reduced the dataset consisting of only structured EHR data to 1,187 features, and the combined dataset to 3,690 features. XGBoost trained with only structured EHR data achieved highest AUROC, AUPRC, and F1. All classification models had lower performance in terms of AUROC if trained on structured EHR data and text.

Table 3. Model performance comparison of different machine learning algorithms data input. LR, logistic regression

Model	Data	AUROC	AUPRC	F1-Macro	Accuracy
XGBoost	Structured EHR	**0.72**	**0.44**	**0.65**	0.80
Random Forest		0.69	0.39	0.62	**0.83**
LR (L1-regularized)		0.69	0.34	0.60	0.69
XGBoost	Structured and text EHR	0.70	0.39	0.62	0.77
Random Forest		0.65	0.28	0.58	0.75
LR (L2-regularized)		0.68	0.35	0.59	0.70

To explain the predictions made by the best performing model based on AUROC, we analyzed the predictions using SHAP (Fig. 2). Coded *White* race had the largest impact on model predictions, increasing the likelihood of case classification. The majority of features with high prediction influence were numerical measurements, comprised of anemia-related, electrolyte and blood count lab values. The coding of specific conditions (diarrhea and gastroesophageal reflux disease) and antibiotic and anti-inflammatory drugs (ciprofloxacin and prednisone) increased the likelihood to classify individuals as at risk for CD diagnosis.

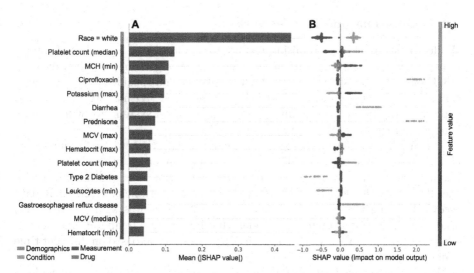

Fig. 2. SHAP values of the XGBoost model (structured EHR only). Mean SHAP value (A) and the SHAP value of individual predictions (B) of the 15 most informative features

4   Discussion

4.1   Clinical and Technical Significance

Using the GI symptom groups developed by Blackwell et al. [4], we can confirm the increased symptom prevalence for abdominal and perianal pain, diarrhea and rectal bleeding in our cohort, at least 180 days prior to the first CD diagnosis. Compared to Blackwell et al., we report higher prevalence of these symptoms for both cases and controls, potentially due to the longer 5 year *observation window* in our study compared to the 12 months. In our best performing CPM, only few of these overrepresented conditions are amongst the 20 most important features.

CPMs with features from both structured EHR and clinical notes often perform better than models that are based on only one of the two data modalities [17]. To enable the identification of yet unknown/unexpected features prior to CD diagnosis, we encoded the clinical text by TF-IDF, an unsupervised method to weight text terms by their appearance frequency in single documents and the whole corpus. The high dimensionality of our input feature matrix when adding TF-IDF-encoded vectors as well as the absence of clinical text in 31% of patients may explain the reduced performance compared to the structured EHR data alone. The advantage of using text information in this study shown for the phenotyping of CD incident cases: Since many CD patients are referred to the MSHS as tertiary care center and are not initially diagnosed on site, we stringently filtered out CD cases with previously diagnosed disease which was not captured sufficiently in the structured data.

4.2   Limitations and Future Work

While our study shows promising results, we acknowledge a number of limitations. First, only the first part of our phenotyping algorithm, the identification of CD patients, has shown to have high sensitivity and specificity [11]. A validation of the CD incident case cohort is further required. With a larger cohort and external validation of our results in a second hospital system, we will be more confident in the generalizability of our results.

Further limitations apply to the nature of the data that we use for our study: clinical research using EHR data is challenging, amongst others due to data quality issues, for instance caused by a data collection bias and missingness in the data, or with regards to accuracy of a patients' ethnicity [1,2]. EHR data recorded during the *observation window* of 1825 days was aggregated in this study. Using advanced prediction models that incorporate temporal information, e.g., recurrent neural networks, or optimizing aggregation based on different time frames might further improve model performance.

We also recognize that the framing of our study would be more applicable to the clinical use case by setting up the prediction model with *left-aligned* patient data, thus having a common prediction time point on a common event across controls and cases [14]. In CD this time point could be defined by the first presentation of GI symptoms. We did not pursue this strategy since only 33% of cases had coded GI symptoms in their structured EHR, resulting in a very small study cohort. To further investigate the magnitude of a potential acceleration of CD diagnosis, predictive modeling with varying lengths of *lead windows* could be explored. Comparing the feature importance between different prediction time points might reveal early identifiable risk features of prospective CD cases.

To further improve the discriminative ability of our models, we are working with the clinical notes in a more supervised manner by extracting and aggregating specific symptoms and conditions. In addition, the aggregation of terms might reduce high dimensionality of our input data. This could be conducted in a supervised manner (e.g., applying the groups described by Blackwell et al. [4]) or by linking terms to biomedical concepts on which hierarchical aggregation could be performed.

5   Conclusion

To our knowledge, this is the first study to describe an EHR-based phenotyping algorithm to identify CD incident cases as well as a diagnostic CPM to predict CD cases prior to their clinical diagnosis. With our best performing ML algorithm, XGBoost, we achieved an AUROC of 0.72 and AUPRC of 0.44, demonstrating the feasibility of this prediction task, though clinical validation of our results is still pending. The high overrepresentation of GI symptoms more than six months prior to the actual diagnosis in our cohort at the MSHS underpins the need to reduce CD diagnostic delay, even at a tertiary care center with a focus on IBD.

Acknowledgements. This work is supported in part through the use of the research platform AI-Ready Mount Sinai (AIR.MS), and through the MSDW resources and staff expertise provided by Scientific Computing and Data at the Icahn School of Medicine at Mount Sinai. The research leading to these results has received funding from the Horizon 2020 Programme of the European Commission under Grant Agreement No. 826117 and by the Joachim-Herz foundation (to SI).

References

1. Beaulieu-Jones, B.K., Lavage, D.R., Snyder, J.W., Moore, J.H., Pendergrass, S.A., et al.: Characterizing and managing missing structured data in electronic health records: Data analysis. JMIR Med. Inf. **6**(1), e11 (2018)
2. Belbin, G.M., et al.: Toward a fine-scale population health monitoring system. Cell **184**(8), 2068–2083 (2021)
3. Bergstra, J., Yamins, D., Cox, D.: Making a science of model search: hyperparameter optimization in hundreds of dimensions for vision architectures. In: Proceedings of the 30th International Conference on Machine Learning, pp. 115–123. PMLR (2013)
4. Blackwell, J., et al.: Prevalence and duration of gastrointestinal symptoms before diagnosis of inflammatory bowel disease and predictors of timely specialist review: a population-based study. JCC **15**(2), 203–211 (2021)

5. Breiman, L.: Random forests. Mach. Learn. **45**(1), 5–32 (2001)
6. Castro, V.M., et al.: Evaluation of matched control algorithms in EHR-based phenotyping studies: a case study of inflammatory bowel disease comorbidities. J. Biomed. Inform. **52**, 105–111 (2014)
7. Chen, T., Guestrin, C.: XGBoost: a scalable tree boosting system. In: Proceedings of the 22nd ACM SIGKDD International Conference on Knowledge Discovery and Data Mining, pp. 785–794 (2016)
8. Danese, S., et al.: Development of red flags index for early referral of adults with symptoms and signs suggestive of Crohn's disease: an IOIBD initiative. JCC **9**(8), 601–606 (2015)
9. Fiorino, G., et al.: Validation of the red flags index for early diagnosis of Crohn's disease: a prospective observational IG-IBD study among general practitioners. JCC **14**(12), 1777–1779 (2020)
10. Goldstein, B.A., Navar, A.M., Pencina, M.J., Ioannidis, J.P.A.: Opportunities and challenges in developing risk prediction models with electronic health records data: a systematic review. J. Am. Med. Inf. Assoc. **24**(1), 198–208 (2017)
11. Ibing, S., Cho, J.H., Böttinger, E.P., Ungaro, R.C.: Second line biologic therapy following tumor necrosis factor antagonist failure: a real world propensity score weighted analysis. CGH (2023, in press)
12. Jayasooriya, N., et al.: Systematic review with meta-analysis: time to diagnosis and the impact of delayed diagnosis on clinical outcomes in inflammatory bowel disease. Aliment. Pharmacol. Ther. **57**(6), 635–652 (2023)
13. Kursa, M.B., Rudnicki, W.R.: Feature selection with the Boruta package. J. Stat. Soft. **36**(11), 1–13 (2010)
14. Lauritsen, S.M., et al.: The framing of machine learning risk prediction models illustrated by evaluation of sepsis in general wards. NPJ Digit. Med. **4**(1), 1–12 (2021)
15. Lundberg, S.M., et al.: From local explanations to global understanding with explainable AI for trees. Nat. Mach. Intell. **2**(1), 56–67 (2020)
16. Nguyen, N.H., et al.: Machine learning-based prediction models for diagnosis and prognosis in inflammatory bowel diseases: a systematic review. JCC **16**(3), 398–413 (2022)
17. Seinen, T.M., et al.: Use of unstructured text in prognostic clinical prediction models: a systematic review. J. Am. Med. Inf. Assoc. **29**(7), 1292–1302 (2022)
18. Torres, J., Mehandru, S., Colombel, J.F., Peyrin-Biroulet, L.: Crohn's disease. Lancet **389**(10080), 1741–1755 (2017)
19. Ungaro, R., Mehandru, S., Allen, P.B., Peyrin-Biroulet, L., Colombel, J.F.: Ulcerative colitis. Lancet **389**(10080), 1756–1770 (2017)

Prognostic Prediction of Pediatric DHF
in Two Hospitals in Thailand

Peter Haddawy[1,2], Myat Su Yin[1(✉)], Panhavath Meth[1], Araya Srikaew[1],
Chonnikarn Wavemanee[1], Saranath Lawpoolsri Niyom[3], Kanokwan Sriraksa[4],
Wannee Limpitikul[5], Preedawadee Kittirat[5], Prida Malasit[7],
Panisadee Avirutnan[6], and Dumrong Mairiang[6,7]

[1] Faculty of ICT, Mahidol University, Nakhon Pathom 73170, Thailand
myatsu.yin@mahidol.edu
[2] Bremen Spatial Cognition Center, University of Bremen, 28359 Bremen, Germany
[3] Faculty of Tropical Medicine, Mahidol University, Bangkok 10400, Thailand
[4] Pediatric Department, Khon Kaen Hospital, Khon Kaen 40000, Thailand
[5] Pediatric Department, Songkhla Hospital, 90100 Songkhla, Thailand
[6] Faculty of Medicine Siriraj Hospital, Mahidol University, Bangkok 10700, Thailand
panisadee.avi@mahidol.edu
[7] National Center for Genetic Engineering and Biotechnology, National Science
and Technology Development Agency, Pathumthani 12120, Thailand
dumrong.mai@biotec.or.th

Abstract. Dengue virus infection is a major global health problem.
While dengue fever rarely results in serious complications, the more
severe illness dengue hemorrhagic fever (DHF) has a significant mortality rate due to the associated plasma leakage. Proper care thus requires
identifying patients with DHF among those with suspected dengue so
that they can be provided with adequate and prompt fluid replacement.
In this paper, we use 18 years of pediatric patient data collected prospec-
tively from two hospitals in Thailand to develop models to predict DHF
among patients with suspected dengue. The best model using pooled data
from both hospitals achieved an AUC of 0.92. We then investigate the
generalizability of the models by constructing a model for one hospital
and testing it on the other, a question that has not yet been adequately
explored in the literature on DHF prediction. For some models, we find
significant degradation in performance. We show this is due to differences
in attribute values among the two hospital patient populations. Possible
sources of this are differences in the definition of attributes and differences in the pathogenesis of the disease among the two sub-populations.
We conclude that while high predictive accuracy is possible, care must
be taken when seeking to apply DHF predictive models from one clinical
setting to another.

Keywords: Dengue Hemorrhagic Fever · Machine learning ·
Prognostic Prediction · Clinical Decision Support · Pediatrics ·
Generalizability

© The Author(s), under exclusive license to Springer Nature Switzerland AG 2023
J. M. Juarez et al. (Eds.): AIME 2023, LNAI 13897, pp. 303–312, 2023.
https://doi.org/10.1007/978-3-031-34344-5_36

1   Introduction

Dengue virus infection is a major global health problem, affecting 390 million people annually [2]. Most symptomatic cases develop into dengue fever, which rarely results in serious complications, but roughly 5% of patients will develop plasma leakage, signifying the more severe illness, dengue hemorrhagic fever (DHF). If DHF patients do not receive adequate and prompt fluid replacement, they may experience hypovolemic shock, which has a significant fatality rate. Given the large number of dengue infections in many countries, determining which patients have a high likelihood of DHF is an essential form of triage for hospital resource management.

In this paper, we develop gradient boosting models for the two-class problem of predicting whether or not a patient will develop DHF, using a dataset of 18 years of short-time series data on pediatric patients with suspected dengue from two hospitals in Thailand. We develop models for two clinical scenarios: a general hospital with full lab facilities and inpatient measurements and a primary care unit without lab facilities. The models achieve sufficient accuracy for clinical use, outperforming previously published results on DHF prediction. An important question for any application of ML to clinical decision support is the extent to which the model performance holds across clinical sites. Almost all previous studies of DHF prediction have used single-site data and no study has adequately addressed this question. To explore this, we develop models using data from one hospital and test them on the other. We find that the accuracy of several models degrades compared to the use of the pooled data. We identify two possible sources of this lack of cross-site generalizability: differences in the definition of attributes and differences in the pathogenesis of the disease among the two sub-populations.

2   Related Work

Phakhounthong et al. [13] built a classification and regression tree (CART) presented for predicting the development of severe dengue fever among pediatric patients with confirmed dengue infection. Their models have a moderate classification accuracy of 64.1%, a sensitivity of 60.5%, and a specificity of 65%. Tan and colleagues [14] presented multiple prediction tools that calculate the risk of developing DHF for adult patients. Their calculators are dynamic, using daily diagnostic measurements from patients to estimate the risk of DHF, as opposed to only using admission data. The AUC of their calculators on the validation set ranges from 0.72-0.79. In [11], the authors developed a simple decision tree to identify adult DHF cases for admission. The Chi-square test was used to decide which variables are most important for classification. Their best decision tree prediction consists of only three variables: clinical bleeding history, serum urea, and total serum protein, but they achieved an accuracy of 48.1% only. Existing studies for the prediction of DHF often include gene expression data and bio-markers rather than simple clinical measures. To differentiate dengue patients from DHF, Gomes and colleagues [7] analyze the expression pattern of 12 genes in peripheral blood mononuclear cells (PBMCs) of dengue patients during acute viral infection. The authors then trained an SVM model using gene expression data of these genes and achieved the highest accuracy of ~85% with leave-one-out cross-validation.

Early diagnosis of dengue cases and determining the severity of dengue cases are essential; however, in resource-limited settings, sophisticated laboratory tests to distinguish the diseases may be unavailable or costly, necessitating separate models and tools to suit the underlying user scenario. In [4], Carrasco and colleagues built predictive tools for adult patients developing severe dengue after hospitalization in well-resourced and resource-limited settings. The authors fitted generalized linear models (GLMs) using demographic, clinical, and laboratory variables at the presentation at the hospital. Their models performed acceptably with optimism-corrected specificity of 29% for a well-resourced setting and 27% for a resource-limited setting, respectively, for 90% sensitivity. The authors then estimated that applying these predictive tools in the clinical setting might reduce unnecessary admissions by 19%, allowing the allocation of scarce public health resources to patients according to the severity of outcomes. Fernandez and colleagues [6] conducted uni- and multi-variable logistic regression analysis to construct a predictive model for severe dengue using demographics and symptoms reported within the first 24 h of admission. In addition, the authors added a dichotomous variable to indicate the type of healthcare facilities, such as a special care facility (for the insured and employed population) or general public health clinics. This was done to account for differences between the two patient populations. Other than the study site variable, the model and analysis were not carried out for separate data sources, and their final model was validated internally using the bootstrap technique. The need for external validation was noted as the limitation.

3 Materials and Methods

3.1 Data

The Institutional Review Board from the Faculty of Medicine Siriraj Hospital, Mahidol University, Bangkok, Thailand, approved this study (approval number 167/2565(IRB3)). The dataset was obtained through a prospective cohort study of 1,845 pediatric patients (age less than 15 years) with suspected dengue at Khon Kaen Hospital (KK-hospital) in Northeast Thailand and Songkhla Hospital (SK-hospital) in Southern Thailand from 2001 up to and including 2017. Since this was a prospective study, the same data collection protocol was followed in both hospitals, although there could be variability in how each feature was measured. The criteria used to define dengue (DF) and dengue hemorrhagic fever (DHF) are in accordance with the World Health Organisation 1997 dengue guidelines. The final diagnosis was confirmed with ELISA and the PCR tests and fell into the following categories: dengue (DF), dengue hemorrhagic fever (DHF), and other febrile illnesses (OFI). The study day data consisted of a short time series (up to 5 days) specific to each patient. As we aim to identify DHF patients from the patients with suspected dengue, we formed the Non-DHF patient group by combining OFI patients with the DF patients since in practice it is impossible to distinguish between them without ELISA and PCR tests. As a result, the final combined-site dataset with no missing values contained 7,243 study days from 1,733 patients with 916 Non-DHF cases and 817 DHF cases. The SK-hospital dataset contained 403 DHF cases and 491 Non-DHF cases, while the KK-hospital dataset included 414 DHF cases and 424 Non-DHF cases.

The features in the dataset can be grouped into four categories: demographics and baseline physical examination/interview (8 features), daily vital signs (9 features), daily physical examination/interview (34 features), and daily laboratory checkup measures (14 features). The demographic and baseline physical examination features were obtained once upon admission; the others were obtained daily. These features were supplemented with some computed ratios, ranges, and average values for the quantitative features. Exploratory data analysis revealed that the quantitative features for DHF and Non-DHF patients differ not only in the values of the features at a given point in time but also in the trajectory of the feature values over time. For each quantitative feature, we thus added estimates of the slope, using the difference between two study day values and the second derivative based on three study day values. We also added the slope and Y-intercept from the least squares fit to three study day values. After generating these additional features, there were 76, 179, and 391 features in the study day 1, 2, and 3 data. Finally, we discretized the continuous variables using entropy-based discretization [8], which selects splits that minimize the entropy of the partitions relative to the target class. Our experiments found this to improve model accuracy. This technique can be helpful when attribute values have very long tails.

3.2 Models and Settings

Classification models to distinguish between Non-DHF and DHF patients were constructed using various algorithms including Random Forest, light gradient-boosting machine (LightGBM) [10] and Support Vector Machine. Since the performance of LightGBM models outperformed the others, we selected LightGBM to carry out the classification task.

We used 10-fold cross-validation to evaluate the models for the combined site dataset. We stratified the data based on the patients to ensure that the data from the same patient did not appear in both training and testing datasets. Our study uses the area under the receiver operating curve (AUC) as the primary measure to evaluate model performance.

In our dataset, the average length of stay at the hospital for admitted patients was three days. The number of patients on study days 4 and 5 was insufficient to train models, so we built three study day models based on the number of days since the patient was admitted to the hospital. The Day 1 model (1-Day) used the admission day data of each patient. Additional study day data was added for Day 2 (2-Day) and Day 3 (3-Day) models so that the number of features available to the models increased with the study days. We developed prognostic models for two clinical settings: resource-limited primary care units (PCU) with only clinical data and well-resourced general hospitals (GH) with clinical, laboratory, and inpatient ward data.

To determine the optimal set of features for each model, we used forward feature selection, adding features in the order of the importance score generated by LightGBM. Features were added incrementally until no improvement in AUC was seen. The numbers of features in the resulting models ranged from 13 to 29, as shown in Table 2.

3.3   Generalizability

Model accuracy may differ for clinical settings or geographic regions beyond those from which the original data came. It is of great importance to understand such variability if we wish to be able to create models that can be widely disseminated. We explored this in our context of DHF prediction by building a model using the data from one of our two hospitals and then evaluating the model using the data from the other hospital. Since the data was collected in a prospective study, with the same data collection protocol followed in both hospitals, any variability would be due only to how features were defined and measured, as well as variations in patient populations.

Table 1. Top 10 features with importance scores from General Hospital (GH) and Primary Care Unit (PCU) scenarios from study Day 1 model.

	1-Day(GH)		1-Day(PCU)	
	Feature	Score	Feature	Score
1	SGOT*/platelet ratio	9971.18	Liver size	4291.92
2	Albumin	2063.3	Abdominal Circumference	2531.99
3	Platelet count	1694.41	Abdominal pain	1721.29
4	Max. daily Hematocrit (fingertip)	1507.28	Height	1368.02
5	Day of fever	1317.77	Avg. daily body temperature	1200.09
6	Abdominal Circumference	1259.02	Weight	943.94
7	Avg. daily body temperature	1080.22	Age	927.79
8	Lymphocyte	1048.64	Cervical	721.89
9	Protein	971.53	Quality Data of Tourniquet test (daily collection)	640.7
10	Hematocrit range	864.13	Max. daily body temperature	608.23

Note *Serum Glutamic-Xaloacetic Transaminase (SGOT) is a liver enzyme also known as Aspartate transaminase (AST).

4   Results

Table 1 shows the top 10 features with their feature importance scores for the 1-Day(GH) and 1-Day(PCU) settings from the combined site dataset. Not surprisingly, the feature SGOT/platelet ratio has by far the highest feature importance score in the GH setting, while the feature Liver size is the top-ranked feature in the PCU setting. The differences in the feature list are expected as the PCU scenario doesn't include the laboratory parameters and features collected in the inpatient ward. The feature lists from other models are not presented due to space limitations. In the 2-Day and 3-Day models, about 25% to 35% of

Table 2. Performance results of the models trained with combined-site data. Note: The column "Sens." refers to Sensitivity, and the column "Spec." refers to Specificity.

	Primary Care Unit (PCU)			Genearl Hospital (GH)			
Model	#Features	AUC	Sens.	Spec.	#Features	AUC	Sens.	Spec
1- Day	15	0.74	0.65	0.66	14	0.87	0.76	0.79
2- Day	23	0.80	0.70	0.74	29	0.90	0.80	0.80
3- Day	13	0.83	0.72	0.75	29	0.92	0.82	0.81

the features in the top 30 are the derived features based on slope and second derivative.

The AUC, sensitivity, and specificity of the 1 - 3 day models trained with the selected features for both PCU and GH settings are shown in Table 2. The AUC values of the models for GH setting in all three study days are higher than PCU since the PCU setting lacks the lab and inpatient ward features. Also, the AUCs of the models improve with increasing days since they have more information available.

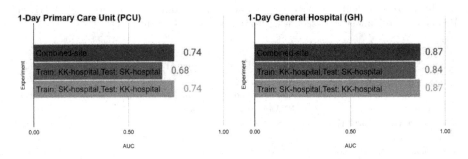

Fig. 1. Accuracy of models trained and tested with the combined-site data, and models trained with data from one hospital and tested on the other.

Figure 1 shows the AUCs of the 1-Day models trained and tested on the multi-site combined dataset and models trained on one hospital's data and then tested on the other. The second bar shows the accuracy of the model trained on KK-hospital data and tested on SK-hospital data and the third bar shows the accuracy of the model trained on SK-hospital data and tested on KK-hospital data. The model trained on the multi-site data and the model trained on SK-hospital and tested on KK-hospital both have equally high accuracy, but the model trained on KK-hospital and tested on SK-hospital has lower accuracy for both settings, with the drop in accuracy largest in the PCU setting. This pattern holds in all models except the 3-Day(PCU) model.

To explore the reason behind the drop in accuracy when applying the model from one hospital to data from another, we examined the features in the models. Table 3 shows the top-ranked features from the models trained with the site-specific datasets in the 1-Day(GH) setting. While the two lists have commonalities, such as SGOT/platelet ratio being top-ranked, they have marked differences. Features highlighted in grey appear in one list of the top 30 features but not in the other. Some other features, such as Abdominal Circumference and Avg. daily body temperature, appear in both lists but with very different rankings. Similar differences exist in the feature rankings for the 2-Day and 3-Day models.

Table 3. Top-ranked features with feature importance scores for the 1-Day(GH) models trained with the two site-specific datasets.

	KK-Hospital		SK-Hospital	
	Feature	Score	Feature	Score
1	SGOT*/platelet ratio	5665.47	SGOT*/platelet ratio	4852.36
2	Abdominal Circumference	1391.13	Platelet count	897.16
3	Albumin	1142.39	Albumin	858.91
4	Platelet count	820.27	Day of fever	812.7
5	Day of fever	729.83	Avg. daily body temperature	765.92
6	Protein	719.08	Protein	703.32
7	Hematocrit range	508.3	Max. daily Hct	661.79
8	Max. daily Hematocrit	508	Hematocrit range	633.74
9	Intake and output diff.	478.76	Max. daily body temperature	624.14
10	Lymphocyte	445.25	PMN*	575.7
11	Avg. daily body temperature	432.74	Abdominal pain	531.54
12	ALT*	423.32	Height	514.44
13	Liver size	402.35	Liver size	513.53
14	...		...	...
16	Age	381.65	Monocyte	372.31
17	...			
21	Bleeding	244.35	Abdominal Circumference	317.56

Note * Serum Glutamic-Xaloacetic Transaminase (SGOT) and Alanine Transaminase (ALT) are liver enzymes, and Polymorphonuclear cell (PMN) is a type of while blood cell also known as neutrophil.

5 Discussion

Existing predictive tools are usually considered for well-resourced settings, but tools for resource-limited settings have gained attention recently [5,9]. The resource-limited setting we considered in our study is the primary care unit, where laboratories, imaging, and genetic data analysis results are not readily available at the point of care. This is an important setting to consider since in many countries where dengue is prevalent, most patients are seen in PCUs. Although the sensitivity of our models in the well-resourced GH setting (78%) (Table 2) and the resource-limited PCU setting (65%) is less than the 90% sensitivity of the generalized linear models using demographic clinical and laboratory variables for adult patients [4], the specificity of our models GH: 78%, PCU: 67% (Table 2) are higher than their well-resourced: 29% and resource-limited: 27% settings. A caveat in this comparison is that many signs and symptoms of dengue manisfest differently in adults and children. A comparison with the ranked features in our models with the WHO 2009 warning signs of severe dengue [15] shows a high degree of consistency. For example, in 1-Day models (1), four of the seven WHO 2009 warning signs — hepatomegaly (`liver size` variable), clinical fluid accumulation (`Intake − output diff` variable) and simultaneous rise in haematocrits (`HCT` variable) with rapid platelet count drop (`plateletcount` variable) are important predictors in our models. The cross-site generalizability of learned medical decision support models is an understudied issue. This is an important issue for the practical deployment of such models since the accuracy of a model may differ for clinical settings or geographic regions beyond those from which the original data came [1,3,12]. Yet, McDermott et al. found that only approximately 23% of ML-based healthcare papers used multiple datasets [12]. Data from different clinical sites can differ due to differences in how clinical features are defined or collected and due to differences in patient populations [16]. Although our data came from a prospective study in which the same data collection protocol was used at both hospital sites, a comparison of the feature rankings of the models for the two hospitals shows significant differences. Examination shows differences mainly in two types of features. The first is features that are somewhat subjective in nature or that are objective, but that may be defined or measured differently at different clinical sites, such as abdominal circumference, bleeding, and abdominal pain. The other is hematological parameters, some of which appear in the top 30 features for SK-hospital but not for KK-hospital. These include `PMN` (rank 10), `Monocyte` (rank 16), and `WBC` (rank 26).

Figure 2 shows the log-values of the `PMN` feature (Polymorphonuclear cell (PMN is a type of while blood cell also known as neutrophil) plotted for two

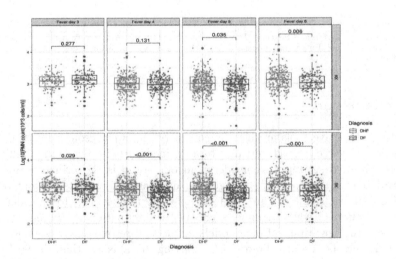

Fig. 2. The log-value of `PMN` between fever days 3 and 6. The p-value of the nonparametric Wilcoxon test between two classes is shown on the bar between the two box plots. A p-value ≤ 0.05 is considered statistically significant.

classes on a range of fever days from each site-specific dataset. For SK-hospital the differences in PMN values are statistically significant for fever days 4–6, while for KK-hospital none of the differences are significant. In addition, we see different patterns, with the PMN value for DHF patients always higher than that for DF patients in SK-hospital, while the value in KK-hospital is lower for fever day 3 and only higher at fever day 6. We hypothesize that these differences may be due to differences in the prevalence of thalassemia blood disorder in the Northeast, where KK-hospital is located, and the South of Thailand, where SK-hospital is located. Around 30–50% of Northeastern Thais have Hemoglobin E type thalassemia, while around 10% of Southern Thais have it. Neutrophils of thalassemia patients have impaired vitality and function. This might explain why the neutrophils of KK-hospital patients did not change much between Non-DHF and DHF patients.

6   Conclusions and Future Work

An early prediction tool to identify the patients who could progress to DHF will improve the utilization of limited hospital resources in dengue-endemic regions. The predictive models developed for the resource-poor setting do not require laboratory-based variables and may be beneficial in resource-limited settings with limited experienced staff. Our models are potentially helpful in guiding outpatient management of fever cases and inpatient monitoring, yet care must be taken regarding their generalizability. Further study of the generalizability of DHF predictive models would be warranted, including how to address variability in clinical data and patient sub-populations. In this study, all necessary data for prediction was readily available; however, in a real clinical setting, prediction of DHF may need to be done at various points in a day before all information is available. Thus, developing models that adequately perform inference with partial information is a topic for future work.

Acknowledgment. This work was partially supported by National Science and Technology Development Agency grant no. P-20-52599, Faculty of Medicine Siriraj Hospital, Mahidol University grant no. R016536004, a grant from the Mahidol University Office of International Relations to Haddawy in support of the Mahidol-Bremen Medical Informatics Research Unit, a Study Group grant from the Hanse-Wissenschaftskolleg Institute for Advanced Study to Haddawy, a fellowship from the Hanse-Wissenschaftskolleg Institute for Advanced Study, and by a Young Researcher grant from Mahidol University to Su Yin.

References

1. Barak-Corren, Y., Chaudhari, P., Perniciaro, J., Waltzman, M., Fine, A.M., Reis, B.Y.: Prediction across healthcare settings: a case study in predicting emergency department disposition. NPJ Digit. Med. **4**(1), 1–7 (2021)
2. Bhatt, S., et al.: The global distribution and burden of dengue. Nature **496**(7446), 504–507 (2013)

3. Burns, M.L., Kheterpal, S.: Machine learning comes of age: local impact versus national generalizability (2020)

4. Carrasco, L.R., et al.: Predictive tools for severe dengue conforming to world health organization 2009 criteria. PLoS Negl. Trop. Dis. **8**(7), e2972 (2014)

5. Chandna, A., et al.: Prediction of disease severity in young children presenting with acute febrile illness in resource-limited settings: a protocol for a prospective observational study. BMJ Open **11**(1), e045826 (2021)

6. Fernández, E., Smieja, M., Walter, S.D., Loeb, M.: A retrospective cohort study to predict severe dengue in Honduran patients. BMC Infect. Dis. **17**(1), 1–6 (2017)

7. Gomes, A.L.V., et al.: Classification of dengue fever patients based on gene expression data using support vector machines. PLoS ONE **5**(6), e11267 (2010)

8. Grzymala-Busse, J.W.: Discretization based on entropy and multiple scanning. Entropy **15**(5), 1486–1502 (2013)

9. Herath, H., et al.: Prediction of plasma leakage phase of dengue in resource limited settings. Clin. Epidemiol. Global Health **7**(3), 279–282 (2019)

10. Ke, G., et al.: LightGBM: a highly efficient gradient boosting decision tree. Advances in Neural Information Processing Systems 30 (2017)

11. Lee, V.J., Lye, D., Sun, Y., Leo, Y.: Decision tree algorithm in deciding hospitalization for adult patients with dengue Haemorrhagic fever in Singapore. Trop. Med. Int. Health **14**(9), 1154–1159 (2009)

12. McDermott, M.B., Wang, S., Marinsek, N., Ranganath, R., Foschini, L., Ghassemi, M.: Reproducibility in machine learning for health research: Still a ways to go. Sci. Transl. Med. **13**(586), eabb1655 (2021)

13. Phakhounthong, K., et al.: Predicting the severity of dengue fever in children on admission based on clinical features and laboratory indicators: application of classification tree analysis. BMC Pediatr. **18**(1), 1–9 (2018)

14. Tan, K.W.: Dynamic dengue Haemorrhagic fever calculators as clinical decision support tools in adult dengue. Trans. R. Soc. Trop. Med. Hyg. **114**(1), 7–15 (2020)

15. World Health Organization and Special Programme for Research and Training in Tropical Diseases and World Health Organization. Department of Control of Neglected Tropical Diseases and World Health Organization. Epidemic and Pandemic Alert: Dengue: guidelines for diagnosis, treatment, prevention and control. World Health Organization (2009)

16. Yang, J., Soltan, A.A., Clifton, D.A.: Machine learning generalizability across healthcare settings: insights from multi-site COVID-19 screening. NPJ Digit. Med. **5**(1), 1–8 (2022)

The Impact of Bias on Drift Detection in AI Health Software

Asal Khoshravan Azar[1], Barbara Draghi[2], Ylenia Rotalinti[2], Puja Myles[2], and Allan Tucker[1(✉)] (iD)

[1] Brunel University London, Uxbridge UB8 3PH, UK
{asal.khoshravanazar,allan.tucker}@brunel.ac.uk
[2] Medicine and Healthcare products Regulatory Agency, London E14 4PU, UK
{barbara.draghi,ylenia.rotalinti,puja.myles}@mhra.gov.uk

Abstract. Despite the potential of AI in healthcare decision-making, there are also risks to the public for different reasons. Bias is one risk: any data unfairness present in the training set, such as the under-representation of certain minority groups, will be reflected by the model resulting in inaccurate predictions. Data drift is another concern: models trained on obsolete data will perform poorly on newly available data. Approaches to analysing bias and data drift independently are already available in the literature, allowing researchers to develop inclusive models or models that are up-to-date. However, the two issues can interact with each other. For instance, drifts within under-represented subgroups might be masked when assessing a model on the whole population. To ensure the deployment of a trustworthy model, we propose that it is crucial to evaluate its performance both on the overall population and across under-represented cohorts. In this paper, we explore a methodology to investigate the presence of drift that may only be evident in sub-populations in two protected attributes, i.e., *ethnicity* and *gender*. We use the BayesBoost technique to capture under-represented individuals and to boost these cases by inferring cases from a Bayesian network. Lastly, we evaluate the capability of this technique to handle some cases of drift detection across different sub-populations.

Keywords: Concept Drift · Data Bias · Healthcare models

1 Introduction

Machine learning technologies in medicine provide many benefits, including automating tasks and analysing big patient data to deliver better healthcare faster. However, different risks have to be considered to ensure reliability and safety. The presence of bias within data [7] is one risk. Bias can manifest in several ways [11] with varying degrees of consequences for the subject group (e.g. biases in online recruitment tools or biases in criminal justice decision-making). This work focuses on two types of algorithmic bias within AI health models: the under-representation of groups of patients (e.g. ethnicity and gender bias).

© The Author(s), under exclusive license to Springer Nature Switzerland AG 2023
J. M. Juarez et al. (Eds.): AIME 2023, LNAI 13897, pp. 313–322, 2023.
https://doi.org/10.1007/978-3-031-34344-5_37

These may be a result of limited available data for the under-represented groups for numerous different reasons [12].

For instance, methods from the Framingham study risk factors [8] have been applied for years to evaluate the probability of developing cardiovascular diseases. However, since the study drew its sample from a non-Hispanic white population that was largely represented, it was prone to misclassification when applied to other subpopulations. Additionally, several pieces of research [9, 10] highlighted that gender bias affects how patients are assessed, diagnosed, and treated within hospitals. When both genders are not offered equal quality treatment and care for the same medical complaints, we can expect patient outcomes to suffer.

Data bias is not the only issue affecting healthcare AI models. Concept drift [6] can cause declining performance over time. This phenomenon refers to how changes within the underlying characteristics of data affect a model's performance when tested on newly available data. For example, it may happen due to the introduction of new technologies which improve measurement accuracy or changes in population demographics over the years. Also, statistical models trained on a domain that does not appropriately generalize to represent the population can potentially pose a risk [4]. For this reason, monitoring and detecting changes within data are fundamental aspects to ensure a reliable and safe algorithm [3].

This paper aims to explore the interaction between bias and concept drift within the context of healthcare models. Although many methodologies investigate concept drift and data bias individually, we are interested in developing a framework to deal with both simultaneously. We demonstrate that drifts within under-represented subgroups might be masked when assessing a model on the whole population. To ensure the deployment of a trustworthy model, we claim that it is crucial to evaluate its performance both on the overall population and across under-represented cohorts.

As healthcare data is often released in batches, a batch-learning-based approach from Rotalinti et al. [13] has been chosen to detect changes within data by monitoring model performance over time. This method updates a model after detecting a drift by boosting data within the training set so that it better represents the current population and, consequently, achieves better performance when the model is tested on new data. We investigate the potential of a novel boosting technique, BayesBoost [5], in handling spurious drifts that are the result of insufficient data for under-represented groups. The resulting models would be up-to-date models that are more representative of the entire population.

2 Methods

2.1 Dataset

The project involved a case study on cardiovascular risks (CVD) based on multiple batches of primary care data from the UK's Clinical Practice Research Datalink (CPRD) [1]. The CPRD is a real-world research service supporting retrospective and prospective public health and clinical studies. It collects anonymized patient data from a network of GP practices across the UK and provides researchers access to high-quality anonymized and synthetic health data that can be used for training purposes or to improve machine learning workflows [14,15].

A high-fidelity synthetic data based on anonymized real primary care data generated from the CPRD Aurum database [2] was used to evaluate the discussed framework. The dataset includes 96709 patients and 16 features correlated with cardiovascular disease (e.g. *smoking behaviour, age, chronic diseases*). Data was collected from 2005 to 2014 and it was split into 9 yearly batches based on the temporal variable representing the date of the event recorded. Data contains 96147 records from which 208 and 468 cases belong to *Black Caribbean* and *Indian* groups respectively.

2.2 Methodology

As mentioned, these experiments have been designed upon two previously developed methodologies i.e., the performance-based drift detection approach from Rotalinti et al. [13] and BayesBoost [5]. The novelty of this piece of work is to design a methodology to explore the presence of drift that may only be evident in sub-populations in two protected attributes, i.e., *ethnicity* and *gender*. Then, the BayesBoost technique is used to capture under-represented individuals and to boost these cases by inferring examples from a Bayesian network.

Performance-Based Drift Detection: The performance-based drift detection approach from Rotalinti et. al. [13] is an algorithm developed to identify drifts within data whose underlying characteristics may change over time. In this paper, the original algorithm is modified to explore whether changes within sub-groups may be masked when assessing drifts on the whole population. Figure 1 summarises the key steps of the framework. In the first step of the analysis, the data is split into batches representing different blocks of time to simulate a real-world scenario (Fig. 1, step 1). A Random forest is learned from a selected training set (Fig. 1, step 2). Then, the post-market deployment of the AI algorithm is simulated by monitoring the performance achieved over time. We name the first batch of data on which the model is tested as the *control set* and the following batch in time as the *next test set*. Ten samples are generated from each subset through bootstrapping techniques to ensure randomness. Depending on the protected attribute chosen (e.g. *ethnicity*), sub-populations (e.g. *Black Caribbean*) are filtered from the control set and the next test set (Fig. 1, step 3). Then, to assess concept drift, performance distributions achieved on subsequent batches of data are compared through a non-parametric statistical test i.e. the Wilcoxon Rank test (Fig. 1, step 4). If the distributions are significantly different, a drift is detected and the model is updated with new batches as the data on which the model was initially trained may be no more representative of the domain. On the other hand, if no drift was detected, the model is retained as is. Furthermore, the framework has been tested on the whole population to assess whether changes within subgroups might be relevant for the analysis. In this paper, the field *ethnicity* and *gender* have been chosen as protected attributes. Additionally, as cardiovascular data used for evaluating the framework is highly imbalanced, we consider recall as a metric to assess model performance.

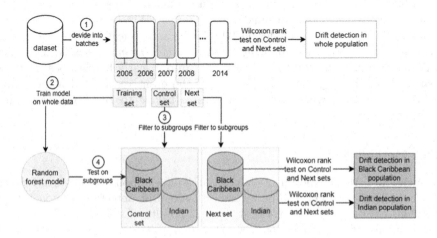

Fig. 1. Performance-based drift detection approach

BayesBoost: BayesBoost [5] is a new technique that combines a Bayesian network synthetic data generator with a boosting approach to identify data biases, correct them and improve classification accuracy. The key steps of the algorithm are presented in Fig. 2. Firstly, it identifies groups of under-represented individuals for a chosen target disease (here *hypertension*) by applying an uncertainty assessment based on the confidence of prediction, Fig. 2a. In this study, 229 and 21 uncertain cases were identified for *Black Caribbean* and *Indian* ethnic groups respectively. To overcome the biases highlighted through the uncertainty analysis, m new synthetic patients are generated for each uncertain case by sampling from a Bayesian Network that is trained on the original dataset, Fig. 2b. The size of the new boosted dataset depends on the value of m. In this study, we have made three attempts similar to the BayesBoost paper by generating different synthetic dataset based on three values of the parameter m. First, m rows are sampled to obtain a dataset with extra boosted cases that are half the size of the original dataset. Second, m is chosen in a way to create a synthetic dataset with an equal number of boosted cases to the original dataset. Lastly, m rows are extracted for each uncertain case to produce a dataset with boosted cases that are twice as large as the original one. The boosted dataset is added to the original dataset (Fig. 2c) generating BB50, BB100, and BB200 respectively.

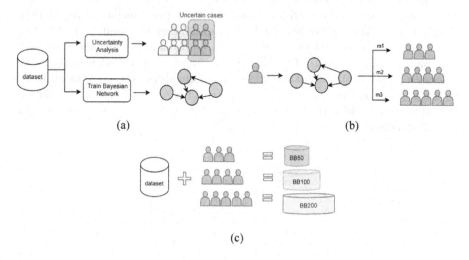

Fig. 2. BayesBoost

In this paper, we developed a framework that combines the two techniques. Firstly, the drift detection algorithm has been implemented by testing model performance on both the whole population and on subsets based on protected characteristics (*ethnicity* and *gender*). Then, experiments are repeated on the boosted datasets (BB50, BB100, BB200) according to BayesBoost methodology. Finally, results obtained are compared and evaluated to assess whether the BayesBoost algorithm could address drift within under-represented subgroups.

3 Results

This section discusses the key findings achieved by implementing the discussed drift detection framework within the subgroups of the chosen protected attributes (*ethnicity* and *gender*) for the original and boosted datasets.

3.1   Protected Attribute: *Ethnicity*

The *ethnicity* attribute within the cardiovascular risks dataset contains 9 different levels i.e., *Bangladeshi, Black African, Black Caribbean, Chinese, Indian, Other, Other Asian, Pakistani, White/no recorded*. Firstly, uncertainty analysis has been implemented to detect under-represented individuals within the original dataset. The *Black Caribbean* population resulted as the most impacted group, followed by the *Indian* group.

Since we are interested in how under-representation affects drift detection, we run experiments on the original dataset focusing on specific subpopulations. Figure 3 shows the results obtained when evaluating the model on the whole population (3a), on the *Black Caribbean* group (3b) and on the *Indian* one (3c). The plots show how drifts detected within the whole population do not coincide with those identified within subgroups, highlighting the importance of our research question. Furthermore, the confidence bounds for both subpopulations investigated are wider than those seen for the whole population, emphasizing model uncertainty.

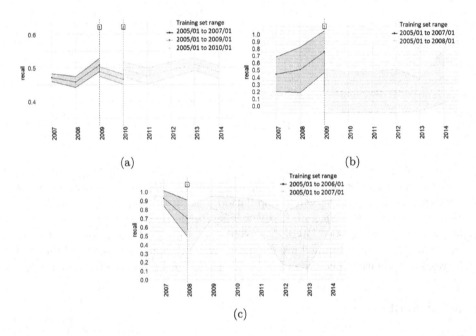

Fig. 3. Protected attribute: *Ethnicity*. Results obtained on the original CVD dataset by evaluating the model's performance with recall metric on the whole population (a), *Black Caribbean* population group (b), and *Indian* group (c).

Experiments have then been repeated on three different datasets (BB50, BB100, BB200) obtained by boosting the original one with different sizes. Figure 4 reveals the results obtained within BB200. The plots show how implementing BayesBoost methods on the original dataset successfully results in removing previously detected drift within the whole population (Fig. 4a) and within the *Black Caribbean* one (Fig. 4b). This finding suggests that drifts detected in the original dataset were presumably caused by the under-representation of individuals within data. On the other hand, new drifts within the *Indian* population are being detected (Fig. 4c). This highlights that any per-turbation of data (including boosting) needs to be applied carefully as it risks to create instabilities. However, it is interesting noting an improvement of per-formance when the model is retrained, suggesting that a more genuine drift in the subpopulation might have been detected.

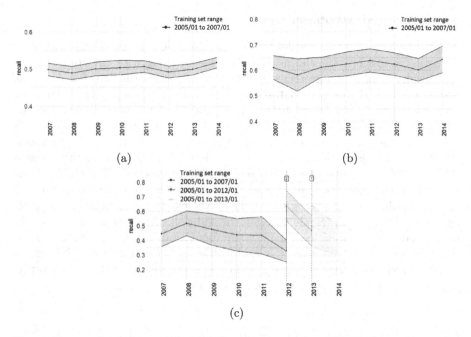

Fig. 4. Protected attribute: *Ethnicity*. Results obtained on BB200 dataset by evalu-ating the model's performance with recall metric on the whole population (a), *Black Caribbean* population group (b), and *Indian* group (c)

Finally, Table 1 summarises the overall number of drifts detected and the recall confidence interval for each under-represented group. It is worth noticing how the correct choice of the size of the boosting m is crucial to shrink the confidence intervals reducing variance and the number of drifts detected.

Table 1. Protected attribute: *Ethnicity*. Results summary on the original CVD, BB50, BB100, and BB200 datasets.

	CVD dataset		BB50 dataset		BB100 dataset		BB200 dataset	
Ethnicity	Drift	Recall range	Drift	Recall range	Drift	Recall range	Drift	Recall range
Black Caribbean	1	0–1	0	0.5–0.7	0	0.6–0.7	0	0.5–0.7
Indian	1	0.1–1	1	0.55–1	1	0.3–0.8	2	0.3–0.75
White	2	0.45–0.5	2	0.35–0.5	0	0.45–0.5	0	0.35–0.45
Other	0	0.25–1	0	0.35–0.8	0	0.55–0.75	0	0.4–0.55
All data	2	0.45–0.5	1	0.45–0.55	1	0.45–0.5	0	0.5–0.55

3.2   Protected Attribute: *Gender*

As previously mentioned, gender bias in data can result in an AI algorithm that perpetuates this in its decisions. For this reason, along with the ethnic group, we consider *gender* as a protected attribute to analyse. Drift detection has been assessed within the original dataset on both the male group (Fig. 5a) and the female one (Fig. 5b). Results show that although drift has been detected within the same batches of data (i.e. batch 2008 and 2010), the trajectory of the model's performance is entirely different as well as the overall bounds (0.2–0.4 for males, 0.6–0.8 for females). When drift detection has been assessed on the boosted dataset BB200, results show a completely different picture (Fig. 6). Applying BayesBoost methods to the original dataset resulted in removing previously detected drifts when assessing the model on the whole population (6a), on male subgroup (6b) and on female one (6c). In addition, it results in improving the overall model's performance. Finally, we present a table summarising the results achieved for each sub-population over the different boosted datasets, Table 2. As was found when on the ethnic group study, an accurate selection of the boosting parameter m is crucial to address spurious drifts within data.

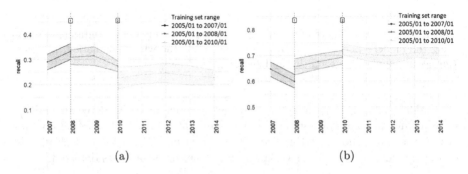

(a)   (b)

Fig. 5. Protected attribute: *Gender*. Results obtained on the original CVD dataset by evaluating the model's performance with recall metric on the male group (a) and the female population group (b)

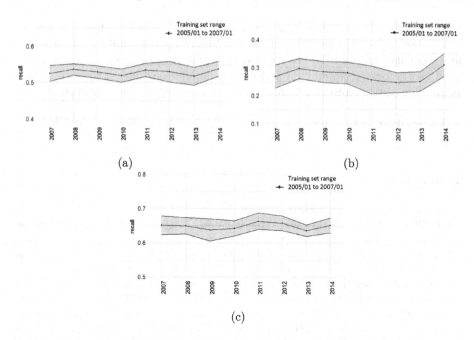

Fig. 6. Protected attribute: *Gender*. Results on BB200 dataset evaluating recall on whole population (a), male population (b), and female population (c)

Table 2. Protected attribute: *Gender*. Results on the original CVD, BB50, BB100, and BB200 datasets.

Gender	CVD dataset		BB50 dataset		BB100 dataset		BB200 dataset	
	Drift	Recall range	Drift	Recall range	Drift	Recall range	Drift	Recall range
Male	2	0.2–0.3	2	0.2–0.4	0	0.2–0.3	0	0.2–0.35
Female	2	0.6–0.75	0	0.55–0.7	0	0.6–0.7	0	0.6–0.7
All data	2	0.45–0.5	1	0.45–0.55	1	0.5–0.55	0	0.5–0.55

4   Conclusions

Artificial intelligence in medicine has an increasingly prominent role in improving diagnosis and treatments. However, it can pose risks such as replicating and even amplifying data biases. Moreover, changes within data over time represent another risk, possibly resulting in obsolete models. Although work has been carried out investigating data bias and data drifts independently, we propose a framework that explores how the two effects interact simultaneously. We highlight how changes within under-represented subgroups may be masked when assessing drifts on whole populations. A framework that combines two novel techniques has been designed to investigate whether drifts within sub-populations may be overcome by reducing biases within data. We show how the application of boosting techniques can resolve drifts caused by under-representation of subgroups. We also show that drifts arising from reasons other than underrepresentation need further investigations. Finally, a key recommendation is that any guidelines for assessing AI algorithms require that performance is evaluated both on the overall population and on sub-populations due to the risk of subpopulation performance being masked by general population performance.

Acknowledgements. This work was funded by NHSX and the Innovate UK Regulatory Pioneer Fund.

References

1. Clinical practice research datalink. https://cprd.com
2. CPRD cardiovascular disease synthetic dataset. https://cprd.com/cprd-cardiovas cular-disease-synthetic-dataset
3. Software and AI as a medical device programme. https://www.gov.uk/governme nt/publications/software-and-ai-as-a-medical-device-change-programme
4. Ditzler, G., Roveri, M., Alippi, C., Polikar, R.: Learning in nonstationary environments: a survey. IEEE Comput. Intell. Mag. **10**(4), 12–25 (2015)
5. Draghi, B., Wang, Z., Myles, P., Tucker, A.: Bayesboost: identifying and handling bias using synthetic data generators. In: Third International Workshop on Learning with Imbalanced Domains: Theory and Applications, pp. 49–62. PMLR (2021)
6. Gama, J., Žliobaitė, I., Bifet, A., Pechenizkiy, M., Bouchachia, A.: A survey on concept drift adaptation. ACM Comput. Surv. (CSUR) **46**(4), 1–37 (2014)
7. Gianfrancesco, M.A., Tamang, S., Yazdany, J., Schmajuk, G.: Potential biases in machine learning algorithms using electronic health record data. JAMA Internal Med. **178**(11), 1544–1547 (2018)
8. Gijsberts, C., et al.: Race/ethnic differences in the associations of the Framingham risk factors with carotid IMT and cardiovascular events. PLoS ONE **10**(7), e0132321 (2015)
9. Hamberg, K.: Gender bias in medicine. Womens Health **4**(3), 237–243 (2008)
10. Krieger, N., Fee, E.: Man-made medicine and women's health: the biopolitics of sex/gender and race/ethnicity. Women's Health, Politics, and Power: Essays on Sex/Gender, Medicine, and Public Health, pp. 11–29 (2020)
11. Mehrabi, N., Morstatter, F., Saxena, N., Lerman, K., Galstyan, A.: A survey on bias and fairness in machine learning. ACM Comput. Surv. (CSUR) **54**(6), 1–35 (2021)
12. Parikh, R.B., Teeple, S., Navathe, A.S.: Addressing bias in artificial intelligence in health care. JAMA **322**(24), 2377–2378 (2019)
13. Rotalinti, Y., Tucker, A., Lonergan, M., Myles, P., Branson, R.: Detecting drift in healthcare AI models based on data availability. In: Machine Learning and Principles and Practice of Knowledge Discovery in Databases, pp. 243–258. Springer Nature Switzerland AG (2022). https://doi.org/10.1007/978-3-031-23633-4_17
14. Wang, Z., Myles, P., Tucker, A.: Generating and evaluating synthetic UK primary care data: preserving data utility & patient privacy. In: 2019 IEEE 32nd International Symposium on Computer-Based Medical Systems (CBMS), pp. 126–131. IEEE (2019)
15. Wolf, A., et al.: Data resource profile: clinical practice research datalink (CPRD) aurum. Int. J. Epidemiol. **48**(6), 1740–1740g (2019)

A Topological Data Analysis Framework for Computational Phenotyping

Giuseppe Albi[1]([✉]), Alessia Gerbasi[1], Mattia Chiesa[2,3], Gualtiero I. Colombo[2], Riccardo Bellazzi[1], and Arianna Dagliati[1]

[1] Department of Electrical, Computer and Biomedical Engineering, University of Pavia, via Ferrata 5, 27100 Pavia, Italy
{giuseppe.albi01,alessia.gerbasi01}@universitadipavia.it,
{riccardo.bellazzi,arianna.dagliati}@unipv.it
[2] Centro Cardiologico Monzino IRCCS, Milan, Italy
{mattia.chiesa,gualtiero.colombo}@cardiologicomonzino.it
[3] Department of Electronics, Information and Biomedical engineering, Politecnico di Milano, Milan, Italy

Abstract. Topological Data Analysis (TDA) aims to extract relevant information from the underlying topology of data projections. In the healthcare domain, TDA has been successfully used to infer structural phenotypes from complex data by linking patients who display demographic, clinical, and biomarker similarities.

In this paper we propose *pheTDA*, a TDA-based framework to assist the computational definition of novel phenotypes. More in details, the *pheTDA* (i) guides the application of the Topological Mapper algorithm to derive a robust data representation as a topological graph; (ii) identifies relevant subgroups of patients from the topology; (iii) assess discriminative features for each subgroup of patients via predictive models.

We applied the proposed tool on a population of 725 patients with suspected coronary artery disease (CAD). *pheTDA* identified five novel subgroups, one of which is characterized by the presence of diabetic patients showing high cardiovascular risk score. In addition, we compare the results obtained with existing clustering algorithms, showing that *pheTDA* obtains better performance when compared to spectral decomposition followed by k-means.

Keywords: Topological Data Analysis · Computational phenotyping · Coronary artery disease · TDA Mapper

1 Introduction

Topological data analysis (TDA) [1] is a field of mathematics that uses qualitative geometric features to extract information from the shape of data. Unlike clustering, TDA can recognise related subgroups, providing better insights on how individual samples relate to the whole population [2]. These properties make TDA successful for the analysis of high-dimensional biomedical datasets [3]. However none of the available open-source tools is developed with the aim of solving biomedical tasks.

J. M. Juarez et al. (Eds.): AIME 2023, LNAI 13897, pp. 323–327, 2023.
https://doi.org/10.1007/978-3-031-34344-5_38

We present *pheTDA*, a TDA-based framework for the computational definition of novel patients' subgroups. Starting from a given sub-phenotype and
a dataset of patients characteristics, *pheTDA* uses a deductive framework to
stratify the individuals into new subgroups, highlighting the most relevant features characterizing novel sub-populations. The pipeline is built with open-source
Python libraries and publicly available[1].

We tested *pheTDA* on a population of patients with suspected coronary
artery disease (CAD). Several scores have been proposed to estimate sex-specific
individual 10-year risk of developing cardiovascular disease, but accurately identifying at-risk individuals is still a challenge [4]. Therefore, there is the need
for new strategies to integrate clinical information and accurately identify CAD
subgroups at an early stage.

2   Materials and Methods

pheTDA provides novel computational definitions of patient subgroups starting by an initial sub-phenotype definition. For each individual $pat = 1, ..., n$
with n being the dimension of the study population, m characteristics $x =
\{x_1, x_2, ..., x_m\}$ are available, constituting a patient matrix $X \in R^{n \times m}$. A label
$\Phi = \{l_1, ..., l_p\}$, representing an initial, clinically defined subgroup is associated
to each patient, with p being the number of initial subgroups.

The framework exploits the TDA Mapper algorithm implemented in the
KeplerMapper package [5] to represent the dataset X with a graph $G = (V, E)$,
where V and E represent the set of k nodes and j edges of the graph, respectively.
To infer the graph G, *pheTDA* performs a grid search over the Mapper's parameters, guided by the initial patient stratification Φ. Then, it applies the Mapper
to project the dataset X in a lower dimensional space $Z \in R^2$ and divides the
projection space in a series of overlapping bins in which subsequently performs
a clustering step. Each cluster constitutes a node v in the final graph representation, while an edge e expresses that two nodes share at least one sample. TDA
Mapper requires the definition of a distance metric d and a lens function f. For
both parameters several solutions are implemented in the proposed tool. Among
the proposed d, we included Gower distance [6], appropriate for datasets with
mixed input data types. Other f are t-Stochastic Neighbor Embedding (t-SNE),
Uniform Manifold Approximation (UMAP) and parametric UMAP.

A community's detection algorithm is then applied on the graph G, in order
to define q novel patient subgroups $\Phi' = \{l'_1, ..., l'_q\}$ where $\forall l'_i \in \Phi'$ with $i =
1, ..., q$. At this point, the framework applies a machine learning model in a
one-vs-rest binary classification setting to predict the probability of belonging
to the new subtypes for each patient. In this way, we obtain a novel subset of
relevant features $x'_{l_i} \subset x$ characterizing novel subgroups. Three machine learning

---

[1] https://github.com/Sep905/pheTDA.

models are compared: Elastic Net (EN) logistic regression, Random Forest and XGBoost. The hyperparameters of the selected model are tuned by maximizing the mean AUC with a grid search using K-cross validation.

2.1   Dataset and Experimental Setup

The study population includes 725 adults with negative history of acute cardiovascular events, undergoing coronary computed tomography angiography (CCTA) for suspected coronary artery disease. Each patient is associated with a CAD label assigned by the physicians via CCTA images. Baseline characteristics were collected for each patient at the first visit, including medical history and examinations. We considered a total of 77 clinical variables, discarding those with at least 10% of missing values. We imputed missing values with missForest [7] and computed sex-specific 10-year cardiovascular risk of occurrence of the first major cardiovascular event by using the CUORE project score [8].

We defined initial sub-phenotypes Φ as NOATH patients without coronary stenosis, and ATH patients with 1% or more coronary stenosis degree. We tested *pheTDA* to discover novel subgroups Φ' of CAD patients, considering as input the initial groups $\Phi = \{NOATH, ATH\}$ and the dataset of baseline clinical features. We split the dataset into training and test set (70/30%), using stratified sampling according to the prevalence of Φ. Once the *pheTDA* pipeline learn the parameters on the training set, we applied the model on the test set to reassign each patient to the novel Φ' predicted subgroups. We compared the new partition with standard clustering methods with the Calinski-Harabasz index.

3   Results

According to the CAD label, the initial subpopulation was composed by 287 NOATH and 438 ATH patients.

pheTDA selected UMAP as the best projection lens, with *minimum distance* = 0.9 and $n°$ *of neighbours* = 50 since it leads to the lowest graph entropy of 0.657. The training set projected in two dimensions using the selected lens and divided by CAD class, is presented in Fig. 1A.

We chose *resolution* = 18, *gain* = 0.5 and DBSCAN as clustering algorithm, with *epsilon* = 0.5 and *minimum samples* = 2. This configuration leads to a connected graph, with a mean degree size and node size of 6.9 and 7.2, respectively. The graph is composed by 275 nodes and 1535 edges and is showed in Fig. 1B.

The modularity community algorithm identified five subgroups $\Phi' = \{\alpha', \beta', \gamma', \delta', \epsilon'\}$. We report in Fig. 1C the graph partitioned according to Φ', coloured by the proportion of the ATH class. In addition, the Sankey diagram presented in Fig. 1D shows the composition of the patients in the training set according to the CAD class and the new subgroups stratification Φ'. Subgroups γ' and ϵ' show higher prevalence of ATH patients, while subgroup δ' is mainly composed of NOATH patients. Subgroup α' is composed of ATH patients only.

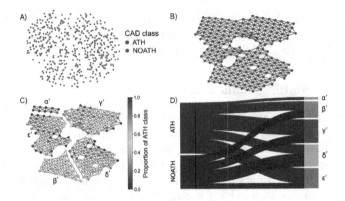

Fig. 1. Training set 2D projections obtained with UMAP after the first step of the grid search, coloured by CAD class (A). Patients graph generated after the second step of the Mapper parameters grid search (B), and partitioned with the identified communities, coloured by the proportion of ATH class (C). Sankey diagram showing the new stratification Φ' introduced in the training set (D).

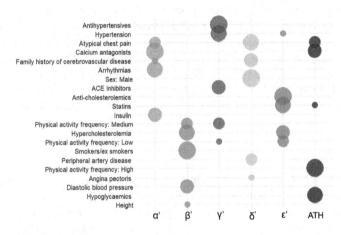

Fig. 2. Bubble plot showing top five features for each subgroup and for the ATH group, according to EN logistic regression. Bubble size is proportional to feature's ranking position. Features are ordered by their importance sum.

EN logistic regression is the model achieving the highest value of AUC for each of the new subgroups. We plot the five features with the highest importance for each subgroup with a bubble plot in Fig. 2.

We compared the *pheTDA* results with agglomerative and spectral clustering. Our method shows a higher Calinski-Harabasz index (6.14) when compared to spectral clustering (0.59), and a lower score when compared to agglomerative clustering with complete linkage (14.1). The same conclusion can be obtained by first applying the UMAP projection lens, with the optimal combination of parameters found in the first step of the grid search, and then by performing the clustering step.

4   Conclusion and Future Works

We present *pheTDA*; a deductive framework to define novel patients' subgroups, given an initial clinical definition, that we want to enhance, and a dataset of clinical features. We tested it using a dataset of coronary artery disease (CAD) patients, since it is known from the clinical practice that some individuals initially classified as *low-risk* patients may experience acute events. This suggests that several individual factors, not considered during the first clinical assessment, may contribute to the onset of acute disease-related manifestations. With *pheTDA* we identified five novel subgroups of CAD patients. The smaller subgroup consists only of ATH patients who represent a subpopulation with diabetes and the highest cardiovascular risk. The other subgroups are more heterogeneous with respect to the proportion of CAD class, but through our framework, we have identified the most discriminating variables in each of them. When applied on an external test set, our method introduces a more robust partition if compared to spectral clustering.

Since the data used to validate this tool is part of an ongoing project, follow-ups data concerning the occurrence of acute events, is currently being collected. In the future we will investigate how the newly introduced stratification impacts cardiac event prediction. It would be of interest to test the proposed framework in different clinical scenarios in order to further assess its generalizability.

References

1. Lum, P., et al.: Extracting insights from the shape of complex data using topology. Sci. Rep. 3, 1236 (2013)
2. Singh, G., et al.: Topological methods for the analysis of high dimensional data sets and 3D object recognition. EPBG@Eurographics (2007)
3. Nicolau, M., et al.: Topology based data analysis identifies a subgroup of breast cancers with a unique mutational profile and excellent survival. PNAS, USA. 108, 7265–7270 (2011)
4. Ge, Y., Wang, T.: Circulating, imaging, and genetic biomarkers in cardiovascular risk prediction. Trends Cardiovasc. Med. 21, 105–112 (2011)
5. Veen, H., et al.: Kepler Mapper: a flexible Python implementation of the Mapper algorithm. JOSS 4, 1315 (2019)
6. Gower, J.: A general coefficient of similarity and some of its properties. Biometrics 27, 857–871 (1971)
7. Stekhoven, D., Bühlmann, P.: MissForest-non-parametric missing value imputation for mixed-type data. Bioinformatics 28, 112–118 (2011)
8. Palmieri, L., et al.: CUORE project: implementation of the 10-year risk score. Euro. J. Cardiovasc. Prev. Rehabil. 18, 642–649 (2011)

Ranking of Survival-Related Gene Sets Through Integration of Single-Sample Gene Set Enrichment and Survival Analysis

Martin Špendl[1], Jaka Kokošar[1], Ela Praznik[1], Luka Ausec[2],
and Blaž Zupan[1(✉)]

[1] Faculty of Computer and Information Science, University of Ljubljana, Ljubljana,
Slovenia
{martin.spendl,jaka.kokosar,ela.praznik,blaz.zupan}@fri.uni-lj.si
[2] Genialis Inc., 177 Huntington Ave Ste 1703, Boston, Massachusetts, USA

Abstract. The onset and progression of a disease are often associated with changes in the expression of groups of genes from a particular molecular pathway. Gene set enrichment analysis has thus become a widely used tool in studying disease expression data; however, it has scarcely been utilized in the domain of survival analysis. Here we propose a computational approach to gene set enrichment analysis tailored to survival data. Our technique computes a single-sample gene set enrichment score for a particular gene set, separates the samples into an enriched and non-enriched cohort, and evaluates the separation according to the difference in survival of the cohorts. Using our method on the data from The Cancer Genome Atlas and Molecular Signatures Database Hallmark gene set collection, we successfully identified the gene sets whose enrichment is predictive of survival in particular cancer types. We show that the results of our method are supported by the empirical literature, where genes in the top-ranked gene sets are associated with survival prognosis. Our approach presents the potential of applying gene set enrichment to the domain of survival analysis, linking the disease-related changes in molecular pathways to survival prognosis.

Keywords: Gene set ranking · Survival analysis · Censored data · Survival curve · Gene expression · Single-sample gene set enrichment scoring

1 Introduction

The onset of diseases and the prediction of their progression are commonly associated with variations in the expression of genes that control specific molecular pathways. Such variations are often more informative and interpretable when

Supported by the Slovenian Research Agency grants P2-0209 and L2-3170.

J. M. Juarez et al. (Eds.): AIME 2023, LNAI 13897, pp. 328–337, 2023.
https://doi.org/10.1007/978-3-031-34344-5_39

considering groups of biologically related genes rather than individual genes alone [12]. Bioinformatics has developed various computational techniques that identify which gene sets are relevant to changes in phenotypes. One such technique is gene set enrichment analysis (GSEA) [22], which calculates an enrichment for each gene set based on the mRNA expression profile of samples and their binary phenotypes, such as tumour type, response, or exposure to a drug. In contrast to binary phenotypes, survival data includes information on the time until a target event and may include censored cases where the event has not yet occurred. We cannot use GSEA or similar enrichment approaches for such data.

Censored data is common in the clinical setting, and hence there is a need for approaches for survival-based gene set scoring methods, which are currently, at best, scarce. Those few reported in the literature rely on assigning gene set-specific scores to a single patient in the dataset. The gene set activity score (GSAS) algorithm [25] relates a score with the expression of transcription factors according to the BASE algorithm [4]. An immune-based prognostic score for ovarian cancer (IPSOV) [20] uses a similar approach to evaluate the association between characteristics and overall survival. Both were used for curating gene sets but not for ranking based on their prognostic power. Similar methods for single-cell RNA-seq were developed [6].

Here, we report on a technique that can rank gene sets based on their survival prognostic ability. Our method relies on calculating sample-based gene-set scores using a single-sample extension of the GSEA [3]. ssGSEA is a method that can assign a gene set enrichment score to each sample individually, thus not requiring a phenotype label. We use ssGSEA to order expression profiled samples based on the expression enrichment of a gene set. Using the median as a splitting criterion, we create two cohorts of equal size: an enriched and non-enriched cohort. The extent of enrichment corresponds to the overexpression of genes in the gene set compared to the average gene expression. Thus, enriched and non-enriched cohorts relate to mostly above and below-average expression of genes, respectively. We then evaluate the importance of a gene set for patient survival using a log-rank test between the cohorts on a Kaplan-Meier survival plot: the more significant the difference in survival characteristics, the higher the importance of a gene set. We rank gene sets according to their log-rank p-value and correct those using Benjamini-Hochberg FDR correction.

2 Methods

Given a gene set, our scoring method for survival data consists of three steps. First, we normalize gene expression values using a common normalization procedure. Second, we rank samples based on their gene set enrichment using a single-sample gene set scoring method [3]. And third, we split samples into two equal-sized cohorts and evaluate the difference in survival using the log-rank test. We repeat the procedure for all gene sets in the relevant gene set database and rank the gene sets according to their score. The ranked list is then subject to interpretation and further investigation by a molecular biologist. We showcase the implemented method on cancer-related data sets and use a standard curated gene set database.

2.1 Data

We collected cancer tissues from The Cancer Genome Atlas Program (TCGA) uploaded to the GEO portal (GSE62944) [17]. Samples are organised in data sets based on their tissue of origin. We collected mRNA sequencing data represented as a gene expression matrix. Different datasets vary in sample size; thus, we included only datasets with more than 100 samples (total of 20). We extracted the sample's survival time and event occurrence from clinical metadata. Survival time is the last known date when a patient is still alive. If a patient dies of cancer, we consider the event has occurred. Otherwise, if its status is unknown or it dies of unrelated death, its event status is censored. Datasets have varying sample sizes and ratios of censored data (Table 1).

Table 1. TCGA project statistics about censored data. The N is the number of samples in the dataset, and the Censored is the ratio of censored samples.

TCGA	CESC	HNSC	KIRC	LAML	LGG	LUAD	READ	SKCM
N	306	504	542	178	532	541	167	472
Censored	0.807	0.675	0.707	0.348	0.846	0.769	0.940	0.661

Gene expression is stored as transcripts per million (TPM); thus, all expression values for each sample sum up to a million. We use a standard procedure of log-transforming each gene expression with pseudo count 1 and z-score normalization for each gene across samples in a dataset. That is, we normalize the columns of the expression matrix.

Gene sets are sets of genes that encode proteins acting together in some biological process. Biologists create and curate them to better understand their function and interactions. We have considered a set of 50 curated gene sets called Hallmark gene sets from the Molecular Signature Database (MSigDB) [11,22], where gene sets represent states and processes in human cells (see Table 2).

Table 2. Example of three Hallmark gene sets and a few genes. N is the number of total genes, and gene names are from HUGO Gene Nomenclature Committee.

Hallmark gene set	N	corresponding genes
ANGIOGENESIS	36	APOH, FGRF1, ITGAV, LPL, VEGFA, ...
APOPTOSIS	161	BAX, BCL10, CASP1, ERBB2, MADD, ...
GYCOLYSIS	200	EGFR, G6PD, GALK1, LDHA, SOD1,

2.2   Single-Sample Gene Set Enrichment Analysis

Single-sample gene set enrichment analysis (ssGSEA) is a single-sample extension of the GSEA algorithm [3]. It assigns the enrichment score to a single sample based on the gene expression profile of a sample. This differs from the original GSEA algorithm, which computes the gene set's enrichment score based on the entire data set. The score represents the gene set's degree of enrichment in a sample in a given dataset. In simplified terms, gene sets with highly expressed genes will have a high enrichment score. Gene expression values of a sample are rank normalized, standardized, are sorted in decreasing order based on their rank r. Genes in the gene set form a probability mass function (PMF) with probabilities $|r|^{\alpha}$, while genes outside form a PMF with genes having equal probability. The enrichment score of a sample is then represented as the difference of cumulative density functions for those PMFs. In essence, the enrichment score of a sample describes the degree of above-average expression of genes in a gene set.

2.3   Gene Set Ranking for Survival Analysis

We aim to evaluate the utility of a gene set in separating samples into enriched and non-enriched cohorts based on their gene set enrichment score. We abstract the approach with the following procedure:

Algorithm 1. Gene Set Ranking for Survival Analysis

1: $data \leftarrow$ samples with normalized expression values
2: $geneSets \leftarrow$ Hallmark gene sets
3: $enrichemntScores \leftarrow$ ssGSEA(data, geneSets)
4: **for each** $score \in enrichemntScores$ **do**
5:     $cohorts \leftarrow$ split sample by median of enrichment score
6:     $p \leftarrow$ log-rank test between cohorts

The literature suggests the median as the least biased approach to split the data into two cohorts [2]. Our null hypothesis is that both cohorts have the same hazard function. We test the null hypothesis using a standard log-rank test, a form of χ^2 test with one degree of freedom (line 6). The 95% confidence intervals of the χ^2 test statistic are calculated using bootstrap without recalculating ssGSEA enrichment scores. We repeat the protocol for other gene sets and correct p-values for the false-discovery rate with the Benjamini-Hochberg procedure.

2.4   Robustness Estimation

We evaluate the robustness of the proposed enrichment scoring in three steps. First, we perform bootstrap sampling 1000 times to estimate the 95% confidence interval of the χ^2 test statistic without recalculating enrichment scores. Recalculating scores for samples on a bootstrapped dataset provides only marginally different 95% CI but requires much more computation. Our method estimates the CI using the same sample enrichment scores as in the original data set.

In the second step, we evaluate how individual genes in the gene set influence the results compared to random genes. We incrementally remove a subset of genes and compare the statistic with the original gene set. Additionally, we remove genes from the gene set and replace them with randomly selected genes. By repeating the procedure 100 times, we calculate 95% confidence intervals of evaluation. The third approach evaluates the robustness in terms of sample size. We downsample the original dataset incrementally to 50% of the original size and compare the χ^2 test statistic. Each downsampling is performed 100 times to estimate the 95% confidence intervals.

3   Results

With the proposed approach, we could find Hallmark genesets that characterize cohorts with significantly different survival characteristics for six of our study's twenty TCGA cancer datasets. In Table 3, we report each data set's top three gene sets and their corresponding test statistics. We would find a gene set significant if the FDR corrected p-value is below 0.01. The table also includes a reference for each gene set that confirms our findings in the existing literature; for brevity, we only include the most relevant articles that have already reported the relation between genes in a Hallmark gene set and their prognostic power in a cancer type.

Table 3. Up to three significant top-ranked Hallmark gene sets for each of the six TCGA cancer types. We report the most relevant reference if p-values are below 0.01. A complete list of literature references is available on our GitHub repository (see Conclusion). CESC - Cervical squamous cell carcinoma and endocervical adenocarcinoma, HNSC - Head and Neck squamous cell carcinoma, KIRC - Kidney renal clear cell carcinoma, LGG - Brain Lower Grade Glioma, LUAD - Lung adenocarcinoma, SKCM - Skin Cutaneous Melanoma.

TCGA	Hallmark	χ^2	p_{value}	References
CESC	UV_RESPONSE_DN	16.2	2.63e-03	[8]
	ANGIOGENESIS	14.7	2.63e-03	[24]
	PROTEIN_SECRETION	14.3	2.63e-03	[15]
HNSC	GLYCOLYSIS	26.0	1.70e-05	[10]
	MTORC1_SIGNALING	17.5	7.25e-04	[21]
	XENOBIOTIC_METABOLISM	15.0	1.78e-03	[14]
KIRC	HEME_METABOLISM	26.1	1.63e-05	[7]
	FATTY_ACID_METABOLISM	18.1	5.28e-04	[5]
	ANDROGEN_RESPONSE	16.7	7.35e-04	[27]
LGG	EMT	19.4	2.87e-04	[23]
	ANGIOGENESIS	18.6	2.87e-04	[16]
	COAGULATION	18.3	2.87e-04	[18]
LUAD	MTORC1_SIGNALING	14.4	4.02e-03	[13]
	HYPOXIA	14.2	4.02e-03	[19]
	GLYCOLYSIS	12.1	8.29e-03	[26]
SKCM	INTERFERON_GAMMA_RESPONSE	15.0	5.36e-03	[1]
	INTERFERON_ALPHA_RESPONSE	13.3	6.78e-03	[9]

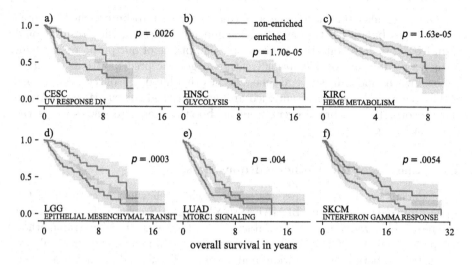

Fig. 1. Kaplan-Meier survival curves for the best-performing gene set. a) CESC-UV response down, b) HNSC-Gylcolysis, c) KIRC-Heme metabolism, d) LGG-Epithelial Mesenchymal Transition, e) LUAD - MTORC1 signalling, f) SKCM-IFN-γ response.

We find literature support for all top-ranked gene set-cancer type pairs. For example, in the case of cervical squamous cell carcinoma and endocervical adenocarcinoma (CESC), the cancer occurrence is highly linked to the infection with human papillomavirus (HPV) infection. This virus produces proteins that transform a human cell into a cancer cell by degrading the main tumour suppressor protein, p53. They also inhibit DNA damage repair in response to UV exposure, leading to extensive mutations. The tumour becomes more invasive by inducing angiogenesis and forming new blood vessels. Our method finds those samples with enriched scores in Hallmark gene sets related to those pathways have lower survival curves.

We observe the survival difference in cohorts suggested by the top-ranked gene sets for TCGA datasets. We observe that the enriched cohort is linked to worse survival compared to non-enriched in 4 out of 6 cases (Fig. 1a,b,d,e), whereas linked to a better prognosis for the other two (Fig. 1c, f).

4   Discussion

The results from the TCGA datasets suggest that our proposed method can pinpoint the relevant gene sets and that the ranking can identify those best related to the phenomena represented in the corresponding dataset. Gene set scoring produces a ranked list with FDR-corrected p-values, but the process is just a hypothesis generation. We should consider significant results with caution. However, a large body of literature confirming our case studies findings suggested that our results are not a result of chance.

There are also Hallmark gene sets that we expect to be enriched in some cancer types but did not appear to be significant. These missing results could stem from our data collection process or assumptions of our proposed method. Namely, we did not consider prior treatment, genetic predispositions, or other diseases when modelling survival time. On the other hand, one of the assumptions is that the ratio between the enriched and non-enriched cohorts is equal. As we show below, this is a broad overstatement, but the literature suggests it is the least biased [2].

4.1   Analysis of the Method's Robustness

We comment on the robustness of the approach by showing an example of the highest ranked gene set HALLMARK_GYLCOLYSIS on the Head and Neck squamous cell carcinoma (HNSC) dataset (Fig. 2). We use the bootstrap method to evaluate the 95% CI of the test statistic. Bootstrap confidence intervals for higher test statistic values are wider and normally distributed, while lower values have a more skewed distribution towards 0. We observe the number of unique samples in a dataset does not affect the size of 95% CI.

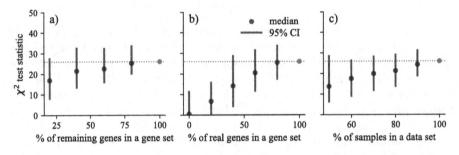

Fig. 2. Robustness of the method. We used the TCGA-HNSC dataset and HALL-MARK_GLYCOLYSIS gene set while comparing χ^2 test statistic. a) Reducing the number of genes in the gene set, b) Replacing gene set genes with random ones, c) Reducing the number of samples in a data set.

We observe how the number of genes in a gene set affects performance (Fig. 2a). Removing as much as half of genes from the gene set has a marginal effect on calculated test statistics. The following shows how scoring is not dependent on any single gene, but their effect is combined in sample ranking. The redundancy of genes in biological pathways and gene sets is known. In contrast, when replacing genes in a gene set with random genes, we observe a clear shift of the test statistic towards lower values (Fig. 2b). Adding noise to the enrichment calculation impacts sample ranking and, thus, the method's performance. Confidence intervals of the mean over multiple runs also become smaller due to the relative distance from zero.

Removing samples from the dataset results in slowly decreasing values in the χ^2 test statistic. Confidence intervals become wider due to the variation in possible cohort combinations. Even when removing 50% of all samples, the gene set is enriched with statistical significance. This suggests that we can use this method even on smaller sample sizes.

4.2 Varying Splitting Threshold

Our cohort formation assumes equally-sized cohorts. Instead of using the median score for splitting, we could search for the score threshold that maximizes the log-rank statistics and find gene set enriched and non-enriched cohorts of different sizes. Figure 3 shows that such threshold search for the HALL-MARK_ADIPOGENESIS gene set on the KIRC dataset improves the results. When using the default median value as a threshold, the log-rank statistic of 8.48 is substantially smaller than 25.37 for the split where 75% of the samples are placed in the enriched cohort. We have observed similar benefits of threshold search for other gene sets and data sets.

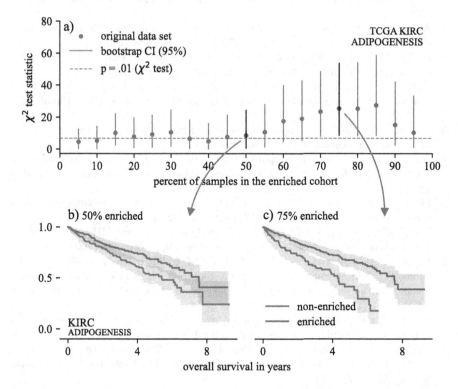

Fig. 3. Varying cohort split selection on HALLMARK_ADIPOGENESIS and TCGA KIRC dataset. a) Varying split threshold between 5 - 95%, b,c) Comparison of Kaplan-Meier plots for 50% and 75% per cent of samples in an enriched cohort.

5   Conclusions

The abundance of censored data and molecular fingerprints in clinical settings encourages the development of methods that can shed light on biological processes that govern disease progression. We propose a survival-related gene set ranking method based on single-sample enrichment scoring. An application of our method on publicly available data sets where the results match those from the literature confirms that our approach produces meaningful results with relevant implications to prognosis. The simplicity of the proposed method also leaves room for additional improvements, such as choosing different splitting criteria for cohort formation. The code and datasets used are available on GitHub (https://github.com/biolab/AIME-2023-paper) and archived on Zenodo [28].

References

1. Alavi, S., Stewart, A.J., Kefford, R.F., Lim, S.Y., Shklovskaya, E., Rizos, H.: Interferon signaling is frequently downregulated in melanoma. Front. Immunol. **9**, 1414 (2018)
2. Altman, D.G.: Prognostic models: a methodological framework and review of models for breast cancer. Cancer Invest. **27**(3), 235–243 (2009)
3. Barbie, D.A., et al.: Systematic RNA interference reveals that oncogenic KRAS-driven cancers require TBK1. Nature **462**(7269), 108–112 (2009)
4. Cheng, C., Yan, X., Sun, F., Li, L.M.: Inferring activity changes of transcription factors by binding association with sorted expression profiles. BMC Bioinf. **8**(1), 1–12 (2007)
5. Du, W., et al.: HIF drives lipid deposition and cancer in CCRCC via repression of fatty acid metabolism. Nat. Commun. **8**(1), 1–12 (2017)
6. Dwivedi, B., Mumme, H., Satpathy, S., Bhasin, S.S., Bhasin, M.: Survival genie, a web platform for survival analysis across pediatric and adult cancers. Sci. Rep. **12**(1), 3069 (2022)
7. Frezza, C., et al.: Haem oxygenase is synthetically lethal with the tumour suppressor fumarate hydratase. Nature **477**(7363), 225–228 (2011)
8. Jackson, S., Storey, A.: E6 proteins from diverse cutaneous HPV types inhibit apoptosis in response to UV damage. Oncogene **19**(4), 592–598 (2000)
9. Kirkwood, J.M., Strawderman, M.H., Ernstoff, M.S., Smith, T.J., Borden, E.C., Blum, R.H.: Interferon alfa-2b adjuvant therapy of high-risk resected cutaneous melanoma: the eastern cooperative oncology group trial EST 1684. J. Clin. Oncol. **14**(1), 7–17 (1996)
10. Kumar, D.: Regulation of glycolysis in head and neck squamous cell carcinoma. Postdoc J.: J. Postdoctoral Res. Postdoctoral Affairs **5**(1), 14 (2017)
11. Liberzon, A., Subramanian, A., Pinchback, R., Thorvaldsdóttir, H., Tamayo, P., Mesirov, J.P.: Molecular signatures database (MSigDB) 3.0. Bioinformatics **27**(12), 1739–1740 (2011)
12. Maleki, F., Ovens, K., Hogan, D.J., Kusalik, A.J.: Gene set analysis: challenges, opportunities, and future research. Front. Genet. **11**, 654 (2020)
13. Marinov, M., Fischer, B., Arcaro, A.: Targeting mTOR signaling in lung cancer. Crit. Rev. Oncol. Hematol. **63**(2), 172–182 (2007)

14. Namani, A., Rahaman, M., Chen, M., Tang, X., et al.: Gene-expression signature regulated by the keap1-nrf2-cul3 axis is associated with a poor prognosis in head and neck squamous cell cancer. BMC Cancer **18**(1), 1–11 (2018)
15. Noordhuis, M.G., et al.: Expression of epidermal growth factor receptor (EGFR) and activated EGFR predict poor response to (chemo) radiation and survival in cervical cancerthe EGFR pathway in advanced-stage cervical cancer. Clin. Cancer Res. **15**(23), 7389–7397 (2009)
16. Plate, K.H., Risau, W.: Angiogenesis in malignant gliomas. Glia **15**(3), 339–347 (1995)
17. Rahman, M., Jackson, L.K., Johnson, W.E., Li, D.Y., Bild, A.H., Piccolo, S.R.: Alternative preprocessing of RNA-sequencing data in the cancer genome atlas leads to improved analysis results. Bioinformatics **31**(22), 3666–3672 (2015)
18. Rong, Y., Post, D.E., Pieper, R.O., Durden, D.L., Van Meir, E.G., Brat, D.J.: PTEN and hypoxia regulate tissue factor expression and plasma coagulation by glioblastoma. Can. Res. **65**(4), 1406–1413 (2005)
19. Salem, A., et al.:: Targeting hypoxia to improve non-small cell lung cancer outcome. JNCI: J. Nat. Cancer Instit. **110**(1), 14–30 (2018)
20. Shen, S., et al.: Development and validation of an immune gene-set based prognostic signature in ovarian cancer. EBioMedicine **40**, 318–326 (2019)
21. Simpson, D.R., Mell, L.K., Cohen, E.E.: Targeting the PI3K/AKT/mTOR pathway in squamous cell carcinoma of the head and neck. Oral Oncol. **51**(4), 291–298 (2015)
22. Subramanian, A., et al.: Gene set enrichment analysis: a knowledge-based approach for interpreting genome-wide expression profiles. Proc. Natl. Acad. Sci. **102**(43), 15545–15550 (2005)
23. Tao, C., Huang, K., Shi, J., Hu, Q., Li, K., Zhu, X.: Genomics and prognosis analysis of epithelial-mesenchymal transition in glioma. Front. Oncol. **10**, 183 (2020)
24. Tomao, F., et al.: Angiogenesis and antiangiogenic agents in cervical cancer. Onco. Targets. Ther. **7**, 2237 (2014)
25. Varn, F.S., Ung, M.H., Lou, S.K., Cheng, C.: Integrative analysis of survival-associated gene sets in breast cancer. BMC Med. Genomics **8**(1), 1–16 (2015)
26. Zhang, L., Zhang, Z., Yu, Z.: Identification of a novel glycolysis-related gene signature for predicting metastasis and survival in patients with lung adenocarcinoma. J. Transl. Med. **17**(1), 1–13 (2019)
27. Zhao, H., Leppert, J.T., Peehl, D.M.: A protective role for androgen receptor in clear cell renal cell carcinoma based on mining TCGA data. PLoS ONE **11**(1), e0146505 (2016)
28. Špendl, M., Kokošar, J.: biolab/aime-2023-paper: Version 1.0 (2023). https://doi.org/10.5281/zenodo.7572951

Knowledge Representation and Decision Support

Supporting the Prediction of AKI Evolution Through Interval-Based Approximate Predictive Functional Dependencies

Beatrice Amico[(✉)] and Carlo Combi[iD]

Department of Computer Science, University of Verona, Verona, Italy
{beatrice.amico,carlo.combi}@univr.it

Abstract. In this paper, we focus on the early prediction of patterns related to the severity stage of Acute Kidney Injury (AKI) in an ICU setting. Such problem is challenging from several points of view: (i) AKI in ICU is a high-risk complication for ICU patients and needs to be suitably prevented, and (ii) the detection of AKI pathological states is done with some delay, due to the required data collection. To support the early prediction of AKI diagnosis, we extend a recently-proposed temporal framework to deal with the prediction of multivalued interval-based patterns, representing the evolution of pathological states of patients. We evaluated our approach on the MIMIC-IV dataset.

1 Introduction

The Intensive Care Unit (ICU) provides a significant example of a clinical scenario where modern monitoring systems continuously store a large amount of data Through a continuous monitoring of pathological states over time, it is possible to anticipate sudden changes in patient's health status, thus preventing possible complications. In this paper, we will focus on the *prediction* task of Acute Kidney Injury (AKI) [4], a multi-organ failure clinical event, a common complication during ICU stays. The early detection and the prompt treatment are elements that likely provide benefits for the patient, possibly avoiding even temporary support from a dialysis machine.

In this paper, we propose an approach to mine ICU data for the early prediction of AKI stage patterns. We extend the 3 window-based framework for deriving Approximate Predictive Functional Dependencies (APFDs) [1], to deal with the prediction of multi-valued pathological state patterns that hold over time intervals. Our analysis is based on real clinical data from the MIMIC-IV dataset. Through a detailed analysis, we performed a set of experiments, considering meaningful clinical aspects and the different severity stages, to find which elements are possibly predictive of the temporal evolution of AKI severity.

J. M. Juarez et al. (Eds.): AIME 2023, LNAI 13897, pp. 341–346, 2023.
https://doi.org/10.1007/978-3-031-34344-5_40

2   Predicting the Evolution of Pathological States

According to the internationally KDIGO criteria [4], a patient receives the diagnosis of AKI if there is (i) an increase in serum creatinine by ≥ 0.3 mg/dl ($\geq 26.5 \, \mu mol/l$) within 48 h, or (ii) an increase in serum creatinine to ≥ 1.5 times baseline within the previous 7 days, or (iii) a urine volume ≤ 0.5 ml/kg/h for six hours. Thus, such criteria have different temporal features which have to be considered. First of all, they allow one to identify AKI with some *implicit delay*. Indeed, criterium (i) considers a window of 48 h, criterium (ii) may be quite immediate but needs previous data for 7 days, while criterium (iii) specifies AKI with a delay of six hours. Furthermore, AKI pathological state has to be re-checked according to some periodicity, e.g. every hour, every three hours, and so on.

Moving to a finer definition of AKI severity stages after the identification of an AKI state, Table 1 specifies the different criteria, considering either serum creatinine concentration or urine output. Even in this case, severity has to be periodically checked and the severity of pathological states may be identified with a delay of even 24 h. Thus, there is the need to (i) predict the evolution of possibly changing pathological states, which have to be periodically re-evaluated; (ii) allow the early prediction of such states, as they are known after some (possibly varying) delay, to reduce the overall risk for patients and to allow the right time for preparing the suitable therapies/interventions.

Table 1. AKI is staged for severity according to the following criteria.

Stage	Serum creatinine	Urine output
1	1.5–1.9 times baseline OR	<0.5 ml/kg/h for 6–12 h
	≥ 0.3 mg/dl (≥ 26.5 mmol/l) increase	
2	2.0–2.9 times baseline	<0.5 ml/kg/h for ≥ 12 h
3	3.0 times baseline OR	<0.3 ml/kg/h for ≥ 24 h OR
	Increase in serum creatinine to ≥ 4.0 mg/dl (≥ 353.6 mmol/l) OR	Anuria for ≥ 12 h
	Initiation of renal replacement therapy	

In [1], we proposed a 3-window framework composed of an Observation Window (OW) where we collect the data, a Waiting Window (WW) as a black hole without any observation, and the Prediction Window (PW) where we collect the predicted pathological states. The WW embodies a central role for two different reasons. As we said before, anticipating some actions to prevent a future event could represent a primary goal in medicine. For this reason, the WW represents a way to create a detachment from the OW and the time window where the predicted pathological state will occur, in order to have to react and, if possible, perform actions to prevent a possible worsening of the patient state.

2.1   Extending APFDs to Deal with Interval-Based Predictions

Let us recall the concept of *functional dependency*. Let r be a relation over the relational schema $R(U)$: let $X, Y \subseteq R$ be sets of attributes of U. r fulfills the functional dependency $X \to Y$ (written as $r \models X \to Y$) if for all the couples of tuples t and t' showing the same value(s) on X, the corresponding value(s) on Y are identical.

The proposed framework in [1] derives *approximate predictive functional dependencies* (APFDs). Their antecedent is characterized by a set of attributes representing the evolution of different patient features, while the consequent consists of the predicted attribute, usually related to a pathological state. Attribute values for antecedent and consequent are evaluated in OW and PW, respectively. It may be worthwhile to explore predicted attributes representing some temporal patterns, over time intervals.

As a first extension of [1], here we propose that the relation containing the values of the predicted attribute, the *target expression*, is interval-based, i.e., its valid time represents the interval during which the considered value is true. According to this new *target expression*, we can extend the definition of *K-State Prediction Expression* (KSPE), a multi-temporal (i.e., with many valid times) relation we will use for mining APFDs. It allows, for any patient, the representation of different attribute values in different valid times, within OW, and of the following pattern value of the target relation, within PW.

Thus, an extended K-State Prediction Expression (eKSPE) has schema $R^{eKSPE}(Z\overline{U}^0\overline{U}^1..\overline{U}^k\dot{B} \cup \{\overline{VT}^0, \overline{VT}^1, .., \overline{VT}^k, \dot{V}T_{start}, \dot{V}T_{end}\})$, where Z is the attribute set identifying the patients we are considering, attributes $\overline{U}^i$ represent properties holding at valid time $\overline{VT}^i$ for $0 \le i \le k$, and for each $t \in R^{eKSPE}$ it holds $t[\overline{VT}^i] < t[\overline{VT}^{(i+1)}]$ for $0 \le i \le k-1$, $\dot{V}T_{start}, \dot{V}T_{end}$ represent the start and the end of valid time, with $t[\overline{VT}^k] < t[\dot{V}T_{start}]$, and $\dot{B}$ is the attribute containing the (pattern) values to be predicted. Moreover, all the valid times $t[\overline{VT}^i]$ of any tuple t are contained in OW, while $t[\dot{V}T_{start}]$ and $t[\dot{V}T_{end}]$ are within PW.

Definition 1 (Interval-based Approximate Predictive Functional Dependency).

Given an eKPSE $R^{eKPSE}(Z\overline{U}^0\overline{U}^1..\overline{U}^k\dot{B} \cup \{\overline{VT}^0, \overline{VT}^1, .., \overline{VT}^k, \dot{V}T_{start}, \dot{V}T_{end}\})$, *an interval-based approximate functional dependency (IAPFD) is expressed as*

$$\overline{X}^h\,\overline{S}^i\,...\overline{W}^j \xrightarrow{\varepsilon} \dot{B} \ with \ 0 \le h < i... < j \le k$$

where $\varepsilon = <\varepsilon_g, \varepsilon_h, \varepsilon_j>$ *and* $\overline{X}^h \subseteq \overline{U}^h, \overline{S}^i \subseteq \overline{U}^i, \overline{W}^j \subseteq \overline{U}^j$

Given an instance w of R^{eKPSE}, w fulfills the IAPFD (written as $w \models \overline{X}^h\,\overline{S}^i\,...\overline{W}^j \xrightarrow{\varepsilon} \dot{B}$), if a relation $s \subseteq w$ exists such that $s \models \overline{X}^h\,\overline{S}^i\,...\overline{W}^j \to \dot{B}$ with $g_3 \le \varepsilon_g \wedge h_3 \le \varepsilon_h \wedge j_3 \le \varepsilon_j$. In other words, $\varepsilon_g, \varepsilon_h, \varepsilon_j$ are the maximum acceptable errors defined by the user for g_3, h_3, and j_3, respectively. Indeed, the proposed approximation considers three different kinds of errors: (i) the percentage of tuples to delete (g_3) (ii) the percentage of patients having the complete deletion of their data (h_3), and (iii) the percentage of tuples to delete per patient (j_3), for obtaining a relation satisfying the given dependency.

3   Experiments: Dataset and Data Transformation

We evaluated the use of APFDs in the context of the Intensive Care Unit (ICU) with Acute Kidney Injury (AKI). As a data source for the analyses, we use the Medical Information Mart for Intensive Care IV (MIMIC-IV) [3].

We transform the MIMIC-IV raw data into a form useful for mining the APFDs, using PostgreSQL and PL/pgSQL, obtaining 73.729 subjects between 18 and 90. Exploiting the KDIGO criteria, we label each patient from admission to discharge from the ICU, with different stages of severity every 3 h. We select the features through a detailed analysis with clinicians, identifying which drugs are particularly weakening for kidneys, studying nephrotoxic and diuretic drugs, and how the level of systolic pressure and exams with contrast radiography fluids influence the severity stage. After this preprocessing part, we use an algorithm inspired by TANE [2]. We compute all the minimal APFDs, considering the three errors, g_3, h_3, j_3. Given a KSPE instance w and the predicted attribute $\dot{B}$, our approach is mainly based on the following steps: (i) derive s by TANE such that $g_3 \leq \varepsilon_g$; (ii) check on s that $h_3 \leq \varepsilon_h$; (iii) if the previous check is fine, check $j_3 \leq \varepsilon_j$.

Table 2. Value combinations of different KSPEs.

$\overline{Nephrotoxic}^1$	$Diuretic^2$	AKI
Acetaminophen	Bumetanide	000
Acetaminophen	Spironolactone	000
Acyclovir	Furosemide	000
Acetaminophen	Chlorothiazide	222
Acetaminophen	Chlorthalidone	111
Acetaminophen	Metolazone	333
Acyclovir	Torsemide	333
Allopurinol	Furosemide	333

(a) Value combinations for
$\overline{Nephrotoxic}^1, \overline{Diuretic}^2 \to AKI$

$\overline{Systolic}^0$	$\overline{Nephrotoxic}^1$	$\overline{Diuretic}^2$	AKI
High	Acetaminophen	Chlorthalidone	111
High	Acetaminophen	Spironolactone	000
High	Acetaminophen	Eplerenone	000
High	Acetaminophen	Bumetanide	333
High	Acetaminophen	Metolazone	333
High	Acyclovir	Furosemide	000
High	Allopurinol	Furosemide	333
High	Aspirin	Furosemide	333

(b) Value combinations for
$\overline{Systolic}^0, \overline{Nephrotoxic}^1, \overline{Diuretic}^2 \to AKI$

3.1   Results

We report the results regarding two different 3-window frameworks, selected relying on the clinical expertise and the KDIGO criteria: 48 h (OW), 12 h (WW), 72 h (PW), and 48 h (OW), 36 h (WW), 48 h (PW). We consider a diagnosis pattern of 9 h, with a different diagnosis every 3 h.

We generated two different K-state Evolution Expressions (KSEs): (i) a KSE with three temporal states $\overline{VT}^0 < \overline{VT}^1 < \overline{VT}^2$, which considers contrast radiography procedures, systolic pressure, and nephrotoxic drugs; (ii) a KSE with three temporal states where $\overline{VT}^k < \overline{VT}^{k-1} + 3$ for $k = 1, 2$ which considers systolic pressure, nephrotoxic drugs, and diuretic drugs.

Starting from analyzing the second KSE (ii) with the first 3-window framework, we obtained a court of 3128 patients and 50 different diagnosis patterns. One of APFDs extracted is $\overline{Nephrotoxic}^1, \overline{Diuretic}^2 \rightarrow A\dot{K}I$ ($g_3 =$ 55%, $h_3 =$ 0%, $j_3 =$ 25%). In Table 2(a) we show some of the different combinations extracted from the database. Regarding the second 3-window framework, we record a reduction in the number of patients (2448). Because of the increase in hours of the waiting window, the patients present at the beginning of the previous prediction window, now are the in the waiting window, therefore out of the considered period for the diagnosis. In Table 2(b), we report different value combinations extracted from the database for the APFD $\overline{Systolic}^0, \overline{Nephrotoxic}^1, \overline{Diuretic}^2 \rightarrow A\dot{K}I$ ($g_3 =$ 53.8%, $h_3 =$ 0%, $j_3 =$ 40%), which involves all the temporal states.

It is easy to see that the most common value combinations are related to homogeneous patterns, like 333, 222, 111, 000. From a clinical point of view, to study the temporal evolution of a diagnosis is more interesting to observe patterns where there is at least a change. Hence, we report another analysis regarding the first KSE (i) with the first 3-window framework, considering only the value combinations for the dis-homogeneous patterns. In Table 3, we report some value combinations of the APFD $\overline{Contrast}^0, \overline{Nephrotoxic}^2 \rightarrow A\dot{K}I$ ($g_3 =$ 56.8%, $h_3 =$ 0%, $j_3 =$ 10%).

Table 3. Value combinations for $\overline{Contrast}^0, \overline{Nephrotoxic}^2 \rightarrow A\dot{K}I$

$\overline{Contrast}^0$	$\overline{Nephrotoxic}^2$	$A\dot{K}I$
Contrast Portal Phlebography	Acetaminophen	033
Contrast Portal Phlebography	Tacrolimus	322
Contrast Phlebography (other)	Vancomycin	322
Contrast Phlebography (other)	Acetaminophen	322
Contrast Portal Phlebography	Vancomycin	122
Contrast Intra-Abdominal Phlebography	Vancomycin	133
Contrast Femoral Phlebography	Acetaminophen	200
Contrast Femoral Phlebography	Clopidogrel	330

4   Conclusions

In this paper, we introduced and discussed an interval-based extension of the 3-window framework for mining the APFDs. We analyzed the possibility to study the evolution of a diagnosis over time intervals, exploring the potential predictiveness of the APFDs. We focus on the temporal aspects of the diagnosis and the criteria to define it. Through a detailed analysis, we reported a set of experiments, considering meaningful clinical features and the different severity stages of AKI, to find which elements are possibly predictive of this disease.

References

1. Amico, B., Combi, C.: A 3-window framework for the Discovery and Interpretation of Predictive Temporal Functional Dependencies. In: Michalowski, M., Abidi, S.S.R., Abidi, S. (eds.) AIME 2022, pp. 299–309. Springer, Cham (2022). https://doi.org/10.1007/978-3-031-09342-5_29
2. Huhtala, Y., Kärkkäinen, J., Porkka, P., Toivonen, H.: TANE: an efficient algorithm for discovering functional and approximate dependencies. Comput. J. **42**(2), 100–111 (1999)
3. Johnson, A., Bulgarelli, L., Pollard, T., Horng, S., Celi, L.A., Mark, R.: Mimic-iv. version 0.4). PhysioNet (2005). https://doi.org/10.13026/a3wn-hq05
4. Khwaja, A.: KDIGO clinical practice guidelines for acute kidney injury. Nephron Clin. Practi. **120**(4), c179–c184 (2012)

Integrating Ontological Knowledge with Probability Data to Aid Diagnosis in Radiology

Charles E. Kahn Jr.[⊠] [iD]

Department of Radiology and Institute for Biomedical Informatics, University of Pennsylvania,
Philadelphia PA 19104, USA
ckahn@upenn.edu

Abstract. Radiological diagnosis requires integration of imaging observations with patient factors, such as age, sex, and medical history. Imaging manifestations of a disease can be highly variable; conversely, one imaging finding may suggest several possible causes. To account for the inherent uncertainty of radiological diagnosis, this report explores the integration of probability data with an ontology of radiological diagnosis. The Radiology Gamuts Ontology (RGO) incorporates 16,839 entities that define diseases, interventions, and imaging observations of relevance to diagnostic radiology. RGO's 55,564 causal ("may cause") relationships link disorders and their potential imaging manifestations. From a cohort of 1.7 million radiology reports on more than 1.3 million patients, the frequency of individual RGO entities and of their pairwise co-occurrence was identified. These data allow estimation of conditional probabilities of pairs of entities. A user interface enables one to traverse the ontology's network of causal relations with associated conditional-probability data. The system generates Bayesian network models that integrate an entity's age and sex distribution with its causally related conditions.

Keywords: Ontology · Knowledge Representation · Probability · Radiological Diagnosis · Data Mining

1 Introduction

1.1 Ontologies

An ontology represents knowledge of a domain of interest as a set of logical statements, or axioms, that relate the ontology's concepts to one another. Ontologies have the advantage of expressing knowledge in a form that computers can process and humans can read [1]. Thus, an ontology allows both humans and computers to describe and reason about the concepts in a domain. Most ontologies employ the subsumption ("is a") relation and its inverse ("has subtype") to define hierarchical relationships between more general and more specific entities, which allows for abstract reasoning. More than 800 biomedical ontologies are available via the National Center for Biomedical Ontology

© The Author(s), under exclusive license to Springer Nature Switzerland AG 2023
J. M. Juarez et al. (Eds.): AIME 2023, LNAI 13897, pp. 347–351, 2023.
https://doi.org/10.1007/978-3-031-34344-5_41

(NCBO) BioPortal site [2]. Various ontologies relevant to radiological diagnosis have been developed, including RadLex, Systematized Nomenclature of Medicine Clinical Terms (SNOMED CT), Disease Ontology, and Human Phenotype Ontology [3].

Ontologies encode knowledge as a set of logical predicates, or axioms. However, relationships between diseases and imaging observations may not follow IF- THEN rules. The imaging manifestations of a disease can be highly variable; conversely, one imaging finding may suggest several possible causes. This report describes an effort to integrate ontological knowledge for imaging diagnosis with a mechanism to incorporate probabilistic information acquired from patient-care data.

1.2 Radiology Gamuts Ontology

The Radiology Gamuts Ontology (RGO) encodes knowledge of diagnostic radiology through a formal representation of the relationships between diseases and imaging findings [4]. It offers "computable knowledge" to aid in radiological diagnosis. RGO includes 16,839 disorders (e.g., "Neiman-Pick disease"), interventions (e.g., "steroid therapy"), and imaging observations (e.g., "splenomegaly"), along with their synonyms and abbreviations. The ontology incorporates knowledge across organ systems, patient populations (adults and children), and imaging modalities; it includes both common and rare diseases.

RGO's distinguishing feature is its 55,564 causal relationships that link disorders and their imaging manifestations. For example, RGO axioms posit "rheumatoid arthritis may cause acro-osteolysis" and "scleroderma may cause achalasia." The "may cause" relation is not as strong as logical implication: "**A** may cause **B**" suggests that $P(\mathbf{B}|\mathbf{A}) \geq P(\mathbf{B})$. This approach offers a probabilistic framework to express causality [5].

2 Methods

2.1 Ethics Statement

The study was approved by the Institutional Review Board and was compliant with the U.S. Health Insurance Portability and Accountability Act (HIPAA); informed consent was waived.

2.2 Patient Cohort

All patients who underwent a radiology examination at the University of Pennsylvania Health System from 1 January 2015 through 31 December 2016 were included; the cohort included 1,702,462 consecutive reports of 1,396,293 patients. Natural language processing (NLP) was applied to identify the presence in the reports of each of RGO's 16,839 disorders, interventions, and imaging findings. In addition to synonyms defined within RGO itself, the search incorporated synonyms from linked terminologies such as RadLex and SNOMED CT.

A natural language processing system with negative-expression filtering (Nuance mPower, Microsoft, Redmond, WA) was applied to the collection of reports. In addition to synonyms defined within RGO itself, equivalent terms were applied from previously established mappings to other biomedical ontologies and vocabularies, including RadLex, Systematized Nomenclature of Medicine Clinical Terms (SNOMED CT), Disease Ontology, Human Phenotype Ontology, and Orphanet Rare Disease Ontology [3, 4]. Reports were aggregated by patient to account for imaging findings or disorders that might appear in different examinations, such as the presence of a cerebral artery aneurysm in a patient with autosomal-dominant polycystic kidney disease.

2.3 Occurrence and Co-occurrence of Entities

An occurrence was defined as positive mention of an RGO entity in a patient's radiology report. For example, a report that indicated "A small pericardial effusion is present" was considered as an occurrence of "pericardial effusion." Negative mentions (e.g., "No splenomegaly") and speculative mentions (e.g., "Rule out AVN") were excluded. Data were aggregated by patient; entities that occurred in reports of the same patient were considered to co-occur, regardless of chronological order. Of RGO's 16,839 entities, 6287 (37.3%) occurred in at least one patient.

Conditional probability of entity **A** given the presence of **B** was estimated as:

$$P(B) = (N_{AB} + \alpha)/(N_B + n\alpha) \tag{1}$$

where N_{AB} is the number of patients in which **A** and **B** co-occurred, N_B is the number of patients in which **B** occurred, Laplace smoothing factor $\alpha = 1$, and n is the number of distinct causes of **A**.

3   Results

3.1   Web Interface

A web interface (https://gamuts.net/probability) was developed to view probability data for RGO entities. The site displays the selected entity's prevalence and age/sex distribution in the dataset. Causally related entities—both those that cause and those caused by the index condition—are listed with their conditional probability value, given the specified entity. An example is presented for cardiomyopathy (Fig. 1). The web interface allows users to traverse the network of causally related entities.

3.2   Bayesian Network

The system combined the ontology's causal relationships with the conditional probability values to generate a Bayesian network model that incorporated the specified entity, its age and sex distribution, and its four most frequent causes and effects. The current work applies a "naïve Bayes" approach, in which age and sex influence the specified entity, which in turn influences all other nodes. The Bayesian network model for cardiomyopathy is shown using the GeNIe platform [7] (Fig. 2).

cardiomyopathy

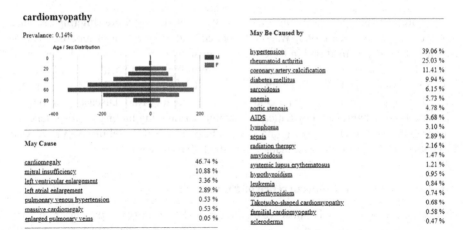

Prevalence: 0.14%

Age / Sex Distribution

■ M
■ F

May Cause

cardiomegaly	46.74 %
mitral insufficiency	10.88 %
left ventricular enlargement	3.36 %
left atrial enlargement	2.89 %
pulmonary venous hypertension	0.53 %
massive cardiomegaly	0.53 %
enlarged pulmonary veins	0.05 %

May Be Caused by

hypertension	39.06 %
rheumatoid arthritis	25.03 %
coronary artery calcification	11.41 %
diabetes mellitus	9.94 %
sarcoidosis	6.15 %
anemia	5.73 %
aortic stenosis	4.78 %
AIDS	3.68 %
lymphoma	3.10 %
sepsis	2.89 %
radiation therapy	2.16 %
amyloidosis	1.47 %
systemic lupus erythematosus	1.21 %
hypothyroidism	0.95 %
leukemia	0.84 %
hyperthyroidism	0.74 %
Takotsubo-shaped cardiomyopathy	0.68 %
familial cardiomyopathy	0.58 %
scleroderma	0.47 %

Fig. 1. Prevalence, age and sex distribution, and causally related entities are displayed. For example, given cardiomyopathy, the probability of mitral insufficiency is 10.88%. The figure does not include all causes of cardiomyopathy.

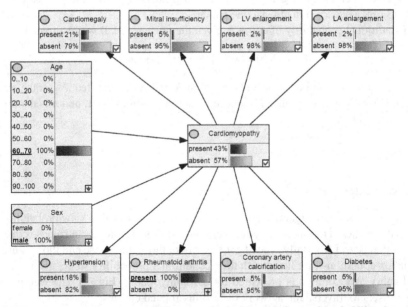

Fig. 2. Bayesian network constructed from ontological knowledge and probability data. The model includes edges between entities for which the Radiology Gamuts Ontology specifies a causal relationship. In this example, for a 60-year-old male with rheumatoid arthritis, the probability of cardiomyopathy is 43%.

4 Discussion

Probability values acquired from mining of radiology reports were combined successfully with ontology-based knowledge of causal relationships. The system's web interface facilitates navigation through the ontology to explore the prevalence and conditional probabilities of causally related entities. The system generates Bayesian network models that integrate the ontology's causal knowledge with empirical probabilistic data to enable decision support for radiological diagnosis.

The current work demonstrates the ability to integrate probability data with ontological knowledge, and our group's current research will address several limitations of the present report. The limitecd number of nodes in the BN model was for demonstration; it is trivial to extend the Bayesian network model to all causally related entities. The naïve Bayes approach simplifies the Bayesian network model's construction but limits its power. Causal and subsumption relationships among all nodes need to be included: for example, in the example shown, left ventricular (LV) enlargement is a form of cardiomegaly; they are not probabilistically independent as suggested. Ongoing work will extend the complexity of the model to incorporate each entity's age and sex distribution and to incorporate all causal and subsumption relationships using noisy-OR or noisy-MAX functions [8]. Current investigations include the analysis of a substantially larger corpus of radiology reports to detect known causal associations and to discover ones not currently encoded in the Radiology Gamuts Ontology.

References

1. Bodenreider, O.: Biomedical ontologies in action: role in knowledge management, data integration and decision support. Yearb Med Inform 67–79 (2008). https://www.ncbi.nlm.nih.gov/pmc/articles/pmid/18660879/
2. Noy, N.F., et al.: BioPortal: ontologies and integrated data resources at the click of a mouse. Nucleic Acids Res **37**, W170-173 (2009). https://doi.org/10.1093/nar/gkp440
3. Filice, R.W., Kahn, C.E.: Biomedical Ontologies to Guide AI Development in Radiology. J. Digit. Imaging **34**(6), 1331–1341 (2021). https://doi.org/10.1007/s10278-021-00527-1
4. Budovec, J.J., Lam, C.A., Kahn, C.E., Jr.: Radiology Gamuts Ontology: differential diagnosis for the Semantic Web. Radiographics **34**, 254–264 (2014). https://doi.org/10.1148/rg.341135036
5. Kleinberg, S., Hripcsak, G. A review of causal inference for biomedical informatics. J Biomed Inform 44(6):1102–12 (2011) https://doi.org/10.1016/j.jbi.2011.07.001
6. Filice, R.W., Kahn, C.E. Jr.: Integrating an ontology of radiology differential diagnosis with ICD-10-CM, RadLex, and SNOMED CT. J Digit Imaging 32, 206–210 (2019). https://doi.org/10.1007/s10278-019-00186-3
7. Druzdzel, M.J.: SMILE: Structural Modeling, Inference, and Learning Engine and GeNIe: a development environment for graphical decision-theoretic models. In: AAAI Proceedings, pp. 902–903. AAAI, Washington, DC (1999). https://www.aaai.org/Papers/AAAI/1999/AAAI99–129.pdf
8. Díez, F.J.: Parameter adjustment in Bayes networks. The generalized noisy OR-gate. In: Uncertainty in Artificial Intelligence In: Proceedings of the Ninth Conference, pp. 99–105. Morgan Kaufmann, San Mateo, CA (1993).https://doi.org/10.1016/B978-1-4832-1451-1.50016-0

Ontology Model for Supporting Process Mining on Healthcare-Related Data

José Antonio Miñarro-Giménez[1]([envelope])[iD], Carlos Fernández-Llatas[2],
Begoña Martínez-Salvador[3], Catalina Martínez-Costa[1], Mar Marcos[3],
and Jesualdo Tomás Fernández-Breis[1]

[1] Dept. Informática y Sistemas, Universidad de Murcia,
CEIR Campus Mare Nostrum, IMIB-Arrixaca, Murcia, Spain
`jose.minyarro@um.es`
[2] SABIEN-ITACA Universitat Politècnica de València, València, Spain
[3] Dept. of Computer Engineering and Science, Universitat Jaume I,
Castelló de la Plana, Spain

Abstract. In the field of Medicine, Process Mining (PM) can be used to analyse healthcare-related data to infer the underlying diagnostic, treatment, and management processes. The PM paradigm provides techniques and tools to obtain information about the processes carried out by analysing the trace of healthcare events in the Electronic Health Records. In PM, workflows are the most frequent formalism used for representing the PM models. Despite the efforts to develop user-friendly tools, the understanding of PM models remains problematic. To improve this situation, we target the representation of PM models using ontologies. In this paper, we present a first version of the Clinical Process Model Ontology (CPMO), aimed at describing the sequential structure and associated metadata of PM models. Finally, we show the application of the CPMO to the domain of prostate cancer.

Keywords: Clinical Process Ontology · Process Mining · Electronic Health Record

1 Introduction

Process mining (PM) is a promising field in many domains, including Medicine [8]. Healthcare institutions generate a large amount of information and data about the management of patients and their treatments, which are stored in Electronic Health Records (EHRs). These data can be analysed with PM techniques to potentially improve their organisational and clinical processes.

In PM, workflows are the most frequent formalism used to represent the inferred models. Despite the efforts dedicated to the development of user-friendly tools, the readability of PM models by physicians remains difficult. The semantic representation of information is a recurring topic in healthcare and life sciences, domains in which collecting, linking and sharing data is key to developing better

J. M. Juarez et al. (Eds.): AIME 2023, LNAI 13897, pp. 352–356, 2023.
https://doi.org/10.1007/978-3-031-34344-5_42

biomedical applications [3]. In PM, ontologies could be used in order to improve the readability of PM models, e.g. by annotating elements with formally defined concepts. Besides, the application of ontologies to EHR data allows specifying implicit knowledge, improving the information available and resulting in better PM models. In this sense, [2] presents a framework that extracts event log data and publishes it as a knowledge graph to support PM algorithms. The knowledge graph is based on an ontological representation of the legacy information system of the organisation. In [1] an ontological representation of the Business Process Model and Notation (BPMN) model is provided, extended with concepts related to clinical pathways.

In this paper, we present a first version of the Clinical Process Model Ontology (CPMO) that aims at describing the sequential structure and associated metadata of PM models generated from healthcare events in EHRs. Unlike [2], the CPMO does not focus on improving the event data to support PM algorithms, but to improve the resulting PM models and ultimately to support methods that can leverage a semantic representation thereof. Moreover, the CPMO distinguishes a layer of sequencing information from a layer about clinical processes and data, by reusing concepts from top-level ontologies, in contrast to [1] which focuses on extending the BPMN notation to include concepts related to clinical pathways. As a use case, we have applied the CPMO to an excerpt of a PM model obtained from synthetic data about patients with prostate cancer.

2   Process Mining on Electronic Health Records

PM techniques use the event logs associated with the healthcare process to infer the model that the patient actually follows, which can be represented graphically as a workflow [12]. During the diagnosis and treatment of patients under certain clinical circumstances, the data and event logs of the processes and treatments carried out are recorded in the EHR.

Using Interactive Process Mining (IPM) techniques [5], the expert can provide information that can create more detailed models for better discovery of healthcare processes. This paradigm takes advantage of the explicit mediation of the human expert for enhancing the system by co-creation of the models in collaboration with the PM expert and the IT personnel of hospitals. In this line, PMApp [11] is the only tool specifically designed for interactive PM analysis in the healthcare domain. This tool is devised to be customizable and upgradable, to allow its adaption to the application domain, and to interoperate with other systems available in health institutions. PMApp uses the Timed Parallel Automaton (TPA) [4] formalism as the base representation for PM models. The TPA has a high expressivity which is comparable to Safe-Petri Nets, and it is possible to translate it to other representations like graphs or BPMN models.

3   Clinical Process Model Ontology

The CPMO provides classes to represent PM models and clinical data, allowing their interrelated representation. Thus, in the development of the CPMO dedicated to PM models, elements are represented using concepts from the BPMN specification. In [6] the ontological formalization of BPMN was investigated. We reuse concepts from BPMN, such as Activity, Task, Sub Process or Gateway, as annotations to describe procedural information of the PM models. BPMN also provides a description of how to classify sequential information, *Exclusive Gateway*, *Parallel Gateway* and *Inclusive Gateway*, and, we link them to the CPMO classes *Sequential transition*, *Multiple transition* and *Synchronous transition*.

Clinical data in the CPMO are represented using as reference framework BioTopLite2 (BTL2) [9], an upper-level ontology for the biomedical domain. BTL2 is linked with BFO [7], which was recently adopted as an ISO standard to support the interchange of information among heterogeneous information systems. BTL2 provides very general categories and relations (e.g., categories like Process or Material entity, and relations like isBearerOf or hasOutcome) that allow standardizing the ontology creation according to a rigorous ontological commitment. The CPMO extends BTL2 by defining informational aspects and integrating them with clinical entities from SNOMED CT as the main reference ontology for the medical domain. This requires the alignment of SNOMED CT main concepts and attributes with BTL2 [10].

4   Application of the CPMO in Prostate Cancer Use Case

To gain a first insight into the CPMO, we used data from theSimulacrum dataset[1], which contains synthetic data provided by the National Cancer Registration and Analysis Service from England.

In this work, we selected the prostate cancer tumour data, given our previous experience with clinical guidelines for this disease. For the sake of simplicity, we chose those patients with one single prostate cancer tumour who had been treated with a Docetaxel regimen.

We applied the PMApp tool to a set of tables containing demographic data and data related to the tumour (such as TNM staging, grade and location). Additionally, the tables contain a number of time-stamped data, among which we selected the following data/events: diagnosis date (`Diagnosis`), surgery date (`Surgery`), date of prescription of the chemotherapy regimen (`presDocetaxel`), and date of administration of each regimen cycle (`cycleDocetaxel`). As a result, we obtained a TPA model (an excerpt is shown at the top of Fig. 1).

The application of the CPMO to the resulting TPA model provides an ontological representation illustrated at the bottom of Fig. 1. We define the instances of the *Node* class, with numbers 1 through 6, to represent each node of the TPA. The transitions between nodes are represented as instances of the *Sequential Transition* class (numbers 7 through 13), as they only have one start node and one end node. Additionally, we associate an instance of a clinical process (numbers 14 through 19) to each instance of the *Node* class. Thus, the CPMO differentiates between the nodes of the TPA model, as objects of information with

---

[1] See https://simulacrum.healthdatainsight.org.uk/.

BPMN annotation, and their corresponding clinical process, as processes from BTL2. To describe the clinical processes of the use case, we included in CPMO more specific subclasses of clinical processes, such as the *Docetaxel Adminis-tration Process* as subclass of the *Medication Administration Process*, with the *hasParticipant* relationship connected to the *Docetaxel* class. Finally, the *Doc-etaxel Administration Process* class is annotated with the SNOMED CT con-cept 18629005| Administration of drug or medicament|. In this way, the CPMO makes explicit the clinical knowledge not represented in the TPA model.

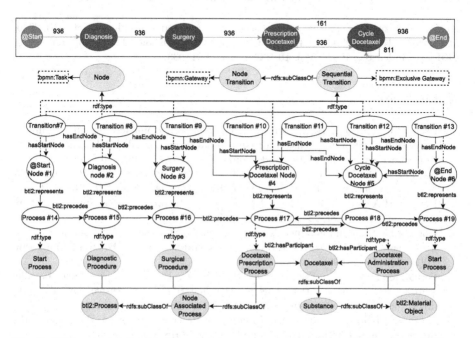

Fig. 1. TPA associated with the prostate cancer use case for patients related to *Doc-etaxel* regimen and its representation using the CPMO.

5 Conclusion and Future Work

The CPMO allows an ontological representation of the resulting TPA model gen-erated with the PMApp tool from events in EHRs. The CPMO includes classes for representing procedural and clinical data in a way that can be independently queried. In this version of the CPMO we have included the classes for diagnosis procedure, medication prescription process, medication administration process, and surgical procedure. Besides, we have applied the CPMO to represent PM model nodes and transitions, and associated metadata. The use of BTL2 and SNOMED CT facilitates the interoperability of the CPMO with other biomedical ontologies and enables enriching resulting PM models.

The CPMO is not designed to represent data at the patient level but statistical data related to the set of patients that follow the same clinical path. This is a design decision due to how TPA models are generated.

Future work must be done to extend the clinical processes that can be represented in the CPMO. This work is planned to include also classes related to representation languages for clinical guidelines. The definition of the CPMO is the first step to develop Knowledge Graph (KG) repositories. Therefore, we have planned to investigate the application of graph-based methods to further exploit the semantic representation of PM models from their associated KGs.

Acknowledgments. This research has been supported through projects PID2020-113723RB-C21, PID2020-113723RB-C22, RTI2018-099039-J-I00 and RYC2020-030190-I funded by MCIN/AEI/10.13039/501100011033.

References

1. Braun, R., Schlieter, H., Burwitz, M., Esswein, W.: Bpmn4cp: design and implementation of a bpmn extension for clinical pathways. In: 2014 IEEE International Conference on Bioinformatics and Biomedicine (BIBM), pp. 9–16 (2014)
2. Calvanese, D., Montali, M., Syamsiyah, A., van der Aalst, W.M.P..: Ontology-Driven Extraction of Event Logs from Relational Databases. In: Reichert, M., Reijers, H.A. (eds.) BPM 2015. LNBIP, vol. 256, pp. 140–153. Springer, Cham (2016). https://doi.org/10.1007/978-3-319-42887-1_12
3. Cheung, K.H., Prud'hommeaux, E., Wang, Y., Stephens, S.: Semantic web for health care and life sciences: a review of the state of the art. Briefings Bioinform. **10**(2), 111–113 (2009)
4. Fernandez-Llatas, C., Pileggi, S.F., Traver, V., Benedi, J.M.: Timed Parallel Automaton: a Mathematical Tool for Defining Highly Expressive Formal Workflows. In: 5th Modelling Symposium (AMS), pp. 56–61. IEEE (2011)
5. Fernandez-Llatas, Carlos (ed.): Interactive Process Mining in Healthcare. HI, Springer, Cham (2021). https://doi.org/10.1007/978-3-030-53993-1
6. Natschläger, C.: Towards a bpmn 2.0 ontology. In: BPMN (2011)
7. Otte, J.N., Beverley, J., Ruttenberg, A.: BFO: Basic formal ontology. Appl. Ontology **17**, 17–43 (2022)
8. Rojas, E., Munoz-Gama, J., Sepúlveda, M., Capurro, D.: Process mining in healthcare: A literature review. J. Biomed. Inform. **61**, 224–236 (2016)
9. Schulz, S., Boeker, M.: BioTopLite: an upper level ontology for the life sciences evolution, design and application. In: GI-Jahrestagung, pp. 1889–1899 (2013)
10. Schulz, S., Martínez-Costa, C.: Harmonizing snomed ct with BioTopLite: An exercise in principled ontology alignment. Stud. Health Technol. Inform. **216**, 832–836 (2015)
11. Valero-Ramon, Z., et al.: Analytical exploratory tool for healthcare professionals to monitor cancer patients' progress. Front. Oncol. **12** (2022)
12. Van Der Aalst, W.: Process mining: data science in action, vol. 2. Springer (2016) https://doi.org/10.1007/978-3-662-49851-4

Real-World Evidence Inclusion in Guideline-Based Clinical Decision Support Systems: Breast Cancer Use Case

Jordi Torres[1]($\boxtimes$) iD, Eduardo Alonso[1,2], and Nekane Larburu[1,3] iD

[1] Vicomtech Foundation, Basque Research and Technology Alliance (BRTA), 20009 Donostia-San Sebastián, Spain
{jtorres,ealonso,nlarburu}@vicomtech.org
[2] Department of Computer Architecture and Technology, University of the Basque Country UPV/EHU, 20018 Donostia—San Sebastián, Spain
[3] Biodonostia Health Research Institute, eHealth Group, 20014 Donostia-San Sebastián, Spain

Abstract. Adopting Clinical Decision Support Systems (CDSS) in clinical practice has shown to benefit both patients and healthcare providers. These CDSS need to be updated when new evidence, data, or guidelines arise since up-to-date evidence directly impacts physician acceptance and adherence to these systems. To this end, in previous studies, methodologies have been developed to update CDSS content by taking advantage of machine learning (ML) algorithms. Modifications in the domain knowledge require a reviewing and validation process before being implemented in clinical practice. Hence, this paper presents a methodology for including real-world evidence in an evidence-based CDSS for breast cancer use case. Decision trees (DT) algorithms are used to suggest modifications based on the analysis of retrospective data, which clinical experts review before being implanted in the CDSS. This way, our methodology allows to combine clinical knowledge from both guidelines and real-world data and enrich the domain clinical knowledge with real-world evidence.

Keywords: Clinical Practice Guidelines · Clinical Decision Support Systems · Real-World Evidence · Decision Trees · Breast Cancer

1 Introduction

Clinical Decision Support Systems (CDSS) help professionals by assisting in patient diagnosis or management, generation medication alerts, etc. [1]. Independently of the type of CDSS, one identified problem that affects the success of these systems in clinical practice is the need to update clinical knowledge. Outdated recommendations reduce clinicians acceptance and adherence to the CDSS [2, 3], and may also put patients at risk or affect their quality care. New knowledge/evidence can be obtained from updates in clinical guidelines or extracting clinical knowledge from real-world clinical data. In this regard, updates from guidelines require clinical experts involvement in guidelines revision, which consumes time and resources, whereas AI-obtained knowledge is

J. M. Juarez et al. (Eds.): AIME 2023, LNAI 13897, pp. 357–361, 2023.
https://doi.org/10.1007/978-3-031-34344-5_43

less human-dependent, but its lack of transparency and interpretability can constrain its acceptance [4, 5].

Therefore, efficient and comprehensive updating of CDSS with clinical knowledge from guidelines or real-world evidence (RWE) extracted from retrospective datasets [6] is crucial. In this sense, methodologies combining AI's potential to extract RWE and update clinical knowledge have been developed. In [7, 8], decision trees (DTs) that model existent CPGs are extended with evidence derived from clinical data. The approach followed in these two studies is similar, a DT is built based on the CPG and validated with clinicians' help. Later, new paths or sub-trees are generated from retrospective data mining and added to the DT, expanding it. Although both studies improve the outcome of the initial knowledge-based CDSS, clinicians are not involved in assessing the changes proposed by data mining techniques. Given the potential impact that any change can cause, a revision step should take place before accepted.

This paper presents a methodology for including RWE from clinical datasets in a supervised manner. Previously we developed a tool that allowed clinicians to enter and edit clinical rules in a rule-based CDSS [9]. This tool has been expanded with the inclusion of AI techniques that suggest modifications to previously formalized rules, but including a revision of the changes in the existent CDSS by end-users. The developed methodology is presented in a use case centered in breast cancer.

2   Methodology

In this work we expand the previously presented platform, eHealth Software Toolkit (eHST) to suggest modifications based on the analysis of data using AI techniques [9]. This is done using a set of different specialized modules, each with a web user interface (UI), that work in conjunction to generate, edit and execute the CIG. All the data is stored in a non-structured database that contains the data collected during the generation and maintenance of CIG, as well as the evaluated data, generated recommendations and applied procedures.

The process of including RWE starts with the analysis of stored retrospective data: the CDSS recommended treatment, the administered treatment, and the users' data. From this data, decision trees (DTs) are generated. Each path from the DT root node to each leaf is used to form a rule, which must meet a minimum quality of evidence based on the number of cases that followed that path for a new AI rule to be created.

Rules derived from decision trees (*empirical rules*) are split into two subsets based on their recommendation being the same or different from each formalized rule (*reference rule*). These two subsets are then used in the similarity analysis to generate modification requests. The similarity between the *reference* and *empirical rules'* conditions is computed with the *Jaccard index*, dividing the size of their intersection with the size of their union. To prepare for this, a preprocessing step is taken for the variables of each rule, and a binary array of length N is generated for each condition set, where N is the total number of possible values in the data model. The values in the binary array are determined by whether a variable is present in the condition set, with numerical variables categorized into different ranges.

After computing Jaccard index, rules are ordered based on the index. To avoid selecting rules that are too dissimilar, a purity condition threshold is set by clinicians. All rules below the threshold are automatically discarded. Next, the analysis is split into two steps: Step 1 suggests modifications to the condition part of the rule for those rules that have the same recommendation as the *reference rule*, while Step 2 suggests modifications to the recommendations for those rules that have a different recommendation from the *reference rule*.

Step 1. Condition modifications.
The first case returns modification requests that suggest the i) *insertion*, ii) *edition* or iii) *deletion* of conditions by comparing the binary arrays of the *reference rule* and *empirical rule*. The applied logic, represented in Fig. 1, determines that new conditions are *inserted* when a class of the data model is present in the *empirical rule* (blue), but not in the *reference* one, and, in the opposite case, a *deletion* is suggested (red). In the case that a variable is present in both, but with different values, an *edition* is suggested (orange). All resulting modification present the conditions affected, and after a reviewing process by clinicians, they are either accepted or rejected.

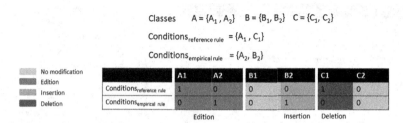

Fig. 1. Generation of modification requests example given 2 condition sets.

Step 2. Recommendation modifications.
The second analysis is similar to *Step 1* but suggesting modifications in the recommendations. The most similar *empirical rule* is selected, and their treatments are compared. Two modification requests are generated to include the new treatment. The treatment can be added in combination with the one from the *reference rule*, or as a separate treatment in a new rule (the *empirical rule*). Clinicians decide the most appropriate modification by reviewing the suggestions made by the system.

3   Breast Cancer Use Case

This use case presents the methodology to gather RWE from breast cancer patients' retrospective data using AI techniques, and include it in a NCCN Breast Cancer[1] guidelines-based CDSS. The NCCN guideline was formalized in collaboration with oncologists from the Hospital Universitario de Torrejón. The clinical scenario presented is the treatment of operated, invasive, triple negative breast cancer (TNBC), where no neoadjuvant treatment is administered. The recommended adjuvant treatment by the guideline is the chemotherapy drug Olaparib.

---

[1] //www.nccn.org/professionals/physician_gls/pdf/breast.pdf.

Applying the similarity analysis, the *reference rule* and the *empirical rules* extracted from the dataset are compared. *Empirical rules* modeling the same clinical scenario but not recommending Olaparib are evaluated, and the most similar rule is selected. The system suggests including the empirical treatment i) in combination with the reference treatment in the *reference rule*, or ii) adding the *empirical rule* in the CDSS, both options generating a modification request.

To assess the validity of these modifications, a clinical reviewer uses the AT validating functionalities (see Fig. 2a). The reviewer can compare the conditions and treatments of the reference rule (left) and the modified rule (right). Colors indicate the type of the modifications made, i) green for new insertions (risk associated conditions), ii) dark blue for editions (Stop treatment instead of Olaparib) and iii) red for deletions. Since the recommendations are opposite a new rule is generated, expanding the CDSS with real-world evidence. As shown in Fig. 2 (**b**), chemotherapy is not recommended for patients with thromboembolisms and low platelet.

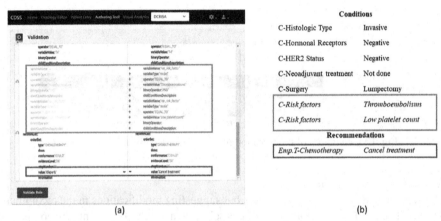

| (a) | (b) |

Fig. 2. AT validator screen; **b)** Schema of the added *empirical* rule. New conditions highlighted in green, treatment edition highlighted in dark blue. (Color figure online)

4   Conclusions and Future Work

In this paper we have presented a methodology that uses AI techniques to gather RWE for the maintenance of rule-based CDSS. The conceived methodology's functioning has been shown with a use case related to breast cancer, where RWE in the form of treatment contraindications has been extracted and included in the CDSS.

Although the proposed methodology can enhance the flexibility of CDSS, limitations have been identified. Possible ambiguities suggesting modification types (insert a condition or edit an existent one), can occur. This could be partially solved by also considering the relational operators that link these conditions. The similarity computation with numerical variables can be improved also considering the distribution of values that it can take. Lastly, the value of the proposed methodology should be assessed with clinical data by performing a comparative analysis of the recommendations given by the CDSS, the suggestions that are generated, and its added value with respect to the original knowledge.

As future work, in addition to addressing the identified constraints, we plan to extend the reasoning approaches of the eHST platform by including ontology-based reasoning functionalities, which will extend the scope of applications of our platform. Furthermore, the methodology to gather RWE and include it in the CDSS will be enhanced by considering the patient outcomes when suggesting changes in the rules, as well as for validating its impact.

Acknowledgements. The authors want to thank Bilbomática and Magda Palka-Kotlowska (from Ribera Salud) for the provided aid during the formalization of the CPG used in the presented use cases, as well as for providing the clinical scenario with which to present the developed methodology.

References

1. S Berner, E.: Clinical decision support systems: theory and practice. Springer Berlin Heidelberg, New York, NY (2016) https://doi.org/10.1007/978-3-319-31913-1
2. Cabana, M.D., et al.: Why Don't Physicians Follow Clinical Practice Guidelines? A Framework for Improvement. JAMA **282**, 1458–1465 (1999)
3. Davis, D.A., Taylor-Vaisey, A.: Translating guidelines into practice: a systematic review of theoretic concepts, practical experience and research evidence in the adoption of clinical practice guidelines. CMAJ **157**, 408–416 (1997)
4. Stiglic, G., Kocbek, P., Fijacko, N., Zitnik, M., Verbert, K., Cilar, L.: Interpretability of machine learning-based prediction models in healthcare. WIREs Data Min. Knowl. Discovery **10**, e1379 (2020)
5. Petch, J., Di, S., Nelson, W.: Opening the Black Box: The Promise and Limitations of Explainable Machine Learning in Cardiology. Can. J. Cardiol. **38**, 204–213 (2022)
6. Sutton, R.T., Pincock, D., Baumgart, D.C., Sadowski, D.C., Fedorak, R.N., Kroeker, K.I.: An overview of clinical decision support systems: benefits, risks, and strategies for success. npj Digit. Med. 3, 1–10 (2020)
7. Bai, Y., et al.: Construction of a Non-Mutually Exclusive Decision Tree for Medication Recommendation of Chronic Heart Failure. Front Pharmacol. **12**, 758573 (2022)
8. Zhao, W., Jiang, X., Wang, K., Sun, X., Hu, G., Xie, G.: Construction of Guideline-Based Decision Tree for Medication Recommendation. Stud Health Technol Inform. (2020)
9. Torres, J., Artola, G., Muro, N.: A Domain-Independent Semantically Validated Authoring Tool for Formalizing Clinical Practice Guidelines. Stud Health Technol Inform. **270**, 517–521 (2020)

Decentralized Web-Based Clinical Decision Support Using Semantic GLEAN Workflows

William Van Woensel[1]($\boxtimes$) (ID), Samina Abidi[2] (ID), and Syed Sibte Raza Abidi[3] (ID)

[1] Telfer School of Management, University of Ottawa, Ottawa, Canada
wvanwoen@uottawa.ca
[2] Medical Informatics Faculty of Medicine, Dalhousie University, Halifax, Canada
[3] NICHE Research Group, Dalhousie University, Halifax, Canada

Abstract. Decentralized clinical decision support (CDS), using the Web browser as a local application platform, fully decouples the CDS from vendor-specific EMR, removes reliance on server infrastructure, and does not require custom software. Using GLEAN, a clinical workflow can be loaded within a Web browser to provide decentralized and specialized CDS at a point-of-care. To that end, GLEAN workflows include all knowledge needed for their local execution; the standards-based and secure data sharing with EMR, if needed; and detection of multimorbidity conflicts. This specialized CDS will execute all decision logic locally in the Web browser; using SMART-on-FHIR, locally entered data can be securely submitted to a FHIR-compliant EMR, and remote data can be retrieved. In such a decentralized setting, clinicians can securely collaborate on multimorbidity patients: (1) by sharing workflow traces, i.e., progression of their local workflows, other clinicians can keep appraised of their decision making; and (2) by leveraging medical online knowledge sources, conflicts (e.g., drug-drug, drug-interaction) between multimorbidity decisions can be detected and resolved.

Keywords: Clinical Decision Support · Multimorbidity · Semantic Web · Notation3

1 Introduction

This paper proposes a decentralized setup for Clinical Decision Support (CDS) at points-of-care, which relies on the Web browser as a local application platform. A fully local CDS removes reliance on server infrastructure, and improves data privacy and security, as manually entered patient data may never have to leave the local computer. A GLEAN (GuideLine Execution and Abstraction using N3) workflow can be loaded into a Web browser to provide special-purpose CDS for diagnosis and/or treatment. To that end, GLEAN workflows are semantically annotated with all knowledge needed for their local execution. If needed, protocol sets such as SMART-on-FHIR [1] can be utilized to securely exchange patient data with an online FHIR-compliant EMR.

To execute GLEAN workflows, we previously defined an extensible Finite State Machine (FSM) execution semantics [2] using a set of state transition rules. GLEAN

J. M. Juarez et al. (Eds.): AIME 2023, LNAI 13897, pp. 362–367, 2023.
https://doi.org/10.1007/978-3-031-34344-5_44

workflows and their execution semantics are represented using the Notation3 (N3) Semantic Web (SW) rule language. This has multiple advantages: it allows integration with the FHIR [3] standard, with its RDF representation (N3 is a superset of RDF); semantic annotation with clinical ontology terms (e.g., SNOMED); and expressive decisional criteria, utilizing a range of built-in operators. Importantly, the N3 rule language can be executed within the Web browser using the eye-js reasoner [4] or by compiling into JavaScript [5]. By leveraging in-browser execution support, GLEAN workflows can be executed at the client-side using an FSM execution semantics.

In case of multimorbidity (i.e., multiple co-occurring illnesses), a patient may interact with multiple clinicians, each individually deciding on treatment for an illness within their specialization. Knowledge-based CDS, such as GLEAN, rely on the computerization of workflows from Clinical Practice Guidelines (CPG), which are known to contain redundant or conflicting advice for multimorbidity illnesses (e.g., contra-indicated drugs) [6]. When decisions on multimorbidity patients are made in isolation, i.e., without knowledge of concomitant treatment, such conflicts may endanger patient safety. In our decentralized setting, we propose a loosely coupled mechanism for collaboration on multimorbidity patients, which involves (a) sharing the GLEAN workflows and their traces between multiple clinicians, together with (b) knowledge-driven tools to detect conflicting advice between workflow artefacts.

In this work, we describe our progress towards a decentralized GLEAN architecture. We previously described a centralized prototype of the GLEAN implementation [7].

2   Methods

GLEAN follows a model-driven engineering paradigm, where executable artefacts are generated from high-level models including workflows and FHIR resources. We briefly summarize the generation, deployment, and execution process below.

(1) *Generation of Executable Workflows.*
    A *semantic GLEAN workflow* represents the overall process flow, decision points, and decisional criteria needed to provide CDS. Based on such a semantic GLEAN workflow, a *self-contained, executable GLEAN workflow* can be generated: to implement decision points, this workflow includes a set of HTML5 forms for manual data entry based on FHIR resources [8], whereas decisional constraints are compiled into JavaScript for execution in the browser (see below).
(2) *Deploying and Executing GLEAN workflows at points-of-care.*
    The *glean-js* JavaScript library loads, visualizes, and runs the executable GLEAN workflows in the Web browser. The library offers multiple visualization modes: an *interactive visual workflow*, which visually explains recommendations in terms of followed workflow paths, useful for education; and a *data-entry form*, which facilitates data-entry and provides textual recommendations. The FSM execution semantics, defined using N3 state transition rules, have been pre-compiled into a JavaScript FSM.

Execution. Following manual data entry (HTML5 form) at a decision point, the input data is packaged as a FHIR record. Next, the input data is checked against the decisional criteria, possibly progressing the workflow as per the FSM. For instance, when the criteria of a certain branch are met, its target task is activated; the other branch's targets are moved to the discarded state; and the decision point will be marked as complete.

2.1 Decentralized Collaborating on Multimorbidity Patients

For multimorbidity patients, multiple clinicians (referred to as *collaborating clinicians*) will typically make individual choices on treatment within their specialty. To avoid conflicts between collaborating clinicians, in our decentralized setting, we propose a loosely coupled mechanism that involves sharing GLEAN workflows and their *workflow traces*, which capture workflow progression and clinicians' choices by recording (a) current task states (e.g., active, completed), and (b) input data that led the workflow down certain paths. We show two example workflows and their traces below:

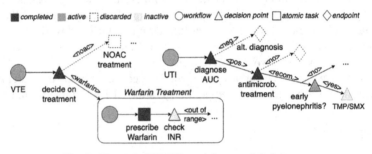

Fig. 1. Example GLEAN workflows and their traces.

In the *VTE* (*venous thromboembolism*) workflow (left), the clinician has decided on Warfarin, since the decision point is marked completed; in the *UTI* (*urinary tract infection*) workflow, suitable treatment is still being decided, as the last decision point is marked active. For each decision point, the trace includes relevant input data as well.

By sharing GLEAN workflows and their latest traces (collectively called *workflow artefacts*), clinicians are kept up-to-date of overall multimorbidity treatment progress; and may track and understand others' decision making in terms of input data (incl. Both patient data and clinician choices) and associated decisional criteria. To concretely share workflow artefacts, either a centralized or decentralized method may be utilized:

Using a Decentralized SOLID Pod. SOLID (Social Linked Data) is a decentralized vision for sharing personal data on the Web [9]. Each person is in charge of a SOLID pod with their personal data; using fine-grained access controls, access can be granted to authorized parties. This effectively decouples online services from personal data storage, obviating privacy concerns of third-party data ownership. SOLID is complementary to our decentralized CDS setting: leveraging the SOLID pod of a multimorbidity patient, workflow artefacts can be stored and shared between collaborating clinicians.

Using a Centralized FHIR-Compliant EMR. It may be practical to share workflow artefacts using an existing, centralized EMR at a point-of-care. this is an option when the

EMR allows linking to third-party applications, such as ours, using standards-based data sharing and secure authentication. An example is the SMART-on-FHIR specification, which is supported by Veradigm (previously AllScripts), Epic, and Cerner. To that end, GLEAN workflow artefacts can be represented and shared using FHIR; using a *Plan-Definition* to represent a workflow, its constituent tasks, decision points and criteria; and *FHIR Tasks* to represent workflow traces, including task states and input data.

2.2 Conflict Detection Between Multimorbidity Workflows

To detect conflicts between multimorbidity GLEAN workflows, tasks are semantically annotated using the SNOMED ontology (namespaces omitted for brevity):

```
:prescribe_TMP_SMX a :Task ;
    :involves [ a snomed:Prescription ; snomed:drugUsed snomed:32792001 ] .
```

This code states that the prescribe_TMP_SMX workflow task involves a snomed:Prescription with drug snomed:32792001 (i.e., product containing TMP). This simple annotation already brings us a long way: e.g., we can check whether different tasks from multimorbidity workflows involve an identical drug, by comparing their snomed:drugUsed properties. Secondly, we can leverage *Medical Linked Open Data (MLOD)* [10], which semantically represents adverse drug interactions from DrugBank and DrugCentral:

```
drug:tmp_smx dv:external-id snomed:32792001 ; dv:drug-interactor dv:conflict:868 .
drug:warfarin dv:external-id snomed:372756006 ; dv:drug-interactor dv:conflict:868 .
dv:conflict:868 a dv:Drug-Drug-Interaction .
```

To query for these adverse interactions, our decentralized approach queries MLOD as a separate online source using secure HTTPS. In particular, we match the local drug identifiers (snomed:drugUsed) to drug identifiers in MLOD (dv:external-id). We further leverage the GLEAN *workflow trace* to flag only relevant conflicts; if a task is in the discarded state (see Fig. 1), i.e., it will not be executed, a conflict will not be flagged.

Once adverse interactions are detected, we aim to provide automated support for collaborating clinicians to find a safe multimorbidity treatment. We previously proposed a set of integration policies [6] to replace, discard, or delay conflicting tasks: e.g., an *Event Conditional Replacement Policy* can reduce Warfarin dosage during and after TMP/SMX treatment, until the INR (International Normalized Ratio) returns to normal. In a distributed setting, we foresee these integration policies being applied in two ways:

1) A clinician treating UTI shares their workflow artefacts with a collaborating clinician, treating the patient's VTE. The latter party detects a conflict between the two workflows; i.e., TMP/SMX and Warfarin. The clinician applies the integration policy on their own workflow, reducing Warfarin dosage to cope with TMP/SMX treatment.
2) A clinician treating VTE shares their workflow artefacts with a collaborating clinician treating the patient's UTI. The latter party similarly detects a conflict, and, in this case, suggests an integration policy for VTE workflow to the collaborating clinician. This is done by sharing the policy in the same way as workflow traces; if needed, the integration policy can be encoded as a *FHIR CarePlan*. An ensuing process to reach consensus on the multimorbidity treatment takes place outside of the GLEAN method.

3   Discussion and Conclusion

This paper presents our progress towards a decentralized GLEAN architecture. We have implemented an important first step, namely, the utilization of a Web browser as local application platform for CDS. To cope with multimorbidity, we outlined our vision towards a decentralized clinician collaboration, which relies on a loosely-coupled sharing of workflow artefacts, automated conflict detection based on online knowledge sources, and applying and/or sharing conflict resolutions. Bottrighi et al. [11] similarly assume multimorbidity treatment will be distributed amongst specialized clinicians; noting this is in line with desiderata from hospital clinicians. To resolve conflicts, the authors propose a comprehensive "distributed but coordinated" protocol: workflows are suspended in case of conflicts until the involved clinicians resolve the conflict. We foresee an approach where workflow execution remains autonomous and uninterrupted, in line with the ultimately autonomous nature of clinician decision making. Instead, we focus on informing collaborating clinicians of overall multimorbidity decision making.

Future work involves fully implementing and evaluating the decentralized GLEAN architecture, including a comparison of response times with a server-side approach.

References

1. HL7 International: SMART on FHIR. http://www.hl7.org/fhir/smart-app-launch/. Accessed 30 Jan 2023
2. Van Woensel, W., Abidi, S., Tennankore, K., et al.: Explainable decision support using task network models in Notation3: computerizing lipid management clinical guidelines as interactive task networks. In: Michalowski, M., Abidi, S.S.R., Abidi, S. (eds.) AIME 2022, vol. 13263, pp. 3–13, pp. Springer, Cham (2022). https://doi.org/10.1007/978-3-031-09342-5_1
3. HL7 International: HL7 Fast Health Interop Resources (FHIR) (2014). https://www.hl7.org/index.cfm
4. de Roo, J., Wright, J.: Eye-JS: Distributing the EYE reasoner for browser and node using WebAssembly (2023).https://github.com/eyereasoner/eye-js
5. Van Woensel, W.: Generating Imperative Code from Knowledge Graphs (N3 rules, OWL ontology) (2022). https://github.com/william-vw/blockiot-cds/tree/main/tool
6. Van Woensel, W., Abidi, S.S.R., Abidi, S.R.: Decision support for comorbid conditions via execution-time integration of clinical guidelines using transaction-based semantics and temporal planning. Artif. Intell. Med. **118**, 102127 (2021)
7. Van Woensel, W., Abidi, S., Tennankore, K., et al.: Clinical guidelines as executable and interactive workflows with FHIR-compliant health data input using GLEAN. In: Michalowski, M., Abidi, S.S.R., Abidi, S. (eds.) AIME 2022, vol. 13263, pp. 421–425. Springer, Cham (2022). https://doi.org/10.1007/978-3-031-09342-5_43
8. Van Woensel, W., Abidi, S.R., Abidi, S.S.R.: Towards model-driven semantic interfaces for electronic health records on multiple platforms using Notation3. In: 4th International Workshop on Semantic Web Meets Health Data Management (SWH'21), New York, NY, USA (2021)
9. Berners-Lee T SOLID. https://solidproject.org/. Accessed 30 Jan 2023
10. Van Woensel, W., Abidi, S.R., Jafarpour, B., Abidi, S.S.R.: Providing comorbid decision support via the integration of clinical practice guidelines at execution-time by leveraging medical linked open datasets. In: 17th World Congress on Medical and Health Informatics (MEDINFO'19), 26 –30 August 2019, Lyon, France (2019)

11. Bottrighi, A., Piovesan, L., Terenziani, P.: Supporting physicians in the coordination of distributed execution of CIGs to treat comorbid patients. Artif. Intell. Med. **135**, 102472 (2023). https://doi.org/10.1016/J.ARTMED.2022.102472

An Interactive Dashboard for Patient Monitoring and Management: A Support Tool to the Continuity of Care Centre

Mariachiara Savino[1], Nicola Acampora[2], Carlotta Masciocchi[3], Roberto Gatta[4]([✉]), Chiara Dachena[3], Stefania Orini[4,5], Andrea Cambieri[2], Francesco Landi[2], Graziano Onder[2], Andrea Russo[2], Sara Salini[2], Vincenzo Valentini[2], Andrea Damiani[3], Stefano Patarnello[3], and Christian Barillaro[2]

[1] Università Cattolica del Sacro Cuore, Diagnostica per immagini, Radioterapia Oncologica ed Ematologia, Rome, Italy
[2] Fondazione Policlinico Universitario Agostino Gemelli IRCCS, Rome, Italy
[3] Real World Data Facility, Gemelli Generator, Fondazione Policlinico Universitario A. Gemelli IRCCS, Rome, Italy
[4] Department of Clinical and Experimental Sciences, University of Brescia, Brescia, Italy
roberto.gatta@unibs.it
[5] Alzheimer's Unit - Memory Clinic, IRCCS Istituto Centro San Giovanni di Dio Fatebenefratelli, Brescia, Italy

Abstract. In recent years, dashboard utilization in healthcare organizations has emerged as a useful tool to monitor and improve the quality of care. In this work, we present a dashboard specifically designed to support the daily clinical activities of the Continuity of Care Centre (CCA) in Fondazione Policlinico Universitario A. Gemelli IRCCS, Rome. CCA team, composed of multidisciplinary professional figures, plans and optimizes the treatment and discharge pathway for frail hospitalized patients. The team monitors the conditions of a large number of patients, which are spread across hospital departments, and fragmentation of patients' information on electronic health records (EHRs) is one of the greatest issues it has to deal with. The presented dashboard integrates and harmonizes data from several data sources providing physicians with a daily updated and interactive visualization of patient longitudinal history throughout hospitalization. The monitoring activity of CCA is supported through filtering options that allow to manage subgroups of patients individually. Moreover, automatic alerts promptly identify critical clinical conditions and deviations of patient processes from hospital protocols. In order to optimize patient discharge and bed occupancy, the dashboard displays also the intensity of medical care trend during hospitalization and highlights low intensity of care in patients ready to be discharged. The dashboard is currently in use by CCA team as a support tool to monitor and manage the admitted patients.

M. Savino and N. Acampora—These authors contributed equally to this work and share first authorship.

Keywords: Dashboard · Data Integration · Frailty · Conformance Checking · Bed Blockers · KPI · Hospital Dashboard

1   Introduction

The progression of life expectancy has led to an increase in hospital admissions of old patients, often affected by multiple comorbidities, with a high clinical and care complexity that increases the risk of adverse events and prolonged hospital stay. The CCA at Fondazione Policlinico Universitario A. Gemelli IRCCS in Rome (Italy) involves different multidisciplinary professional figures experienced in the management of chronicity and complexity [1]. Main aims of CCA team are: (a) to optimize and plan the discharge pathway for frail patients (b) to minimize delayed hospital discharges, i.e. bed-blockers, in order to optimize hospital beds and resources availability. During clinical practice, CCA team provides consulting for all hospital wards, and a large number of heterogeneous patients need to be monitored every day. Moreover, the fragmentation of patients' information on EHR makes not easy to reconstruct their longitudinal clinical history.

In this work we present a dashboard, developed using Shiny R library [2], that supports CCA team by monitoring patients conditions and clinical pathways during hospitalization. The dashboard is daily updated through a data extraction and processing pipeline, from hospital Data Warehouse (DWH), that allows the visualization of patient's longitudinal history during hospitalization. The dashboard integrates also alerts to promptly identify critical clinical situations and analyses deviations of patient processes from hospital protocols using process mining techniques. Moreover, the dashboard supports CCA team in optimizing bed occupancy with the identification of low levels of medical care in patients close to be discharged.

2   Dashboard Implementation

2.1   Data Integration and Preprocessing

The dashboard integrates several data sources extracted from the hospital's DWH. Data are extracted from views provided by the DWH and processed by Gemelli Generator Real World Data facility [3] through daily scheduled Extract, Transform and Load procedures that produce flat tables with relevant information about hospitalization. In DWH systems the data are not shaped according to a formal Data Model, thus a preprocessing pipeline harmonizes and integrates them. The pipeline integrates the information associated to the same hospitalization and shapes them in the form of Event Log containing information about the hospital stay, emergency room admission, CCA consultations and nursing scales. Moreover, the pipeline splits the longitudinal history of hospitalization and prepares data to be visualized on the dashboard in terms of individual transfers and stays in a ward.

2.2   Intensity of Medical Care

Delayed hospital discharge, also known as bed-blocking, occurs when patients are medically ready to be discharged but remain hospitalized due to reasons related to the bed availability in the chosen discharge setting, domiciliary care activation or social factors [4]. Delayed discharge typically results in reduced levels of medical care intensity, defined as the level of resources and number of clinical activities provided to an inpatient [5]. Therefore, the proactive identification of patients with low intensity of care can help reduce bed-blockers and improve the discharge process. In our work, we estimated for each patient the trend of intensity of care throughout the hospitalization by considering the number of: radiological examinations, laboratory exams, medical consultations and surgical procedures. Estimated trend of intensity of medical care was then smoothed using a 3-day simple moving average. We defined low intensity of care as a value under 1 for at least 3 consecutive days of hospitalization, if the time period doesn't include weekends, 5 days otherwise.

2.3   Conformance Checking

The developed dashboard implements and displays a conformance checking application to evaluate the adherence of patient pathways to a specific hospital procedure for early identification of patients at risk of prolonged hospitalisation or difficult discharge. The procedure requires that all patients admitted to the hospital and transferred to a ward must be evaluated by nursing staff within 24 h using the Blaylock Risk Assessment Screening Score. A Brass score greater than or equal to 20 indicates the need for a CCA consultation within 48h [6]. The conformance checking analysis allows the early identification of deviations from this hospital procedure that could lead to prolonged hospital stay. We employ pMineR software [7], an R based software for process mining in healthcare, that implements Pseudo-Workflow formalism to represent protocols and clinical guidelines [8].

3   Interface Design and Functionality

The developed dashboard implements a sidebar with filtering options and a main panel. The sidebar allows to filter and update the patient cohort displayed in the main panel, on the basis of hospitalization date and specific features (E.R. admission, level of intensity of care, ward, etc.). The main panel consists of 4 tabs:

- The first tab (Fig. 1) shows an interactive table including information related to all transfers of the patient cohort. The table implements some alerts, i.e., yellow color highlights all transfers in which Brass scale is performed after the defined time threshold, whereas orange color highlights Brass scale higher than 20. By clicking on a table row is shown: (1) the hospitalization timeline with also the possibility to visualize the texts of CCA consultations, (2) the intensity of care trend during hospitalization.

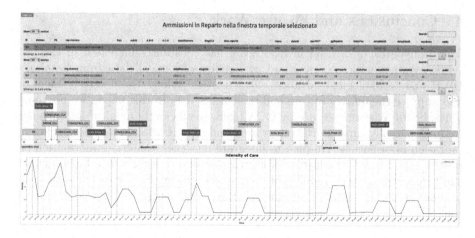

Fig. 1. Fist tab showing information about single transfers, hospitalization longitudinal history and intensity of care trend.

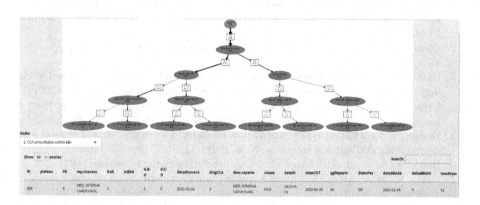

Fig. 2. Conformance Checking tab displaying adherence to the hospital procedure for early identification and management of complex patients. Numbers on the graph indicate patients who activated each transition.

- Second and third tabs show some clinical KPIs related to hospital procedures and some statistics about length of stay in the selected cohort.
- In the last tab the conformance checking analysis is proposed (Fig. 2). The graph shows for each ward transfer the adherence to the hospital procedure. In the first node is displayed the number of total transfers and the set of nodes and edges represents the possible paths that each patient can follow. Moreover, the user can select a single node with a drop-down menu and visualize a summary table with all patients going through the selected node.

4 Conclusions and Future Work

The dashboards utilization in healthcare has constantly increased in the last years, together with the enhanced availability of clinical data from hospital's DWH and technological advances [9]. Their use allows to effectively monitor and improve the quality of care and treatments [10]. In this paper, we proposed a dashboard, currently in use by CCA team, to efficiently monitor and manage complex patients through the integration and visualization of all relevant hospitalization information. We plan to integrate into the dashboard information on therapies and predictive factors of length of stay or potential bed-blockers that can help optimize the planning of discharge pathway. Furthermore, dashboard usability and efficacy will be evaluated after a sufficiently long period of clinical use.

References

1. Giovannini, S., et al.: A new model of multidimensional discharge planning: continuity of care for frail and complex inpatients. Eur. Rev. Med. Pharmacol. Sci. **24**, 13009–13014 (2020)
2. Chang, W., et al.: Shiny: web application framework for R. R package version 1.7.4.9002. (2023)
3. Damiani, A., et al.: Building an artificial intelligence laboratory based on real world data: the experience of gemelli generator. Front. Comput. Sci. **3**, 768266 (2021)
4. Everall, et al.: Patient and caregiver experience with delayed discharge from a hospital setting: a scoping review. Health Expect. 22.5, 863–873 (2019)
5. Beglinger, J.E.: Quantifying patient care intensity: an evidence-based approach to determining staffing requirements. Nurs. Adm. Q. 30.3, 193–202 (2006)
6. Blaylock, A., et al.: Discharge planning: predicting patients' needs (1992)
7. Gatta, R., et al.: pMineR: an innovative r library for performing process mining in medicine. In: ten Teije, A., Popow, C., Holmes, J.H., Sacchi, L. (eds.) AIME 2017. LNCS (LNAI), vol. 10259, pp. 351–355. Springer, Cham (2017). https://doi.org/10.1007/978-3-319-59758-4_42
8. Gatta, R., et al.: A framework for event log generation and knowledge representation for process mining in healthcare. In: 2018 IEEE 30th International Conference on Tools with Artificial Intelligence (ICTAI), pp. 647–654. IEEE (2018)
9. Buttigieg, S.C., et al.: Hospital performance dashboards: a literature review. J. Health Organ. Manage. (2017)
10. Ludlow, K., et al.: Co-designing a dashboard of predictive analytics and decision support to drive care quality and client outcomes in aged care: a mixed-method study protocol. In: BMJ open 11.8, e048657 (2021)

A General-Purpose AI Assistant Embedded in an Open-Source Radiology Information System

Saptarshi Purkayastha[1]([✉]), Rohan Isaac[1], Sharon Anthony[1], Shikhar Shukla[1], Elizabeth A. Krupinski[3], Joshua A. Danish[2], and Judy Wawira Gichoya[3]

[1] Indiana University-Purdue University Indianapolis, Indianapolis, IN 46202, USA
{saptpurk,risaac,shaantho,shikshuk}@iu.edu
[2] Indiana University Bloomington, Bloomington, IN 47405, USA
jdanish@indiana.edu
[3] Emory University, Atlanta, GA 30322, USA
{ekrupin,judywawira}@emory.edu

Abstract. Radiology AI models have made significant progress in near-human performance or surpassing it. However, AI model's partnership with human radiologist remains an unexplored challenge due to the lack of health information standards, contextual and workflow differences, and data labeling variations. To overcome these challenges, we integrated an AI model service that uses DICOM standard SR annotations into the OHIF viewer in the open-source LibreHealth Radiology Information Systems (RIS). In this paper, we describe the novel Human-AI partnership capabilities of the platform, including few-shot learning and swarm learning approaches to retrain the AI models continuously. Building on the concept of machine teaching, we developed an active learning strategy within the RIS, so that the human radiologist can enable/disable AI annotations as well as "fix"/relabel the AI annotations. These annotations are then used to retrain the models. This helps establish a partnership between the radiologist user and a user-specific AI model. The weights of these user-specific models are then finally shared between multiple models in a swarm learning approach.

Keywords: radiology · Human-AI collaboration · few shot learning · swarm learning

1 Introduction

Integration of artificial intelligence (AI) into radiology has been the topic of debates in the last several years - ranging from suggestions to "not train any more radiologists since their work is redundant" to AI demonstrating superhuman performance for various diagnostic tasks [6,11]. The radiologist's job encompasses various aspects, including image interpretation and analysis, report creation, and patient and doctor consultation [5]. The diagnostic process typically includes the detection and characterization of findings in medical images [6].

© The Author(s), under exclusive license to Springer Nature Switzerland AG 2023
J. M. Juarez et al. (Eds.): AIME 2023, LNAI 13897, pp. 373–377, 2023.
https://doi.org/10.1007/978-3-031-34344-5_46

In this context, AI in radiology promises to improve not only diagnostic/interpretive tasks but also workflow, increase efficiency and enhance the quality of service [13]. The realization of this impact of AI remains elusive because the real-world implementation of many AI models is fraught with multiple challenges [8], such as dataset variability [7], lack of adequate testing [14], variation in practices [3], and limited tooling that integrates AI with existing workflows [1,10]. Many have argued that better partnership between human radiologists and AI models is a way to remove these implementation challenges and build end-user trust in AI [2,16]; but few in-situ implementations, if any, can be found in literature [17].

In this paper, we describe updates to LibreHealth Radiology (LH-radiology), a well-regarded standards compliant [4,9], open-source Radiology Information System (RIS), which integrated an AI model service into the full workflow of an RIS in implementation. The LibreHealth AI model service is designed with Human-AI partnership in mind, where a human radiologist can retrain a model, and live updates to the AI model can be shared with other radiologists on the network. The AI Model Service enables two-way communication supporting training and inference between different radiology AI models in the LibreHealth RIS. This allows for the integration of AI-enabled functionality, including disease classification, body organ segmentation, automated study lists, volumetric measurement, image enhancements, and decision support tools into the regular radiologist workflow.

2   LibreHealth RIS Architecture

The AI-integrated LibreHealth RIS consists of two main components: the RIS module and the AI model service. The system is built on an EHR toolkit that provides the foundation for API and data models used by many health IT systems. When a radiologist accesses the study list in the LH-radiology module, the information is stored in the MYSQL database, and the AI model service is called.

2.1   LibreHealth Radiology Module

The LH-radiology system adds the abilities of a Radiology Information System (RIS) into the LibreHealth Toolkit. This adds functionality to the EHR to manage radiology studies, review orders, configure web configuration for PACS and viewer connectivity, support REST API-based communication with the AI model service and the MySQL database. Standards-compliant interfaces with HL7, DICOM, and WADO are some of the main strengths of the RIS.

2.2   LibreHealth Radiology Artificial Intelligence Model Service

The AI model service is a REST API-based service that is used to communicate with AI models from LibreHealth RIS. The service is able to run as a standalone system that can receive DICOM Modality Worklist (MWL) or DICOM images from the PACS system and return AI model output as a JSON file. The AI Model Service also supports the retraining of the models when their outputs are modified by a radiologist.

3   Methods

3.1   OHIF Viewer and Its Extensions

The OHIF viewer communicates with the AI model service using DICOM-SR annotations, which provide a standard way for radiologists to annotate their findings. The viewer is configured with a custom extension that allows 2-way communication, and core extensions from OHIF that are useful for measurement tasks are installed.

The LibreHealth radiology module source is available on GitLab: (https://gitlab.com/librehealth/radiology/lh-radiology).

3.2   AI Model Service

The AI model service is an assistant that works with different AI models and communicates with the OHIF viewer using REST API endpoints. A sample GET request is used to obtain information about AI models and their status.

The viewer's GET request payload returns the AI model inference result, including bounding box coordinates, image annotation, the model applied for inference and its status.

If the radiologist feels the inference is inaccurate, new annotations can be generated by adjusting the bounding box coordinates on the OHIF viewer, which is then used to retrain the model. In this case, the AI model service receives a POST request containing the new bounding box coordinates as payload. The AI model service is modular and can be extended in functionality. Future extensions may include deploying AI results to prioritize studies on the worklist.

The design of the AI model service is modular, allowing for future extensions in its functionality. For example, we have used this functionality to deploy the AI results to prioritize studies on the worklist. The AI model service source code is GitLab: (https://gitlab.com/librehealth/radiology/lh-radiology-aimodel-service).

3.3   Swarm Learning and Few Shot Learning

Customizing AI models based on a radiologist's preferences can lead to bias, as the model may pick up on the radiologist's regular errors or fail to correct them. To reduce bias while maintaining personalization, multiple human-AI models can collaborate through swarm learning, sharing weights via the AI model service to improve accuracy while preserving confidentiality. Swarm learning is utilized in the AI Model Service, where models are built independently on private data for each individual radiologist [15] as shown in Fig. 1. The trained model weights are then shared with other personalized radiologist models via a swarm application interface and merged to create an updated model with updated parameters [12]. The resulting model has accurate inferences equivalent to or better than those trained on a central site on all gathered data.

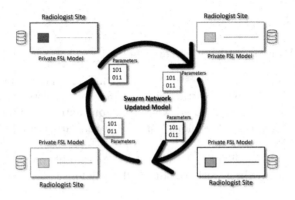

Fig. 1. LH-radiology Swarm Network Architecture

4   Discussion

We present an end-to-end deployment of an AI inference and training service tailored for use by radiologists. Our work is notable for several reasons, including designing and implementing a system that supports multiple modalities and models, personalized AI development through swarm learning, and a standards-based approach for system integration. This platform is versatile and suitable for various use cases, making it an excellent choice for exploring how radiologists can effectively leverage AI technology.

5   Conclusion

This personalized approach could potentially lead to more accurate and reproducible radiology assessments, but further studies are needed. Readers are invited to join the community to test and improve the system to support better infrastructure for validating AI models for multiple end-users.

Acknowledgements. This work was funded by the US National Science Foundation (#1928481) from the Division of Electrical, Communication & Cyber Systems.

References

1. Daye, D., et al.: Implementation of clinical artificial intelligence in radiology: who decides and how? Radiology **305**(3), 555–563 (2022)
2. Do, H.M., et al.: Augmented radiologist workflow improves report value and saves time: a potential model for implementation of artificial intelligence. Acad. Radiol. **27**(1), 96–105 (2020)
3. Dustler, M.: Evaluating ai in breast cancer screening: a complex task. Lancet Digital Health **2**(3), e106–e107 (2020)

4. Gichoya, J.W., Kohli, M., Ivange, L., Schmidt, T.S., Purkayastha, S.: A platform for innovation and standards evaluation: a case study from the openmrs open-source radiology information system. J. Digit. Imaging **31**(3), 361–370 (2018)

5. Hwang, E.J., et al.: Development and validation of a deep learning-based automated detection algorithm for major thoracic diseases on chest radiographs. JAMA Netw. Open **2**(3), e191095–e191095 (2019)

6. Linguraru, M.G., Maier-Hein, L., Summers, R.M., Kahn Jr, C.E.: RSNA-MICCAI panel discussion: 2. leveraging the full potential of ai-radiologists and data scientists working together. Radiology: Artif. Intell. **3**(6) (2021)

7. Maguolo, G., Nanni, L.: A critic evaluation of methods for covid-19 automatic detection from x-ray images. Inf. Fusion **76**, 1–7 (2021)

8. Mazaheri, S., Loya, M.F., Newsome, J., Lungren, M., Gichoya, J.W.: Challenges of implementing artificial intelligence in interventional radiology. In: Seminars in Interventional Radiology, vol. 38, pp. 554–559. Thieme Medical Publishers, Inc. (2021)

9. Moses, D.A.: Deep learning applied to automatic disease detection using chest x-rays. J. Med. Imaging Radiat. Oncol. **65**(5), 498–517 (2021)

10. Omoumi, P., et al.: To buy or not to buy-evaluating commercial ai solutions in radiology (the eclair guidelines). Europ. Radiol. **31**(6), 3786–3796 (2021)

11. Rajpurkar, P., et al.: Chexnet: Radiologist-level pneumonia detection on chest x-rays with deep learning. arXiv preprint arXiv:1711.05225 (2017)

12. Saldanha, O.L., et al.: Swarm learning for decentralized artificial intelligence in cancer histopathology. Nat. Med. **28**(6), 1232–1239 (2022)

13. Strohm, L., Hehakaya, C., Ranschaert, E.R., Boon, W.P., Moors, E.H.: Implementation of artificial intelligence (ai) applications in radiology: hindering and facilitating factors. Eur. Radiol. **30**(10), 5525–5532 (2020)

14. Tariq, A., et al.: Current clinical applications of artificial intelligence in radiology and their best supporting evidence. J. Am. Coll. Radiol. **17**(11), 1371–1381 (2020)

15. Warnat-Herresthal, S., et al.: Swarm learning for decentralized and confidential clinical machine learning. Nature **594**(7862), 265–270 (2021)

16. Wichmann, J.L., Willemink, M.J., De Cecco, C.N.: Artificial intelligence and machine learning in radiology: current state and considerations for routine clinical implementation. Invest. Radiol. **55**(9), 619–627 (2020)

17. Yang, L., Ene, I.C., Arabi Belaghi, R., Koff, D., Stein, N., Santaguida, P.: Stakeholders' perspectives on the future of artificial intelligence in radiology: a scoping review. Europ. Radiol. **32**(3), 1477–1495 (2022)

Management of Patient and Physician Preferences and Explanations for Participatory Evaluation of Treatment with an Ethical Seal

Oscar Raya[1]([✉])[iD], Xavier Castells[2][iD], David Ramírez[2][iD], and Beatriz López[1][iD]

[1] Control Engineering and Intelligent Systems (eXiT) Research Group,
University of Girona, Girona, Spain
{oscar.raya,beatriz.lopez}@udg.edu
[2] Translab Research Group, Department of Medical Sciences,
University of Girona, Girona, Spain
xavier.castells@udg.edu

Abstract. Clinical practice guidelines (CPGs) suffer from several limitations, including limited patient participation. APPRAISE-RS is a methodology for generating treatment recommendations that overcomes this limitation by enabling both patients and clinicians to express their personal preferences about the treatment outcomes. However, patient, and clinical preferences are treated with equal importance, while it seems reasonable/fair to give more importance to clinicians' preferences as they have more experience on the matter. In this work we present APPRAISE-RS-E, which considers different ponderations when including users' preferences based on their experience for the generation of treatment recommendations. Moreover, since users are involved in the decision loop, an explanation of the recommendations is provided. Finally, as APPRAISE-RS-E uses AI methods, it has been evaluated using a set of principles and observable indicators, getting an ethical seal that informs users about the ethical issues involved. The experiments have been carried out in the field of attention deficit hyperactivity disorder (ADHD).

Keywords: Computerized clinical practice guidelines · Clinical decision support systems · Ethical AI · Explainable AI · Participatory medicine · ADHD

1 Introduction

Clinical practice guidelines (CPGs) have become more patient-centered, yet clinicians and patients poorly adhere to them [2]. The paltry consideration of clinicians and patients' preferences in CPG writing could justify this non-compliance [4]. AI research has developed computer-assisted tools to personalize treatment recommendations [6]. One such tool is APPRAISE-RS [5], which proposes to include patient and clinician preferences in the recommendation

© The Author(s), under exclusive license to Springer Nature Switzerland AG 2023
J. M. Juarez et al. (Eds.): AIME 2023, LNAI 13897, pp. 378–383, 2023.
https://doi.org/10.1007/978-3-031-34344-5_47

process. APPRAISE-RS requires participants to have a certain degree of expertise and knowledge of the disease and possible side effects of available treatments. Despite their expertise, clinicians need prior training to ensure a suitable patient's preferences choice. Concurrently, clinicians' influence on treatment selection should prevail over patients' one. We propose addressing these challenges in APPRAISE-RS-E by considering user expertise when ranking preferences, explainability, and ethical considerations. We use multi-criteria decision analysis (MCDA) [8] to consider clinicians and patients' preferences according to their expertise, describe explainability capabilities, and analyze our application with the PIO test [7] ensuring our tool is not only effective but also ethically sound.

The paper is organized as follows: Sect. 2 provides background, Sect. 3 describes our methodological contributions, Sect. 4 explains our achieved results and implications, and Sect. 5 concludes the paper.

2   Background

Our research is based on two past works: APPRAISE-RS and the PIO model. APPRAISE-RS [5] is a computer-assisted tool that proposes to include patient and clinician preferences in the recommendation process, which can improve the validity of clinical practice guidelines (CPGs) and reduce non-compliance. The methodology automates, adapts, extends, and iterates the Grading of Recommendations Assessment, Development and Evaluation (GRADE) working group methodology [1] to formulate automated, up-to-date, participatory, and personalized treatment recommendations from an updated database of clinical studies. The goal of the methodology is to help and not replace clinical judgement.

The PIO model [7] was established by the Observatory of Ethics in Artificial Intelligence of Catalonia (OEIAC) to promote ethical and responsible AI systems. It consists of seven ethical principles that AI systems should abide by: transparency and explainability, justice and equity, security and non-maleficence, responsibility and accountability, privacy, autonomy, and sustainability.

3   Method

Figure 1 summarizes contributions of this paper: MCDA, explanations of the process, and the ethical seal generated by the PIO model. Note that APPRAISE-RS-R is based on APPRAISE-RS which recommends treatments based on medical evidence (clinical studies). APPRAISE-RS-E adds information about the user experience to balance the influence of the preferences in the final ranking.

MCDA is used in APPRAISE-RS-E to consider both clinical and patient preferences in the decision-making process, as shown in Fig. 2. The preferences are represented using a Likert scale and are considered only if they have a value equal to or greater than 7. The interventions recommended by APPRAISE-RS are divided into two groups: those recommended "in favor" and those recommended "against." Two multi-criteria analyses are then performed, one using

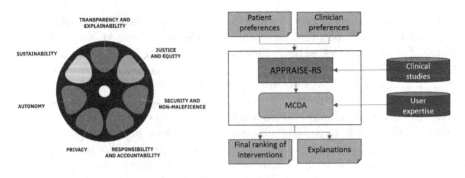

Fig. 1. APPRAISE-RS-E methodology

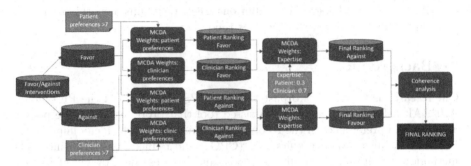

Fig. 2. MCDA in the APPRAISE-RS-E methodology.

the patient's preferences as weights and the other using the clinician's preferences, resulting in four rankings of interventions. These rankings are merged based on a weighting (30/70 for this paper) and the coherence of the "in favor" and "against" rankings is compared to produce the final ranking of interventions. The final ranking indicates the most appropriate intervention for a given situation, taking into account their advantages and disadvantages.

Explainability is ensured by the use of 'white-box' AI techniques such as rule-based systems, meta-analysis from APPRAISE-RS, and MCDA from this work. Clinicians receive explanations on the quality and strength of each recommendation, as well as access to intermediate results for analyzed preferences. Patients also receive explanations from the clinician.

The developers of APPRAISE-RS-E, used the PIO model to obtain an ethics badge for the tool. The badge shows the position on the 7 ethical principles, displayed in a traffic light color schema.

4   Results

APPRAISE-RS-E has been applied for ADHD intervention recommendations, resulting in the TDApp tool (https://tdapp.org), and tested over 28 simulated patients (https://caleta.udg.edu/git/eXiT_Research_Group/TDAH_Data). As a preliminary approach, we use the weighted average [3] in the MCDA.

Table 1. Comparition of APPRAISE-RS-E with 4 CPGs and APPRAISE-RS

Patient	Age	Complex	Number of interventions recommended to be administered					
			CPG Spain	CPG UK	CPG USA	CPG Canada	APPRAISE-RS	APPRAISE-RS-E
1	Child/Teen	No	8	2	16	6	4	1
2	Child/Teen	No	8	2	16	6	5	1
3	Child/Teen	No	8	2	16	6	4	1
4	Child/Teen	No	8	2	16	6	1	1
5	Child/Teen	No	8	2	16	6	1	1
6	Child/Teen	Yes	8	2	16	6	1	1
7	Child/Teen	Yes	8	2	16	6	2	1
8	Child/Teen	Yes	8	2	12	2	0	1
9	Child/Teen	Yes	8	2	16	6	1	0
10	Child/Teen	Yes	8	2	16	6	2	1
11	Child/Teen	Yes	6	2	16	6	0	1
12	Child/Teen	No	6	2	16	6	3	1
13	Child/Teen	No	8	2	16	6	1	1
14	Child/Teen	No	8	2	16	6	3	1
15	Child/Teen	No	8	2	16	6	5	1
16	Child/Teen	Yes	6	2	16	4	4	1
17	Child/Teen	Yes	8	2	6	2	1	1
18	Child/Teen	Yes	6	2	16	4	0	1
19	Adult	Yes	6	2		5	2	1
20	Adult	No	8	2		5	3	1
21	Adult	Yes	6	2		5	1	1
22	Adult	No	2	2		5	2	1
23	Adult	No	2	4		5	2	1
24	Adult	Yes	8	4		3	0	1
25	Adult	No	2	4		2	2	1
26	Child/Teen	No	8	2	16	6	2	1
27	Child/Teen	No	6	2	16	6	0	1
28	Adult	No	2	4		5	1	1

Table 1 compares the results of the APPRAISE-RS-E method to four CPGs and the APPRAISE-RS, that does not consider user expertise. Fewer first-line treatment recommendations were provided, implying more personalized recommendations. Note that APPRAISE-RS could not make a recommendation, in some cases, while APPRAISE-RS-E could.

Regarding the explainability of APPRAISE-RS-E, the system provides transparency and understandability of the decision-making process and allows users to rate their satisfaction with the recommendations, providing valuable feedback data for continuous improvement. Clinicians who used the system provided positive feedback about its explainability feature. However, it may require a certain level of statistical knowledge for complete understanding.

The results of the PIO test show that APPRAISE-RS-E performed superbly in terms of Responsibility and Accountability (100%), Privacy (100%), and Security and Non-Maleficence (89%) and displayed a relatively high performance in Transparency and Explainability (87%) and Autonomy (83%), displayed by the badge on Fig. 1. However, there is still room for improvement in the areas of Justice and Equity (71%) and Sustainability (57%). These findings clearly indicates the strengths and limitations of APPRAISE-RS-E, guiding future developments and improvements regarding the understanding difficulties for patients with low socio-educational backgrounds.

Our results suggest that using MCDA to rank interventions can improve personalized treatment for ADHD patients. However, the limitation of our in silico data requires further study in clinical settings with real patients. A clinical study is currently being conducted to determine the extent of existing differences between patients' and clinicians' preferences.

5　Conclusions

Non-compliance with medical recommendations remains a significant challenge in healthcare, and one of the main reasons for this is the lack of consideration for user' preferences. Our paper proposes a new approach to address this issue by using MCDA, explainability capabilities, and ethical considerations to create a more personalized, patient-centered, and ethically sound treatment recommender system/tool. The proposed methodology, APPRAISE-RS-E, is tested for ADHD. Results show how APPRAISE-RS-E achieves specific recommendations, while explainability, thought useful, assumes clinicians have some statistical knowledge. The process conducted to get the ethical seal has provided several reflections on the design of the tool regarding patients with some vulnerability conditions regarding the use of technology. Therefore, as future work we look for improvements regarding the citizens' accessibility/usability of the tool. Moreover, we will analyze different MCDA alternative methods.

Acknowledgements. This work received joint funding from the European Regional Development Fund (ERDF), the Spanish Ministry of the Economy, Industry and Competitiveness (MINECO), the Carlos III Research Institute (grant no. PI19/00375), and received support from the Generalitat de Catalunya 2021 SGR 01125.

References

1. Alonso-Coello, P., Oxman, A., et al.: GRADE Evidence to Decision (EtD) frameworks: a systematic and transparent approach to making well informed healthcare choices. 2: Clinical practice guidelines. BMJ (Clin. Res. ed.) **353**, BMJ.I2089 (2016). https://doi.org/10.1136/BMJ.I2089
2. Arts, D., Voncken, A., et al.: Reasons for intentional guideline non-adherence: a systematic review. Int. J. Med. Inf. **89**, 55–62 (2016). https://doi.org/10.1016/J.IJMEDINF.2016.02.009
3. Cabral, J.B., Luczywo, N.A., Zanazzi, J.L.: cikit-criteria: Colección de métodos de análisis multi-criterio integrado al stack científico de Python. In: 45JAIIO, pp. 59–66 (2016). http://45jaiio.sadio.org.ar/sites/default/files/Sio-23.pdf
4. Kim, C., Armstrong, M.J., et al.: How to identify, incorporate and report patient preferences in clinical guidelines: a scoping review. Health Expect. **23**(5), 1028–1036 (2020). https://doi.org/10.1111/hex.13099
5. López, B., Raya, O., et al.: APPRAISE-RS: automated, updated, participatory, and personalized treatment recommender systems based on grade methodology. Heliyon (2023). https://doi.org/10.1016/j.heliyon.2023.e13074

6. López, B., Torrent-Fontbona, F., et al.: HTE 3.0: Knowledge-based systems in cascade for familial hypercholesterolemia detection and dyslipidemia treatment. Exp. Syst. **39**, e12835 2022). https://doi.org/10.1111/EXSY.12835

7. OEIAC: Pio self assessment model (2023). https://oeiac.cat/en/pio-self-assessment-model/. Accessed 16 Mar 2023

8. Torra, V., Narukawa, Y.: Modeling decisions : information fusion and aggregation operators. Springer (2010). https://doi.org/10.1007/978-3-540-68791-7

Author Index

Printed in the United States
by Baker & Taylor Publisher Services

Printed in the United States
by Baker & Taylor Publisher Services